P9-DWP-892

114

Advances in Biochemical Engineering/Biotechnology

Series Editor: T. Scheper

Advances in Biochemical Engineering/Biotechnology

Series Editor: T. Scheper

Recently Published and Forthcoming Volumes

Engineering of Stem Cells

Volume Editor: Ulrich Martin

With contributions by

M. Amit · A. Bosio · S. Donath · M. Geraerts · S. Gerecht
I. Gruh · U.A.O. Heinlein · P. Hennemann · V. Huppert ·
J. Itskovitz-Eldor · R.-K. Li · E. Luong · M. Malchow
U. Martin · S. Mitalipov · C.M. Verfaillie · R.D. Weisel
D. Wolf · J. Wu · F. Zeng · R. Zweigerdt

Editor
Professor Dr. Ulrich Martin
Hannover Medical School
Leibniz Research Laboratories for
Biotechnology and Artificial Organs
Carl-Neuberg-Str. 1
30625 Hannover
Germany
martin.ulrich@mh-hannover.de

ISSN 0724-6145 e-ISSN 1616-8542
ISBN 978-3-540-88805-5 e-ISBN 978-3-540-88806-2
DOI: 10.1007/978-3-540-88806-2
Springer Heidelberg Dordrecht London New York

Library of Congress Control Number: 2009933990

Cover design: WMXDesign GmbH, Heidelberg, Germany

Printed on acid-free paper

Springer is part of Springer Science+Business Media (www.springer.com)

Series Editor

Volume Editor

Dr. Ulrich Martin

Hannover Medical School
Leibniz Research Laboratories for
Biotechnology and Artificial Organs
Carl-Neuberg-Str. 1, 30625 Hannover
Germany
martin.ulrich@mh-hannover.de

Editorial Board

Honorary Editors

Advances in Biochemical Engineering/ Biotechnology Also Available Electronically

Advances in Biochemical Engineering/Biotechnology is included in Springer's eBook package *Chemistry and Materials Science*. If a library does not opt for the whole package the book series may be bought on a subscription basis. Also, all back volumes are available electronically.

For all customers who have a standing order to the print version of *Advances in Biochemical Engineering/Biotechnology*, we offer the electronic version via SpringerLink free of charge.

If you do not have access, you can still view the table of contents of each volume and the abstract of each article by going to the SpringerLink homepage, clicking on "Chemistry and Materials Science," under Subject Collection, then "Book Series," under Content Type and finally by selecting *Advances in Biochemical Bioengineering/Biotechnology*

You will find information about the

- Editorial Board
- Aims and Scope
- Instructions for Authors
- Sample Contribution

at springer.com using the search function by typing in *Advances in Biochemical Engineering/Biotechnology*.

Color figures are published in full color in the electronic version on SpringerLink.

Aims and Scope

Advances in Biochemical Engineering/Biotechnology reviews actual trends in modern biotechnology.

Its aim is to cover all aspects of this interdisciplinary technology where knowledge, methods and expertise are required for chemistry, biochemistry, microbiology, genetics, chemical engineering and computer science.

Special volumes are dedicated to selected topics which focus on new biotechnological products and new processes for their synthesis and purification. They give the state-of-the-art of a topic in a comprehensive way thus being a valuable source for the next 3-5 years. It also discusses new discoveries and applications.

In general, special volumes are edited by well known guest editors. The series editor and publisher will however always be pleased to receive suggestions and supplementary information. Manuscripts are accepted in English.

In references *Advances in Biochemical Engineering/Biotechnology* is abbreviated as *Adv. Biochem. Engin./Biotechnol.* and is cited as a journal.

Special volumes are edited by well known guest editors who invite reputed authors for the review articles in their volumes.

Impact Factor in 2008: 2,569; Section "Biotechnology and Applied Microbiology": Rank 48 of 138

Attention all Users of the "Springer Handbook of Enzymes"

Information on this handbook can be found on the internet at springeronline.com

A complete list of all enzyme entries either as an alphabetical Name Index or as the EC-Number Index is available at the above mentioned URL. You can download and print them free of charge.

A complete list of all synonyms (more than 25,000 entries) used for the enzymes is available in print form (ISBN 3-540-41830-X).

Save 15%

We recommend a standing order for the series to ensure you automatically receive all volumes and all supplements and save 15% on the list price.

Preface

I am very pleased to present this volume on engineering stem cells in *Advances in Biochemical Engineering and Biotechnology*. This volume stays abreast of recent developments in stem cell biology and the high expectations concerning the development of stem cell based regenerative therapies.

Regenerative medicine is the focus of current biomedical research, with unique challenges related to scientific, technical and ethical issues of stem cell research, and the potential added value of connecting biomedicine with enabling technologies such as materials sciences, mechanical- and nano-engineering. Research activities in regenerative medicine include strategies in endogenous regeneration of injured or degenerated tissues by means of gene therapy or cell transplantation, as well as complex approaches to replace or reconstruct lost or malformed tissue structures, by applying tissue engineering approaches. In most cases, the specialized functional cell types of interest cannot be isolated from the diseased organ or expanded to a sufficient degree, and various stem and progenitor cell types represent the only applicable cell source.

In almost all cases, stem cells have to be engineered, sometimes for functional improvement, in many cases to produce large numbers of cells, and frequently to achieve efficient and specific differentiation in the cell type(s) of interest. Engineering stem cells can take place on different technological levels including genetic manipulation, treatment with growth factors and small molecules and seeding on functionalized nanostructured surfaces or culture in 3-dimensional matrices, as well as static or continuous mass culture in various types of bioreactors. The mode of engineering depends critically not only on the functional cell type that needs to be generated, but also on the necessary cell number and the stem cell type that is used. Prior to clinical application, GXP-conform protocols have to be developed and potential risks such as tumour formation have to be assessed and minimized.

To prepare this volume, contributions from leading researchers and experts in specific fields of basic and applied stem cell research were assembled. In particular, the technological aspects of stem cell research, including critical discussion of technical limitations, are the focus of most contributions. They address important aspects, including isolation of adult and embryonic stem cells, generation of pluripotent cell sources by means of reprogramming, differentiation of stem cells, purification of stem cells and their derivatives, and the development of large scale culture protocols. Recent developments in stem cell-based tissue engineering round off the volume.

I hope that this volume will be useful not only to stem cell researchers but also to investigators in related fields, including physicians, chemists and engineers, who intend to enter the field of stem cell research. In addition, I anticipate that it will provide basic reading material for students starting their research in the field of stem cell biology and regenerative medicine.

I thank all the authors for their excellent contributions and Springer for implementation of this project. I would especially like to thank Prof. Thomas Scheper and Ulrike Kreusel for their patience and excellent work as production editors.

Hannover, Summer 2009 Ulrich Martin

Contents

Adv Biochem Engin/Biotechnol (2009) 114: 1-21
DOI: 10.1007/10_2008_21

Published online: 17 April 2009

Adult Stem and Progenitor Cells

Martine Geraerts and Catherine M. Verfaillie

Abstract The discovery of adult stem cells in most adult tissues is the basis of a number of clinical studies that are carried out, with therapeutic use of hematopoietic stem cells as a prime example. Intense scientific debate is still ongoing as to whether adult stem cells may have a greater plasticity than previously thought. Although cells with some features of embryonic stem cells that, among others, express Oct4, Nanog and SSEA1 are isolated from fresh tissue, it is not clear if the greater differentiation potential is acquired during cell culture. Moreover, adult more pluripotent cells do not have all pluripotent characteristics typical for embryonic stem cells. Recently, some elegant studies were published in which adult cells could be completely reprogrammed to embryonic stem cell-like cells by overexpression of some key transcription factors for pluripotency (Oct4, Sox2, Klf4 and c-Myc). It will be interesting for the future to investigate the exact mechanisms underlying this reprogramming and whether similar transcription factor pathways are present and/or can be activated in adult more pluripotent stem cells.

Keywords Adult stem cell, Plasticity, Pluripotency

Contents

M. Geraerts (✉) and C.M. Verfaillie (✉)
Interdepartementaal Stamcelinstituut Leuven (SCIL), Katholieke Universiteit Leuven, Herestraat 49 bus 804, 3000 Leuven, Belgium
e-mail: Martine.Geraerts@med.kuleuven.be; Catherine.Verfaillie@med.kuleuven.be

Abbreviations

BM	Bone marrow
BMSC	Bone marrow stem cell
EB	Embryoid body
ESC	Embryonic stem cell
HSC	Hematopoeietic stem cell
iPS	Induced pluripotent stem cell
MAPC	Multipotent adult progenitor cell
MEF	Mouse embryonic fibroblast
MSC	Mesenchymal stem cell

1 Stem Cells: General Concepts

Over the last decade, stem cell research has made significant strides due to important new discoveries both in the embryonic and the adult stem cell field. Stem cells are the most primitive, unspecialized cells in embryonic, fetal or adult tissues. Due to lack of definitive markers, they are generally defined based on three functional properties. First, unlike most specialized tissue-specific cells, stem cells that do not express tissue-specific transcripts, proteins or functions, have the capacity to replicate themselves clonally for many times through symmetrical cell divisions and both daughter stem cells continue to be identical to the unspecialized parent stem cell. Alternatively, in asymmetric stem cell divisions, one of the two daughter cells is identical to the parent stem cell. This proliferating capacity is called long-term self-renewal.

Second, unspecialized stem cells can give rise to specialized cells in general via asymmetric divisions, where one of the two daughter cells undergoes lineage commitment and differentiation under influence of signals inside and outside the cell (cell-intrinsic and -extrinsic factors). The potency of a stem cell is defined based on the number of different specialized cells that can be generated. The zygote and early blastomeres are *totipotent* stem cells that make up a full organism including extraembryonic lineages. A *pluripotent* stem cell can generate all cells of the three germ layers (endodermal, mesodermal and ectodermal layer) as well as the germline, but not the extraembryonic trophoblast. Pluripotent stem cells are present in the inner cell mass of the blastocyst, and can be isolated and cultured in vitro, cells referred to as embryonic stem cells [1, 2]. The more restricted *multipotent* stem cells only give rise to cells of a specific tissue and are often named after the tissue from which they are derived. For example, neural stem cells are self-renewing cells that can differentiate into the two major cell types of the nervous system; neurons and glia. Most of the stem cells from adult tissues are multipotent [3]. Spermatogonial stem cells are an example for *unipotent* adult stem cells, as they can only generate sperm cells [4]. It is well-known that BM, intestine and lung have stem cell populations. Other organs that were thought to be "post-mitotic" and unable to regenerate now have also been shown to contain stem cell populations, including the brain [3], the heart [5] and the kidney [6]. Adult stem cells are essential for continuously renewing tissues such as the BM, blood and intestine, and play an important role in recovery from injury in tissues.

Third, stem cells and their progeny are able to reconstitute functionally a given tissue upon transplantation in vivo. The best characterized adult stem cell for transplantation with proven therapeutic efficacy is without doubt the hematopoietic stem cell [7]. Transplantation of undifferentiated embryonic stem cells mostly results in the formation of teratomas, tumors composed of cells of the three germ layers [2]. This proves their true pluripotency, but suggests that ESC-based therapies will only be possible with purified, differentiated cell populations.

2 Functional Characteristics of Adult BM-Derived Stem Cells

The BM was for many years regarded as the main source of hematopoietic stem cells. Non-hematopoietic stem cells, such as mesenchymal stem cells and endothelial progenitor cells, can also be isolated from the BM compartment. This reflects the complexity of this organ, in which several cell populations cohabit. Intriguingly, during the last 7 years, new, more pluripotent cell populations have been isolated from the BM by several investigators using different experimental strategies. However, it is important to know that there are various ways to prove the true pluripotency of cells. In vitro induced differentiation and analysis of cell-type specific markers is the easiest and most accessible method of analyzing tri-lineage differentiation capacity of cells. However, expression of some more or less tissue specific transcripts or proteins does not prove that the presumed differentiated cells have acquired the same properties of their in vivo counterparts. In general, expression of lineage specific transcripts and proteins can only be seen as a first step to demonstrate lineage specification/differentiation, but demonstrating that the differentiated cells acquired functional characteristics in vitro and more importantly in vivo is required.

As described above, ESC are considered pluripotent. This designation can be demonstrated using different assays. For instance, ESC have the capacity to form embryoid bodies, three-dimensional aggregates that closely resemble the core structure of a post-implantation embryo where spontaneous differentiation into cells of the three germ layers is seen. This is generally regarded as a typical characteristic of pluripotent cells. However, this does not demonstrate that the differentiated cells are functionally equivalent to cells found in tissues of the three germlayers. Likewise, teratomas generated from subcutaneously transplanted ESC do not prove that ESC can promote normal development. The ultimate proof that ESC are pluripotent, i.e., can generate cells of all organs and tissues, can only be obtained by injection of ESC in blastocyst and generation of germ-line competent chimeric mice. The most stringent test for pluripotency is tetraploid complementation: test cells are injected into 4n blastocysts and somatic lineages are only composed of the injected cells, since 4n host cells only form extraembryonic cell types as placental trophoblast [8].

Similar levels of proof for the presence of a classical multipotent stem cell, namely the hematopoietic stem cells (HSC), exist. HSC are characterized by the presence of certain cell surface proteins and transcripts, which is insufficient to demonstrate that the cells in question are indeed functional HSC. HSC can be induced to differentiate in vitro in most, if not all, of the cells in the hematopoietic system. However, no in vitro assay has been developed that can definitively prove that HSC were present. The only method that conclusively demonstrates that cells have HSC characteristics is transplantation and subsequent reconstitution of the hematopoietic system in a lethally irradiated recipient in which no endogenous hematopoietic cells remain.

In this review, we will give an overview of the BM-derived cell populations (Table 1) keeping the remarks listed above for demonstration of cell differentiation and potency of stem cells in mind.

2.1 Hematopoietic Stem Cells

Hematopoietic stem cells (HSC) were first defined in the early 1960s as a population of clonogenic BM cells with the ability to generate myeloerythroid colonies in the spleens of lethally irradiated hosts [9, 10] and reconstitution of all blood cell lineages after injection into secondary hosts [11]. HSC are by far the most extensively studied stem cells, and knowledge gained from these studies has allowed their use in clinical applications for the treatment of hematological disorders and malignancies. HSC can be harvested from BM, peripheral blood or umbilical cord blood. HSC are capable of long-term self-renewal in vivo, and sit atop a hierarchy of progenitors that become progressively restricted to initially multiple and subsequent single blood lineages. Differentiation into fully specialized blood cells of the lymphoid (T,B and natural killer cells) and myeloid lineages (granulocytes (neutrophils, eosinophils and basophils), monocytes-macrophages, erythrocytes, megakaryocytes and mast cells) goes via stepwise differentiation through intermediate, proliferating cell populations that become progressively more restricted in their differentiation potential, which is accompanied by decreased proliferative potential.

More than 40 years of research has yielded great insight into the identity of HSC, but it should be kept in mind that, despite the many studies, many aspects of HSC biology remain to be identified and that, for instance, the HSC from human origin still cannot be isolated to homogeneity. Enrichment for HSC occurs by combining selection based on specific cell surface markers that are expressed on HSC and elimination of cells expressing cell surface markers present on differentiated cells. In the mouse, HSC are enriched as “LSK” cells (lineage negative cells that are Sca1^{+} and c-Kit^{+}) or by using antibodies against the SLAM family (CD150^{+}, CD244^{-} and CD48^{-}) [12, 13]. The expression pattern of surface antigens on HSC differs between species and some markers change depending on the activation state of the cells: mouse HSC are CD34$^{low/-}$, Sca-1^{+}, Thy1$^{+/low}$, CD38^{+}, c-Kit^{+}, Flt3^{-}, lin^{-} and human HSC are CD34^{+}, CD59^{+}, Thy1^{+}, CD38$^{low/-}$, c-Kit$^{-/low}$, lin^{-} (Table 1).

Table 1 Overview of BM-derived stem cell populations

Celltype	Derived from	Cultured in	Potency in vitro	potency in vivo	Surface markers	Transcription factors for pluripotency
mHSC	Peripheral blood, bone marrow	Co-culture with stromal feeders	Hematopoietic cells	Hematopoietic cells	Lin$^-$, Sca$^+$, c-Kit$^+$, CD150$^+$, CD34$^{low/-}$, Thy1$^{+/low}$, CD38$^+$, Flt3$^-$	/
hHSC	Peripheral blood, bone marrow, umbilical cord blood	Co-culture with stromal feeders	Hematopoietic cells	Hematopoietic cells	CD34$^+$, CD59$^+$, Thy1$^+$, CD38$^{low/-}$, c-Kit$^{-/low}$, Lin$^-$	/
mMSC	Bone marrow, almost every tissue	10% FBS	Mesenchymal cells: osteocytes, adipocytes, chondrocytes	Mesenchymal tissue: bone, fat, cartilage	CD10$^+$, CD13$^+$, CD29$^+$, CD44$^+$, CD49a-f$^+$, CD73$^+$, CD90$^+$, CD105$^+$, CD106$^+$, CD140b$^+$, CD166$^+$,SSEA1$^+$	/
hMSC	Bone marrow, almost every tissue	10% FBS	Mesenchymal cells	Mesenchymal tissue	CD10$^+$, CD13$^+$, CD73$^+$, CD90$^+$, CD105$^+$, CD140b$^+$, CD146$^+$, C D271$^+$, CD340$^+$, CD349$^+$, W8B2$^+$, SSEA4$^+$, Stro-1$^+$	/
mMAPC	Bone marrow, muscle, brain	2% FBS, LIF, EGF, PDGF-BB	Mesoderm endoderm neuroectoderm	Low frequency chimeric mice: endothelium, neural cells, hepatocytes, intestinal epithelium, retina, kidney, lung, hematopoietic cells	CD45$^-$, CD34$^-$, MHC II, MHC I^{Low}, SSEA1^{+l}, cKit$^+$, CD9$^+$	Oct4$^+$,Sox2$^-$, Nanog$^-$

(continued)

Table 1 (continued)

Celltype	Derived from	Cultured in	potency in vitro	Potency in vivo	Surface markers	Transcription factors for pluripotency
rMAPC	Bone marrow	2% FBS, LIF-EGF-PDGF-BB	Mesoderm Endoderm Neuroectoderm		$CD45^-$, $CD34^-$, MHC II, MHC I^{low}, $CD31^+$	$Oct4^+$,$Sox2^-$, $Nanog^-$
hMAPC	Bone marrow	2% FBS, EGF, PDGF-BB	Mesodermal Endodermal	Endothelial	$CD45^-$, MHC II^-, MHC I^{low}, $CD44^{low}$	$Oct4^-$, $Sox2^-$
hMIAMI	Bone marrow	2% FBS	Mesenchymal cells Neuroectoderm Pancreatic cells	Not described	$SSEA4^+$, $CD45^-$, $CD34^-$	$Oct4^+$, $Rex1^+$
hBMSC	Bone marrow	17% FBS	Endothelium Hepatocytes Neuroectoderm	Myocardium regeneration	$CD45^-$, MHC I/II^-, c-Kit^{low},$CD90^{low}$, $CD105^{low}$	$Oct4^-$
hUSSC	Umbilical cord blood	Myelocult medium or 30% FCS	Mesenchymal Hematopoietic Immature hepatocytes Neuroectoderm: dopamin ergic neurons	Chondrocytes Neural-like cells Hepatocytes Cardiomyocytes Hematopoeitic cells	$CD45^-$, c-Kit^-, HLA-DR^-, $CD10^{low}$, $Flk1^{low}$	$Oct4^-$
hAFS	Amniotic fluid	15% ES-FBS, 18% Chang B and 2% Chang C	Mesenchymal Endothelial Neuroectoderm Hepatocytes	Bone Neuroectoderm	$cKit^+$, MHC-I^+, MHC-II^{low}, $CD45^-$, $CD34^-$, $CD133^-$, $CD29^+$, $CD44^+$, $CD73^+$, $CD90^+$, $CD105^+$ and $SSEA4^+$	$Oct4^+$

hMASC	Liver, heart, bone marrow	Mesencult medium and subsequent 2% FBS, PDGF-BB, EGF	Neuroectoderm Mesenchymal Hepatocytes	Not described	$CD13^+$, $CD49b^+$, $CD90^+$, $CD73^{low}$, $CD44^{low}$, HLA-ABC^{low}, $CD29^{low}$, $CD105^{low}$, KDR^{low}, $CD49a^{low}$, $CD117^-$, $CD34^-$,$CD133^-$, $CD45^-$, $CD14^-$, $CD38^-$	Oct-4^+, $Nanog^+$,$Sox2^+$
m-maGSC	Neonatal and adult testis	(Stra-8^+ sorted) derived with GDNF, expansion on MEFs with LIF	Trilineage differentiation from embryoid bodies	Teratoma formation germline competent chimeric mice	$SSEA1^+$	Oct-4^+, $Nanog^+$,$Sox2^+$
m-GPR125-MASC	Testis	($GPR125^+$ sorted) derived with GDNF on testicular stroma, expansion on MEFs with LIF	Trilineage differentiation from embryoid bodies	Teratoma formation Chimeric mice	Not described	Oct-4^+, $Nanog^+$,$Sox2^+$
hVSEL	Umbilical cord blood	Not expanded	Not described	Not described	$CXCR4^+$, $AC133^+$, $CD34^+$, Lin^-, $CD45^-$	Oct-4^+, $Nanog^+$
mVSEL	Bone marrow Mobilized peripheral blood	Co-cultures for differentiation (BM) and expansion (C2C12)	Neuroectoderm pancreatic cells cardiomyocytes	Myocytes Capillaries	$CXCR4^+$, $SSEA1^+$, Sca-1^+, $CD45^-$, Lin^-, HLA-DR^-, MHC-I^-, $CD90^-$, $CD29^-$, $CD105^-$	Oct-4^+, $Nanog^+$
mPre-MSC	Bone marrow	Mesencult medium and mMAPC medium after sort	Mesenchymal cells Hepatocytes Endothelial cells Astrocytes	Hematopoietic cells	$SSEA1^+$	Oct-4^+, $Nanog^+$, $Rex1^+$

Flow cytometry sorting using a combination of the KLS or SLAM phenotype with additional cell surface markers, can enrich HSC to near homogeneity in the mouse [12, 14]. It has been estimated that HSC represent only about 1 out of every 100,000 cells in mouse BM. However, in large part due to the absence of good in vivo reconstitution assays from human HSC, the phenotype of human HSC is yet to be fully determined.

During embryologic development, HSC are derived from the ventral mesoderm [15]. A first wave of blood production in mammals occurs in the yolk sac. During this primitive hematopoiesis, mainly red blood cells are generated that help in oxygenating all the growing tissues of the developing embryo. A second wave of hematopoiesis occurs in an area surrounding the dorsal aorta termed the aorta-gonad mesonephros (AGM) region. It is believed that the cells originating in the AGM region subsequently populate the fetal liver, later the fetal thymus, spleen and finally the BM.

The HSC niche, defined as a specialized microenvironment in different tissues capable of housing and maintaining hematopoiesis, is starting to be characterized [12]. In postnatal animals, where HSC are chiefly present in the BM, individual HSC occupy facultative niches scattered over BM sinusoids (specialized blood vessels that allow cells to pass in and out the circulation) and near the vast endosteal surface (interface of bone and marrow) of the trabecular bone [16]. It is not clear, however, whether these two sites represent separate niches or if perivascular, endothelial cells and endosteal osteoblasts/osteoclasts collaborate in a common niche. HSC constantly circulate from one BM compartment to another (for instance from femur to tibia). It has been hypothesized that recirculation of HSC between one facultative niche may be required for the maintenance of the HSC phenotype. Alternatively, this apparent recirculation between possible different niches may simply reflect the passage of HSC through some of these locations during their migration. The spleen and liver, where HSC are present during fetal live, contain only a few HSC under normal conditions. However, in certain hematopoietic malignancies or other stresses, hematopoiesis can be re-established in these organs, demonstrating that facultative niches that support the long-term maintenance of HSC and hematopoiesis can be re-activated in these organs.

CXCL12, previously termed SDF1, and angiopoietin-1 are some of the factors that regulate HSC maintenance and that are produced by multiple cell types within the HSC niche, including osteoblasts, perivascular and endosteal cells in different regions of the BM [17]. The BM microenvironment is a complex system wherein several factors work together in inducing differentiation or maintenance of HSC self-renewal. Despite the many years of investigation, no single cytokine responsible for HSC self-renewal has been identified. As a result, HSC can only be maintained in vitro with a supportive cellular microenvironment of mixed or cloned stromal cells [18]. Several groups have evaluated the expressed gene profile of different stromal feeders that support HSC in vitro. However, this has not yet yielded sufficient information to allow one to develop a culture system wherein HSC can be maintained or expanded in the absence of feeders but solely supplemented with defined proteins generated by such feeders. In fact, the study of hematopoietic niches in vivo has demonstrated that cell–cell based signals, such as for instance the Notch pathway, play a significant role in maintaining HSC undifferentiated. Morphogens such as bone morphogenetic proteins, hedgehogs and Wnts, commonly thought of as factors that govern defined steps in development, are a second class of factors that play a major role in HSC self-renewal [19]. Finally, cell intrinsic factors like the activation of specific transcription factors, such as the homeobox genes Hox-A4 and Hox-A3, are also known to govern self-renewal of HSC [20].

Because the hematopoietic system is so well-studied, it can serve as a model system to be applied to define the phenotype and function of other adult stem cells. Prospective isolation of (subsets of) cells and subsequent analysis in well-defined cell culture systems or after transplantation as has been done for HSC is crucial for the characterization of all stem cells. Only this approach will provide insight in the phenotype and developmental potential of other stem cells.

2.2 *Mesenchymal Stem Cells*

Next to HSC, the BM harbors a second stem cell population that was discovered by the groundbreaking work of Friedenstein in the early 1970s [21–23]. He placed the whole BM into tissue culture flasks, removed the non-adherent cells and characterized the spindle-like adherent colony-forming fibroblast-like cells as rapidly growing cells that can be differentiated by various factors to osteocytes, chondrocytes and adipocytes. Subsequent studies confirmed these findings and demonstrated that these colony forming fibroblasts (CFU-F), as Friedenstein termed them, can, at the clonal level, differentiate to multiple connective tissue types. These cells were then renamed mesenchymal stem cells or marrow stromal cells (MSC). There is no consensus yet concerning their phenotypic and functional characteristics, as preparations of cells generated through adherence and culture differ among species and laboratories.

No single marker or combination of markers is known that can unequivocally identify MSC neither in vitro nor in vivo and there are no quantitative assays to assess the presence of MSC in a given cell population. Currently, MSC are defined by a combination of morphologic, phenotypic and functional properties [24]. Human and rodent MSC are enriched by their preferential ability to adhere to culture plastic. It is hence unavoidable that hematopoietic cells such as macrophages, and endothelial cells or smooth muscle cells, which also adhere to plastic, "contaminate" the cultures. Further enrichment of MSC is obtained by repeated passaging of the mixed cell population, by plating cells at low densities, by exposure to potassium thiocyanate that selectively kills macrophages and other hematopoietic cell types, or by negative selection to exclude hematopoietic cells (CD45, Glycophorin-A) with commercially available columns, flow cytometry sorting or immunomagnetic selection [25–27]. Because culture methods that differ between different investigators, likely select or expand different cell types or sub-populations, the phenotypic expression of culture expanded MSC varies. Thus, several cell surface antigens have been described that would identify cultured MSC, including CD10, CD13, CD29, CD44, CD49a–f, CD63, CD90, CD105, CD106, CD140b and SB-10 (antibody against CD166) [28, 29]. Cultured MSC do not express antigens found on endothelial (progenitor) cells (CD31), although CD105 is found on EPC, and hematopoietic cells (CD45, CD3, CD14, CD11b, CD19, CD38 and CD66b). Number, differentiation potential and maximal life span of MSC declines with age [30]. The frequency of colony forming fibroblasts from the BM is low but can be enriched 100-fold by positive selection with the Stro-1 antibody, as described by Simmons et al. [31]. After this initial paper, various other surface markers have been used for positive selection of MSC, such as Sca-1, SH3/SH4 (antibodies against CD73), SH2 (antibody against CD105), SSEA1/4, MCAM/CD146, GD2, STRO-1 (binds to tissue nonspecific alkaline phosphatase) and CD271 (low-affinity nerve growth factor receptor) (Table 1) [17, 32–36].

To demonstrate multipotency of MSC, one needs to demonstrate that clonally isolated and expanded MSC differentiate to alizarin red positive osteoblasts, oil-red O-positive adipocytes, and alcian blue positive chondrocytes. However, many studies have used non-clonal isolations, which cannot prove multilineage differentiation at the single cell level. It should also be noted that these in vitro assays correlate poorly with in vivo differentiation assays [37]. In contrast to in vivo studies with HSC, no in vivo assays to assess self-renewal and differentiation properties of freshly isolated or culture-expanded MSC at the clonal level have been developed. In vivo analysis of MSC multipotency is mostly carried out by heterotopic transplantation and only approximately 10% of clonal MSC are able to form bone, stroma, and marrow adipocytes. Although some in vitro studies have suggested that MSC can also differentiate into other mesodermal cell types, such as skeletal and cardiac muscle or endothelial cells, this has not been proven at the clonal level after heterotopic transplantation [38]. Assaying self-renewal of MSCs in vitro is based on sustained growth in culture and on the retention of differentiation properties after multiple population doublings. However, after 20–40 population doublings, depending on the isolate, MSC senesce due to progressive telomere shortening [39]. Long-term expansion of human and mouse MSC induces cell transformations and, after transplantation of these cells in immuno-compromised mice, sarcomas are formed [40, 41]. Demonstration of self-renewal of MSC in vitro and in situ relies on persistent expression of cell surface markers thought to identify primitive MSC. In addition, as little is known about the normal physiological role, the exact tissue location and the development of MSC in vivo, identification of primitive MSC markers is crucial.

Recently, Sacchetti et al. [17] identified MCAM/CD146 as an in situ MSC marker in human BM. MCAM marks self-renewing adventitial reticular cells, a stromal cell type in the subendothelial layer of BM sinusoids. MCAM is also expressed on circulating endothelial progenitors [42] and pericytes [43], an elusive cell type originally defined by its morphology and close contact to endothelial cells in the microvasculature of every connective tissue. In this regard, BM adventitial reticular cells might function as pericytes in the BM sinusoids. Moreover, pericytes can differentiate into osteoblasts, chondrocytes, adipocytes, smooth muscle cells, a property shared with MSC. As MSC are not only found in the BM but are present in nearly every organ [44], it has been hypothesized that pericytes and MSCs are one and the same cell, which was highlighted in a study by Covas et al. [43]. Further studies will be needed to define fully the differences and similarities in phenotype and differentiation ability between MSC and pericytes derived from different tissues. Lineage tracing studies have suggested that the first wave of MSC may be derived from Sox-1 + neuroepithelium and not from mesoderm, but that during subsequent steps in organogenesis such neuroepitelium-derived MSC are replaced by MSC from multiple developmental origins [45]. These lineage tracing studies have also suggested that some CFU-Fs in BM are derived from neuroepithelium and neural crest precursors. Many groups have described unexpected differentiation of MSC into neural cells [46], cardiomyocytes [47] and pneumocytes [48]. Although some of these studies may have used not foolproof methods to prove such unexpected differentiation [49], the varied developmental origin of MSC may in part account for these results.

In bone marrow, stromal cells serve two functions: providing a supportive microenvironment for HSC and development/maintenance of the sinusoidal network. These properties together with the ability to form mesenchymal structures like bone and cartilage, resulted currently in a large number of clinical trials with BM-derived cells for organ repair and for tissue engineering applications to treat congenital diseases as Osteogenesis Imperfecta, methachromatic leukodystrophy. It has been shown that MSC aid in engraftment of hematopoietic stem cells, they have an immunosuppressive effect in graft vs host disease and could be beneficial in osteoarthritis and cardiac ischemia [50]. These easily generated, maintained and expanded MSC can in addition be used as vehicles for growth factors or drug delivery.

2.3 Adult Stem Cells with a Greater Potency

Since the late 1990s, several reports described surprising properties of adult stem cells that questioned long-held dogmas that, during development, pluripotent cells were specified to ectoderm, endoderm and mesoderm, and that all adult multipotent stem cells hence belonged to a single germlayer and even more a specific tissue, giving rise only to cells of the tissue they reside in. However, a number of reports described that freshly isolated blood or BM derived cells transplanted in recipient animals were able to differentiate into – aside from the expected hematopoietic lineage, and also into various cell types from endodermal (endocrine pancreas, liver, bile ducts) [51–54], ectodermal (epidermis and neural cells) [55, 56] and mesodermal (endothelium, skeletal and cardiac muscle) origin [57–59]. In most of these experiments whole BM populations were utilized. However, a number of studies also evaluated purified HSC and found that their progeny apparently differentiated into liver, lung, gastrointestinal, and skin epithelium [60, 61]. A number of subsequent papers confirmed these initial unexpected observations, whereas other studies suggested that the degree of lineage switch that occurs in vivo is minimal or non-existent [62, 63]. It should be noted that most studies claiming a possible lineage switch, based this conclusion chiefly on the acquisition of phenotypic characteristics of the new cell type, not acquisition of functional characteristics, nor evidence for real repopulation of an organ different than the hematopoietic system in vivo. One notable exception was the study by Lagasse et al. [61], demonstrating that grafting of as few as 50 wild-type KLS cells in mice with a fatal liver disorder due to a genetic mutation in the FAH gene could correct the liver disease. Subsequent studies have, however, shown that this rescue was not a cell autonomous effect of the HSC, but was due to the fusion between macrophages derived from the HSC and hepatocytes with introduction of a normal copy of the FAH gene in host hepatocytes [64]. Interestingly, the fusion resulted in the partial effacement of the hematopoietic gene program, which is consistent with a cellular reprogramming event. Aside from fusion resulting in an apparent lineage switch of adult cells, transplantation of heterogeneous cell populations comprising two different stem cells, each of which gives rise to the expected tissue, but not to another tissue, may explain at least some of the apparent stem cell plasticity [65–67]. Obviously technical difficulties in proving lineage switch, including false positive immuno-histological assessments, as well as technical difficulties with the use of sex chromosomes to identify donor and host cells, among others, may explain part of the discrepancies between different studies evaluating this phenomenon. However, another possibility is that trans- or de- and re-differentiation can occur. A final possibility is that more pluripotent, less lineage restricted stem cells persist postnatally and are part of the cells that are injected.

A second series of studies also suggests greater potency of adult cells. In these studies, BM cells [68], spermatogonial stem cells [69] or neurospheres [70] are cultured in vitro, and are subsequently shown to have greater differentiation potency. Clarke et al. [71] isolated neurospheres from Rosa mice, cultured the spheres in vitro, and subsequently injected the spheres in the blastocyst. Of the offspring at E11, 12% were partial chimeras in which neurospheres contributed to the CNS, heart, liver, intestine. However no life chimeric offspring was generated and this study could not be replicated by others [72].

In 2002, our group published the isolation via culture of cells we termed multipotent adult progenitor cells or MAPC from mouse and rat BM. We demonstrated that these cells could be expanded without telomere shortening, and could at the clonal level differentiate into mesodermal (endothelium, smooth muscle cells, skeletal muscle and osteoblasts), neuroectodermal and endodermal (hepatocytes) cells [68, 73–77]. Aside from the presence of transcripts and proteins consistent with the specific cell types, functional attribution of the differentiated mesodermal, hepatic and neuroectodermal cells was demonstrated in vitro. Cells were first isolated from human [78] and rodent BM [68], as well as from newborn rodent brain and muscle tissue [79]. The major difference in culture conditions is the need for LIF to isolate and maintain the rodent cells, but not the human cells. Subsequently, MAPC were also isolated from swine [80], and like the human cells, these do not require addition of LIF to the culture. Since the initial description of the isolation methods, improvements have been made to the culture system, where isolations and cell maintenance are now done at 5% O_2. Mouse, rat and human MAPC do not express CD45 and other more mature hematopoietic cell surface antigens, are MHC Class II negative and express low levels of MHC-class I. Mouse MAPC described in 2002 express low levels of SSEA1, whereas those isolated under hypoxic conditions are SSEA1 negative. Recent mouse MAPC isolates are also c-Kit, EpCam, VLA-6 and CD9 positive, but CD34 negative, whereas rat clones are CD31 positive and human clones are $CD44^{low}$. The MAPC population described in 2002 contributed to many somatic tissues after mouse blastocyst injection, although the degree of contribution was low in most chimeric mice and no germ-line transmission was detected [68]. More recent isolates contribute less than 1–5% to E12 mouse embryos, and no significant contribution to life offspring has been detected. Upon transplantation into sub-lethally irradiated NOD-SCID mice, murine MAPC engraft and differentiate to hematopoietic cells, that upon secondary transfer can rescue the hematopoietic system; and in a limited fashion to epithelium of liver, lung and gut [68, 81]. Moreover, undifferentiated human and mouse MAPC contribute to endothelium, smooth muscle and skeletal muscle when grafted in an ischemic limb model, where they also improve limb function via the secretion of trophic factors [74]. A similar trophic effect has also been noted for mouse and swine MAPC grafted in an acute myocardial infarct model [82, 83].

Comparative transcriptome analysis of MAPC, MSC and ESC showed that MAPC cluster closer to ESC and are significantly different from MSC and MSC-like cells (cells isolated under MAPC conditions that do not express detectable levels of Oct4) [84]. The rodent MAPC gene signature was remarkable for the finding that a number of early endodermal transcription factors are expressed, whereas MSC only express mesoderm specific transcripts. Rat and mouse MAPC express Oct-4, a gene known to maintain pluripotency in ESC, at levels between 5% and 20% of murine ESC. Aside from Oct4 and Rex-1, MAPC also express a number of ESC associated genes (Ecats) but they do not express Nanog and Sox2, two other genes known to play a significant role in the maintenance of the pluripotency transcriptional network in ES cells [85, 86]. Of note, when Nanog is suppressed in ESC using shRNA mediated knock-down, a similar expression of endoderm specific transcripts as is seen in MAPC can be detected. Moreover, there is mounting evidence that Nanog expression fluctuates in ESC [87] and lower levels of Nanog may tip the balance towards differentiation rather than staying pluripotent [88]. Although these results suggest that the presence of ESC specific transcripts may be responsible for the greater potency of MAPC than MSC, studies wherein Oct4 is knocked-down will be needed to prove this notion. Gene expression profiling also identified cell surface markers that could be used for prospective isolation of MAPC such as c-Kit and PDGF-Rα.

Since the isolation of MAPC, multiple groups have reported isolation of more pluripotent stem cells not only from rodent and human BM [33, 89–91], but also from heart, liver [92], umbilical cord blood [93–95], dermis [96], hair follicles [97], amniotic fluid [98] and skeletal muscle [99, 100]. Methods used for isolation of these cells were in general relatively similar, even though the O_2 tension in the incubator chamber varied between 3 and 20% O_2, the serum concentration used ranged from 2 to 20%, and in many instances no other growth factors were added apart from serum; cell densities used differed as well. The potential of cells was evaluated by demonstrating acquisition of transcripts and/or proteins of cells from the three germ layers; functional attributes of the differentiated cells in vitro was only assessed in a limited number of studies and in vivo repopulation was seldom tested.

D'Ippolito et al. [89] isolated a population of "pluripotent" cells from BM of people aged 3–72 years, termed marrow-isolated adult multilineage inducible (MIAMI) cells, that differentiate into cells expressing transcripts and proteins found in mesenchymal lineages as well as neural and pancreatic lineages. When maintained at 3% O_2, MIAMI cells could be expanded for more than 50 population doublings. MIAMI cells are SSEA4$^+$, CD45$^-$, CD34$^-$, express telomerase and the transcription factors Oct-4 and Rex-1 (Table 1). Yoon et al. [91] reported the isolation of human BM-derived multipotent stem cells (hBMSC) with the capacity to differentiate into cells expressing transcripts and proteins found in cells of the three germ layers (endothelium, hepatocytes and neuroectoderm) in vitro. hBMSCs may also be able to differentiate into cardiomyocytes in vivo. Single cell clones (CD45$^-$, MHC I/II$^-$, c-Kit^{low},CD90^{low}, CD105^{low}) could be expanded for more than 140 population doublings without loss of telomere length but did not express the transcription factor Oct-4. Kogler et al. [93] isolated similar cells by culturing umbilical cord blood, naming the cells unrestricted somatic stem cells (USSC). USSC are CD45$^-$, c-Kit$^-$, HLA-DR$^-$, CD10^{low} and Flk1^{low}, can be expanded for more than 40 population doublings and were shown to differ from MSC based on their immunophenotype, telomere length, mRNA expression and differentiation capacity. USSCs give rise in vitro to mesenchymal cells (osteoblasts, chondroblasts, adipocytes) and cells with protein expression pattern and some functional attributes of neuroectodermal and hepatic cells. USSCs grafted in utero in pre-immune sheep differentiated into chondrocytes, neuron-like cells, and contributed to a low extent to cardiomyocytes and hematopoietic cells. Similar cells were also isolated by de Coppi et al. [98] by culture of human amniotic fluid cells (AFS cells). C-Kit positive cells were selected from human amniocentesis specimens and can be maintained in culture for more than 250 population doublings without karyotypic instabilities and telomere shortening. Clonal lines express Oct4 and are further MHC-I$^+$, MHC-II^{low}, CD45$^-$, CD34$^-$, CD133$^-$, CD29$^+$, CD44$^+$, CD73$^+$, CD90$^+$, CD105$^+$ and SSEA4$^+$. In vitro, AFS cells were shown to differentiate into mesenchymal, endothelial, neuronal and hepatic lineages. Finally, Beltrami et al. [92] published that multipotent adult stem cells (MASC) could be isolated by culture of human BM, cardiac and liver derived cells. MASC express Oct4, Nanog and Sox2, expand without telomere shortening for 40 population doublings and at the clonal level differentiate into cells with phenotypic and functional attributes of several mesodermal cell types, hepatic cells and neuroectodermal cells.

Another example wherein lineage restricted stem cells gained greater potency are spermatogonial stem cells, cultured in vitro. In 2004, Kanatsu-Shinohara et al. [69] demonstrated for the first time that when neonatal spermatogonial stem cells were cultured for 4–7 weeks in vitro in the presence of bFGF, EGF, LIF and GDNF, approximately 3% of the cells generated ESC-like colonies that could be maintained in ESC conditions (on MEF in medium with 15% FCS and LIF). These multipotent germ stem cells (mGS) express Oct4, Nanog and Rex1 and could form teratomas and germ-line chimeric mice. Subsequently Guan et al. [101] showed that culture of highly purified spermatogonial stem cells (Stra8$^+$) from adult mouse testis with GDNF and subsequently on MEF with LIF yielded cells that had all attributes of ESC: EB formation, teratoma formation, and germ-line competent contribution to chimeric mice. These cells were termed maGSC, multipotent adult germline stem cells. In 2007, the group of Rafii [102] found that spermatoginal stem cells selected based on the expression of GPR125, cultured on mouse testicular stromal cells with GDNF, generate after 2–3 months GPR125$^+$ multipotent adult spermatogonial derived stem cells (GPR-125-MASC), ESC-like cells that can be maintained and expanded in ESC conditions on MEF. GPR-125-MASC express Oct4, Nanog and Sox2 but not other ESC transcripts as Gdf3 and Rex1. GPR-125-MASC form EBs, teratomas and contribute in part to chimeric mice. Hence, these spermatogonial derived multipotent stem cells are the only postnatal derived stem cells with all pluripotency features of ESC.

A third set of studies suggest that cells expressing gene transcripts responsible for the pluripotency of ESC, such as Oct4, Nanog and Sox2, and cell surface antigens found on ESC, such as SSEA1 and SSEA4, may be isolated from fresh BM or umbilical cord blood. Very Small Embryonic-like (VSEL) cells [94] were sorted from human cord blood as well as mouse BM as CXCR4$^+$, AC133$^+$, CD34$^+$, Lin$^-$, CD45$^-$ cells. Murine BM-derived VSEL, express Oct-4, Sox2, Nanog. Cells isolated from human express SSEA-4 and from mouse, SSEA1. Whether human VSEL can be expanded is unknown, although mouse VSEL can be expanded as spheres that maintain Oct4, Nanog and SSEA1 expression in co-cultures over C2C12 cells [103]. For mouse VSEL, differentiation to cells with transcripts consistently found in neuroectoderm, pancreas and cardiomyocytes was shown following co-culture with the respective tissues [90]. VSEL only represent 0.02% of the BM mononuclear cells.

Anjos-Afonso et al. [33] demonstrated that SSEA1$^+$ primitive cells can be sorted from murine BM. SSEA1$^+$ cells could be detected not only in fresh Lineage negative BM but also in BM cells cultured in MesenCult medium for 1–2 passages. SSEA1$^+$ cells sorted from both fresh and cultured BM expressed Oct4, Nanog and Rex-1 transcripts and protein, albeit with levels significantly lower than in ESC. SSEA1$^+$ cells represented the majority of quiescent (G_O) Lineage negative cells and represent only 0.45–0.97% of the total cells. When SSEA1 + cells were plated in Mesencult medium, expression of Oct4, Nanog and Rex1 was lost. By contrast, when cells were maintained in medium also used by Jiang et al. [68] to maintain MAPC, cells retained Oct4, Nanog and Sox2 expression; in fact, the transcript levels of these three transcription factors increased 100-fold. In no instance was Oct4, Nanog or Sox2 expression found when the cultures were initiated with SSEA1 negative cells. One of the clonal cell populations generated under these conditions differentiated in vitro to mesenchymal cell types as well as astrocyte-, endothelial- and hepatocyte-like cells. When grafted intra-femoral, differentiation to osteocytes, adipocytes, cartilage, endothelium as well as hematopoietic cells was noted. As described for MAPC, expression of for instance $Sox17^+$, PDGF-$R\alpha^+$, and c-Kit^+ was found, but in contrast to MAPC a number of mesodermal transcription factors, such as $Brachyury^+$, VE-Cad^+, and $GATA2^+$, were also expressed.

The relationship between the different adult "more multipotent" stem cells derived from non-germline tissues is not clear. Despite differences in cell surface phenotype, they are likely all related cell populations. Collaborative studies wherein the relationship between these cells can be assessed using transcriptome and perhaps proteome analysis, as well as by using standardized differentiation studies in vitro and in vivo, will be needed to define the relationship.

3 General Conclusions: Multipotent or Pluripotent Adult Stem Cells?

It is well-known that Oct4, together with Sox2 and Nanog, is the major transcription factor that allows maintenance of pluripotency of ESC and loss of Oct4 results in loss of pluripotency [104]. In freshly isolated adult somatic cells, high levels of Oct4 can be detected in germline cells and downregulation of Oct4 leads to apoptosis [105]. A pluripotent state can also be induced in adult somatic cells by forced expression of Oct4 together with Sox2, Klf4 and c-Myc in mouse and human cells [106–108] or in combination with Sox2, LIN28 or Nanog in human cells [109]. These induced pluripotent stem cells (iPS) are almost indistinguishable from ESC. Of the genes that need to be introduced to generate iPS cells, Oct4 and Sox2 are the two key transcription factors that allow de-differentiation of differentiated cells to an iPS state [110, 111].

The finding that Oct4 is also detected in some adult somatic cells and in cancer cells, raises the question whether Oct4 can be used as a marker for pluripotency in adult cells [112]. Although some studies have used immunohistochemistry on histological sections to demonstrate that Oct4 positive cells exist in vivo, this notion should be viewed with care as recent studies have shown that false positive staining is possible. In addition, care should be taken when evaluating Oct4 transcripts in isolated cell populations from human origin because many Oct4 and Nanog pseudogenes exist which are expressed in normal somatic cells and poorly designed primers cannot distinguish between pseudogenes and the specific gene [113]. A recent study evaluated whether Oct4 has a physiological role in postnatal life, employing a conditional Oct4 knockout mouse. This study demonstrated that Oct4 is dispensable for both self-renewal and maintenance of somatic cells from several tissues including intestinal epithelium, BM, skin, brain and liver [114]. In addition, this group also used a second genetically modified mouse, wherein IRES-eGFP was knocked-in behind the fifth exon of Oct4. ESC from these animals are eGFP positive; however, analysis of different somatic tissues did not identify eGFP positive cells. This may at first sight be inconsistent with the studies from Kucia et al. [90] and Anjos-Afonso et al. [33], who isolated SSEA1/4 positive cells that express Oct4 transcripts and proteins. However, Oct4 levels found in BM derived SSEA1 + cells in the Anjos-Afonso et al. studies were >100-fold lower than in ESC, and the sensitivity of Oct4-IRES-GFP may be too low to yield GFP positive cells in different tissues. Whether low level expression of Oct4 has physiologically relevance is obviously not known. As we hinted earlier in the chapter, Oct4 expression in somatic cells has in general been reported in cultured cells. As most studies demonstrating greater potency of spermatogonial stem cells, that already express Oct4 in vivo, or adult somatic stem cells, coinciding with presence of Oct4, have demonstrated this potential only in cells cultured ex vivo, the possibility exists that the acquisition of potency and/or Oct4 expression may be a culture-induced phenomenon. Alternatively, the possibility exists that during development primitive cells, perhaps pre-gastrulation stage cells, are left in different tissues, which can be enriched by ex vivo culture.

Of note, in the first publication describing iPS cells, the iPS cell lines were not completely reprogrammed to ESC, as shown by gene expression profiling and methylation studies. This first generation of iPS cells formed embryoid bodies and teratomas but could not give rise to postnatal chimeric mice [106]. When reprogramming was allowed to proceed for a longer period of time, iPS cells were highly similar to ESC, again documented by gene expression and DNA methylation [115, 116]. The latter cells form chimeric mice with germ-line transmission [111, 115, 117]. As has also been shown for epiblast cells, that may be slightly more differentiated compared with ESC, the identification of iPS cells that are reprogrammed to an almost ESC state and iPS cells that are reprogrammed to a state indistinguishable from ESC suggests that there are several stages in pluripotency. It is hence possible that the populations of cells isolated from somatic cell cultures may have different degrees of pluripotency, either inherent to the cells that are selected by the culture, or induced by the culture method. Detailed gene expression, epigenetic and genetic studies will be necessary to determine whether transcription factor pathways essential in pluripotent cells like ES and iPS are also present or can be further induced in adult somatic cells with more pluripotent characteristics like MAPC, USSC, MIAMI cells. Regardless of their origin, adult more pluripotent cells may be very valuable as a model for de-,re- and transdifferentiation and form a source of stem cells for cell therapies and drug screening.

References

1. Thomson JA, Itskovitz-Eldor J, Shapiro SS, Waknitz MA, Swiergiel JJ, Marshall VS, Jones JM (1998) Science 282:1145
2. Rossant J (2008) Cell 132:527
3. Gage FH (2000) Science 287:1433
4. Kanatsu-Shinohara M, Lee J, Inoue K, Ogonuki N, Miki H, Toyokuni S, Ikawa M, Nakamura T, Ogura A, Shinohara T (2008) Biol Reprod 78:681

5. Beltrami AP, Barlucchi L, Torella D, Baker M, Limana F, Chimenti S, Kasahara H, Rota M, Musso E, Urbanek K, Leri A, Kajstura J, Nadal-Ginard B, Anversa P (2003) Cell 114:763
6. Oliver JA, Maarouf O, Cheema FH, Martens TP, Al-Awqati Q (2004) J Clin Invest 114:795
7. Ballester G, Tirona MT, Ballester O (2007) Oncology (Williston Park) 21:1576
8. Denker HW (2006) J Med Ethics 32:665
9. Till JE, Mc CE (1961) Radiat Res 14:213
10. Becker AJ, Mc CE, Till JE (1963) Nature 197:452
11. Siminovitch L, McCulloch EA, Till JE (1963) J Cell Physiol 62:327
12. Kiel MJ, Yilmaz OH, Iwashita T, Terhorst C, Morrison SJ (2005) Cell 121:1109
13. Spangrude GJ, Heimfeld S, Weissman IL (1988) Science 241:58
14. Matsuzaki Y, Kinjo K, Mulligan RC, Okano H (2004) Immunity 20:87
15. Orkin SH, Zon LI (2008) Cell 132:631
16. Morrison SJ, Spradling AC (2008) Cell 132:598
17. Sacchetti B, Funari A, Michienzi S, Di Cesare S, Piersanti S, Saggio I, Tagliafico E, Ferrari S, Robey PG, Riminucci M, Bianco P (2007) Cell 131:324
18. Wagner W, Roderburg C, Wein F, Diehlmann A, Frankhauser M, Schubert R, Eckstein V, Ho AD (2007) Stem Cells 25:2638
19. Blank U, Karlsson G, Karlsson S (2008) Blood 111:492
20. Argiropoulos B, Humphries RK (2007) Oncogene 26:6766
21. Friedenstein AJ, PiatetzkyII S, Petrakova KV (1966) J Embryol Exp Morphol 16:381
22. Friedenstein AJ, Deriglasova UF, Kulagina NN, Panasuk AF, Rudakowa SF, Luria EA, Ruadkow IA (1974) Exp Hematol 2:83
23. Friedenstein AJ, Chailakhyan RK, Latsinik NV, Panasyuk AF, Keiliss-Borok IV (1974) Transplantation 17:331
24. Dominici M, Le Blanc K, Mueller I, Slaper-Cortenbach I, Marini F, Krause D, Deans R, Keating A, Prockop D, Horwitz E (2006) Cytotherapy 8:315
25. Baddoo M, Hill K, Wilkinson R, Gaupp D, Hughes C, Kopen GC, Phinney DG (2003) J Cell Biochem 89:1235
26. Modderman WE, Vrijheid-Lammers T, Lowik CW, Nijweide PJ (1994) Exp Hematol 22:194
27. Tondreau T, Lagneaux L, Dejeneffe M, Delforge A, Massy M, Mortier C, Bron D (2004) Cytotherapy 6:372
28. Benayahu D, Akavia UD, Shur I (2007) Curr Med Chem 14:173
29. Park PC, Selvarajah S, Bayani J, Zielenska M, Squire JA (2007) Semin Cancer Biol 17:257
30. Kern S, Eichler H, Stoeve J, Kluter H, Bieback K (2006) Stem Cells 24:1294
31. Simmons PJ, Torok-Storb B (1991) Blood 78:55
32. Quirici N, Soligo D, Bossolasco P, Servida F, Lumini C, Deliliers GL (2002) Exp Hematol 30:783
33. Anjos-Afonso F, Bonnet D (2007) Blood 109:1298
34. Gang EJ, Bosnakovski D, Figueiredo CA, Visser JW, Perlingeiro RC (2007) Blood 109:1743
35. Gronthos S, Fitter S, Diamond P, Simmons PJ, Itescu S, Zannettino AC (2007) Stem Cells Dev 16:953
36. Martinez C, Hofmann TJ, Marino R, Dominici M, Horwitz EM (2007) Blood 109:4245
37. Bianco P, Kuznetsov SA, Riminucci M, Gehron Robey P (2006) Methods Enzymol 419:117
38. Bianco P, Robey PG, Simmons PJ (2008) Cell Stem Cell 2:313
39. Pittenger MF, Mackay AM, Beck SC, Jaiswal RK, Douglas R, Mosca JD, Moorman MA, Simonetti DW, Craig S, Marshak DR (1999) Science 284:143
40. Li H, Fan X, Kovi RC, Jo Y, Moquin B, Konz R, Stoicov C, Kurt-Jones E, Grossman SR, Lyle S, Rogers AB, Montrose M, Houghton J (2007) Cancer Res 67:10889
41. Rubio D, Garcia S, Paz MF, De la Cueva T, Lopez-Fernandez LA, Lloyd AC, Garcia-Castro J, Bernad A (2008) PLoS ONE 3:e1398
42. Delorme B, Basire A, Gentile C, Sabatier F, Monsonis F, Desouches C, Blot-Chabaud M, Uzan G, Sampol J, Dignat-George F (2005) Thromb Haemost 94:1270
43. Covas DT, Panepucci RA, Fontes AM, Silva Jr WA, Orellana MD, Freitas MC, Neder L, Santos AR, Peres LC, Jamur MC, Zago MA (2008) Exp Hematol 36(5):642–54
44. da Silva Meirelles L, Chagastelles PC, Nardi NB (2006) J Cell Sci 119:2204

45. Takashima Y, Era T, Nakao K, Kondo S, Kasuga M, Smith AG, Nishikawa S (2007) Cell 129:1377
46. Cho KJ, Trzaska KA, Greco SJ, McArdle J, Wang FS, Ye JH, Rameshwar P (2005) Stem Cells 23:383
47. Pittenger MF, Martin BJ (2004) Circ Res 95:9
48. Rojas M, Xu J, Woods CR, Mora AL, Spears W, Roman J, Brigham KL (2005) Am J Respir Cell Mol Biol 33:145
49. Phinney DG, Prockop DJ (2007) Stem Cells 25:2896
50. Togel F, Westenfelder C (2007) Dev Dyn 236:3321
51. Ianus A, Holz GG, Theise ND, Hussain MA (2003) J Clin Invest 111:843
52. Petersen BE, Bowen WC, Patrene KD, Mars WM, Sullivan AK, Murase N, Boggs SS, Greenberger JS, Goff JP (1999) Science 284:1168
53. Theise ND, Badve S, Saxena R, Henegariu O, Sell S, Crawford JM, Krause DS (2000) Hepatology 31:235
54. Kale S, Karihaloo A, Clark PR, Kashgarian M, Krause DS, Cantley LG (2003) J Clin Invest 112:42
55. Brazelton TR, Rossi FM, Keshet GI, Blau HM (2000) Science 290:1775
56. Mezey E, Chandross KJ, Harta G, Maki RA, McKercher SR (2000) Science 290:1779
57. Ferrari G, Cusella-De Angelis G, Coletta M, Paolucci E, Stornaiuolo A, Cossu G, Mavilio F (1998) Science 279:1528
58. Gussoni E, Soneoka Y, Strickland CD, Buzney EA, Khan MK, Flint AF, Kunkel LM, Mulligan RC (1999) Nature 401:390
59. Orlic D, Kajstura J, Chimenti S, Limana F, Jakoniuk I, Quaini F, Nadal-Ginard B, Bodine DM, Leri A, Anversa P (2001) Proc Natl Acad Sci U S A 98:10344
60. Krause DS, Theise ND, Collector MI, Henegariu O, Hwang S, Gardner R, Neutzel S, Sharkis SJ (2001) Cell 105:369
61. Lagasse E, Connors H, Al-Dhalimy M, Reitsma M, Dohse M, Osborne L, Wang X, Finegold M, Weissman IL, Grompe M (2000) Nat Med 6:1229
62. Wagers AJ, Sherwood RI, Christensen JL, Weissman IL (2002) Science 297:2256
63. Alvarez-Dolado M, Pardal R, Garcia-Verdugo JM, Fike JR, Lee HO, Pfeffer K, Lois C, Morrison SJ, Alvarez-Buylla A (2003) Nature 425:968
64. Camargo FD, Finegold M, Goodell MA (2004) J Clin Invest 113:1266
65. Jackson KA, Mi T, Goodell MA (1999) Proc Natl Acad Sci U S A 96:14482
66. McKinney-Freeman SL, Jackson KA, Camargo FD, Ferrari G, Mavilio F, Goodell MA (2002) Proc Natl Acad Sci U S A 99:1341
67. Kawada H, Ogawa M (2001) Blood 98:2008
68. Jiang Y, Jahagirdar BN, Reinhardt RL, Schwartz RE, Keene CD, Ortiz-Gonzalez XR, Reyes M, Lenvik T, Lund T, Blackstad M, Du J, Aldrich S, Lisberg A, Low WC, Largaespada DA, Verfaillie CM (2002) Nature 418:41
69. Kanatsu-Shinohara M, Inoue K, Lee J, Yoshimoto M, Ogonuki N, Miki H, Baba S, Kato T, Kazuki Y, Toyokuni S, Toyoshima M, Niwa O, Oshimura M, Heike T, Nakahata T, Ishino F, Ogura A, Shinohara T (2004) Cell 119:1001
70. Wurmser AE, Nakashima K, Summers RG, Toni N, D'Amour KA, Lie DC, Gage FH (2004) Nature 430:350
71. Clarke DL, Johansson CB, Wilbertz J, Veress B, Nilsson E, Karlstrom H, Lendahl U, Frisen J (2000) Science 288:1660
72. Greco B, Low HP, Johnson EC, Salmonsen RA, Gallant J, Jones SN, Ross AH, Recht LD (2004) Stem Cells 22:600
73. Jiang Y, Henderson D, Blackstad M, Chen A, Miller RF, Verfaillie CM (2003) Proc Natl Acad Sci U S A 100(Suppl 1):11854
74. Aranguren XL, McCue JD, Hendrickx B, Zhu XH, Du F, Chen E, Pelacho B, Penuelas I, Abizanda G, Uriz M, Frommer SA, Ross JJ, Schroeder BA, Seaborn MS, Adney JR, Hagenbrock J, Harris NH, Zhang Y, Zhang X, Nelson-Holte MH, Jiang Y, Billiau AD, Chen W, Prosper F, Verfaillie CM, Luttun A (2008) J Clin Invest 118:505

75. Schwartz RE, Reyes M, Koodie L, Jiang Y, Blackstad M, Lund T, Lenvik T, Johnson S, Hu WS, Verfaillie CM (2002) J Clin Invest 109:1291
76. Aranguren XL, Luttun A, Clavel C, Moreno C, Abizanda G, Barajas MA, Pelacho B, Uriz M, Arana M, Echavarri A, Soriano M, Andreu EJ, Merino J, Garcia-Verdugo JM, Verfaillie CM, Prosper F (2007) Blood 109:2634
77. Ross JJ, Hong Z, Willenbring B, Zeng L, Isenberg B, Lee EH, Reyes M, Keirstead SA, Weir EK, Tranquillo RT, Verfaillie CM (2006) J Clin Invest 116:3139
78. Reyes M, Lund T, Lenvik T, Aguiar D, Koodie L, Verfaillie CM (2001) Blood 98:2615
79. Jiang Y, Vaessen B, Lenvik T, Blackstad M, Reyes M, Verfaillie CM (2002) Exp Hematol 30:896
80. Zeng L, Rahrmann E, Hu Q, Lund T, Sandquist L, Felten M, O'Brien TD, Zhang J, Verfaillie C (2006) Stem Cells 24:2355
81. Serafini M, Dylla SJ, Oki M, Heremans Y, Tolar J, Jiang Y, Buckley SM, Pelacho B, Burns TC, Frommer S, Rossi DJ, Bryder D, Panoskaltsis-Mortari A, O'Shaughnessy MJ, Nelson-Holte M, Fine GC, Weissman IL, Blazar BR, Verfaillie CM (2007) J Exp Med 204:129
82. Pelacho B, Nakamura Y, Zhang J, Ross J, Heremans Y, Nelson-Holte M, Lemke B, Hagenbrock J, Jiang Y, Prosper F, Luttun A, Verfaillie CM (2007) J Tissue Eng Regen Med 1:51
83. Zeng L, Hu Q, Wang X, Mansoor A, Lee J, Feygin J, Zhang G, Suntharalingam P, Boozer S, Mhashilkar A, Panetta CJ, Swingen C, Deans R, From AH, Bache RJ, Verfaillie CM, Zhang J (2007) Circulation 115:1866
84. Ulloa-Montoya F, Kidder BL, Pauwelyn KA, Chase LG, Luttun A, Crabbe A, Geraerts M, Sharov AA, Piao Y, Ko MS, Hu WS, Verfaillie CM (2007) Genome Biol 8:R163
85. Jaenisch R, Young R (2008) Cell 132:567
86. Kim J, Chu J, Shen X, Wang J, Orkin SH (2008) Cell 132:1049
87. Chambers I, Silva J, Colby D, Nichols J, Nijmeijer B, Robertson M, Vrana J, Jones K, Grotewold L, Smith A (2007) Nature 450:1230
88. Mullin NP, Yates A, Rowe AJ, Nijmeijer B, Colby D, Barlow PN, Walkinshaw MD, Chambers I (2008) Biochem J 411:227
89. D'Ippolito G, Diabira S, Howard GA, Menei P, Roos BA, Schiller PC (2004) J Cell Sci 117:2971
90. Kucia M, Reca R, Campbell FR, Zuba-Surma E, Majka M, Ratajczak J, Ratajczak MZ (2006) Leukemia 20:857
91. Yoon YS, Wecker A, Heyd L, Park JS, Tkebuchava T, Kusano K, Hanley A, Scadova H, Qin G, Cha DH, Johnson KL, Aikawa R, Asahara T, Losordo DW (2005) J Clin Invest 115:326
92. Beltrami AP, Cesselli D, Bergamin N, Marcon P, Rigo S, Puppato E, D'Aurizio F, Verardo R, Piazza S, Pignatelli A, Poz A, Baccarani U, Damiani D, Fanin R, Mariuzzi L, Finato N, Masolini P, Burelli S, Belluzzi O, Schneider C, Beltrami CA (2007) Blood 110:3438
93. Kogler G, Sensken S, Airey JA, Trapp T, Muschen M, Feldhahn N, Liedtke S, Sorg RV, Fischer J, Rosenbaum C, Greschat S, Knipper A, Bender J, Degistirici O, Gao J, Caplan AI, Colletti EJ, Almeida-Porada G, Muller HW, Zanjani E, Wernet P (2004) J Exp Med 200:123
94. Kucia M, Halasa M, Wysoczynski M, Baskiewicz-Masiuk M, Moldenhawer S, Zuba-Surma E, Czajka R, Wojakowski W, Machalinski B, Ratajczak MZ (2007) Leukemia 21:297
95. Rogers I, Yamanaka N, Bielecki R, Wong CJ, Chua S, Yuen S, Casper RF (2007) Exp Cell Res 313:1839
96. Toma JG, Akhavan M, Fernandes KJ, Barnabe-Heider F, Sadikot A, Kaplan DR, Miller FD (2001) Nat Cell Biol 3:778
97. Yu H, Fang D, Kumar SM, Li L, Nguyen TK, Acs G, Herlyn M, Xu X (2006) Am J Pathol 168:1879
98. De Coppi P, BartschJr G, Siddiqui MM, Xu T, Santos CC, Perin L, Mostoslavsky G, Serre AC, Snyder EY, Yoo JJ, Furth ME, Soker S, Atala A (2007) Nat Biotechnol 25:100
99. Tamaki T, Okada Y, Uchiyama Y, Tono K, Masuda M, Wada M, Hoshi A, Ishikawa T, Akatsuka A (2007) Stem Cells 25:2283

100. Romero-Ramos M, Vourc'h P, Young HE, Lucas PA, Wu Y, Chivatakarn O, Zaman R, Dunkelman N, el-Kalay MA, Chesselet MF (2002) J Neurosci Res 69:894
101. Guan K, Nayernia K, Maier LS, Wagner S, Dressel R, Lee JH, Nolte J, Wolf F, Li M, Engel W, Hasenfuss G (2006) Nature 440:1199
102. Seandel M, James D, Shmelkov SV, Falciatori I, Kim J, Chavala S, Scherr DS, Zhang F, Torres R, Gale NW, Yancopoulos GD, Murphy A, Valenzuela DM, Hobbs RM, Pandolfi PP, Rafii S (2007) Nature 449:346
103. Kucia M, Wysoczynski M, Ratajczak J, Ratajczak MZ (2008) Cell Tissue Res 331:125
104. Nichols J, Zevnik B, Anastassiadis K, Niwa H, Klewe-Nebenius D, Chambers I, Scholer H, Smith A (1998) Cell 95:379
105. Kehler J, Tolkunova E, Koschorz B, Pesce M, Gentile L, Boiani M, Lomeli H, Nagy A, McLaughlin KJ, Scholer HR, Tomilin A (2004) EMBO Rep 5:1078
106. Takahashi K, Yamanaka S (2006) Cell 126:663
107. Park IH, Zhao R, West JA, Yabuuchi A, Huo H, Ince TA, Lerou PH, Lensch MW, Daley GQ (2008) Nature 451:141
108. Takahashi K, Tanabe K, Ohnuki M, Narita M, Ichisaka T, Tomoda K, Yamanaka S (2007) Cell 131:861
109. Yu J, Vodyanik MA, Smuga-Otto K, Antosiewicz-Bourget J, Frane JL, Tian S, Nie J, Jonsdottir GA, Ruotti V, Stewart R, Slukvin II, Thomson JA (2007) Science 318:1917
110. Nakagawa M, Koyanagi M, Tanabe K, Takahashi K, Ichisaka T, Aoi T, Okita K, Mochiduki Y, Takizawa N, Yamanaka S (2008) Nat Biotechnol 26:101
111. Wernig M, Meissner A, Cassady JP, Jaenisch R (2008) Cell Stem Cell 2:10
112. Lengner CJ, Welstead GG, Jaenisch R (2008) Cell Cycle 7
113. Liedtke S, Enczmann J, Waclawczyk S, Wernet P, Kogler G (2007) Cell Stem Cell 1:364
114. Lengner CJ, Camargo FD, Hochedlinger K, Welstead GG, Zaidi S, Gokhale S, Scholer HR, Tomilin A, Jaenisch R (2007) Cell Stem Cell 1:403
115. Okita K, Ichisaka T, Yamanaka S (2007) Nature 448:313
116. Wernig M, Meissner A, Foreman R, Brambrink T, Ku M, Hochedlinger K, Bernstein BE, Jaenisch R (2007) Nature 448:318
117. Brambrink T, Foreman R, Welstead GG, Lengner CJ, Wernig M, Suh H, Jaenisch R (2008) Cell Stem Cell 2:151

Adv Biochem Engin/Biotechnol (2009) 114: 23-72
DOI: 10.1007/10_2008_38

Published online: 04 April 2009

Isolation and Enrichment of Stem Cells

Andreas Bosio, Volker Huppert, Susan Donath, Petra Hennemann, Michaela Malchow, and Uwe A. O. Heinlein

Abstract Stem cells have the potential to revolutionize tissue regeneration and engineering. Both general types of stem cells, those with pluripotent differentiation potential as well as those with multipotent differentiation potential, are of equal interest. They are important tools to further understanding of general cellular processes, to refine industrial applications for drug target discovery and predictive toxicology, and to gain more insights into their potential for tissue regeneration. This chapter provides an overview of existing sorting technologies and protocols, outlines the phenotypic characteristics of a number of different stem cells, and summarizes their potential clinical applications.

Keywords stem cells, differentiation, tissue regeneration, cancer, magnetic cell separation

Contents

A. Bosio, V. Huppert, S. Donath, P. Hennemann, M. Malchow, and U.A.O. Heinlein (✉)
Miltenyi Biotec GmbH, Friedrich-Ebert-Straße 68, 51429, Bergisch Gladbach, Germany
e-mail: uweh@miltenyibiotec.de

Abbreviations

ACL	Acute lymphoid leukemia
ALDH	Aldehyde dehydrogenase
AML	Acute myeloid leukemia
ASC	Adult tissue-specific stem cell
betaNGF	Beta Nerve growth factor
bFGF	Basic fibroblast growth factor
BMC	Bone marrow cell
BMP-4	Bone morphogenic protein
BTSC	Brain tumor stem cell
CABG	Coronary Artery Bypass Grafting
CC-IC	Human colon cancer-initiating cell
CR	Clinical remission
CSC	Coronary sinus catheter
CXCR4	Alpha chemokine receptor type 4
EFS	Event-free survival
EGF	Epidermal growth factor
EPC	Endothelial progenitor cell
EPCAM	Epithelial cell adhesion molecule
ESA	Epithelial-specific antigen
ESC	Totipotent embryonic stem cell
GBM	Glioblastoma stem cell
HCC	Hepatocellular carcinoma
HEF	Human embryonic fibroblast
hESC	Human embryonic stem cell
HGF	Hematopoietic growth factor
HLA	Human leukocyte antigen
HPSC	Human pluripotent stem cell
HSC	Pluripotent hematopoietic stem cell
iPS	Induced pluripotent stem cell
KIR	Killer inhibitory receptor
LIF	Leukemia inhibitory factor
LRP	Lineage-restricted progenitor cell
LTR-HSC	Longterm repopulatory hematopoietic stem cell
MAPC	Multipotent adult progenitor cell

MEF	Mouse embryonic fibroblast
mESC	Mouse embryonic stem cell
MMFD	Mismatched related family donor
MSC	Mesenchymal stem cell
MUD	Matched unrelated donor
NK	Natural killer
PSA-NCAM	Polysialic acid-neural cell adhesion molecule
RMS	Rostal migratory stream
SLAM	Slow as molasses
SP	"Side population" phenotype
SPC	Sphingosylphosphorylcholine
SSC	Spermatogonial stem cell
SSEA-1	Stage-specific antigen 1
TGF-beta1	Transforming growth factor beta 1
TH-EGFP	Tyrosine Hydroxylase- Enhanced green fluorescent protein

1 Introduction

Stem cells have the potential to revolutionize tissue regeneration and engineering. The hematopoietic stem cells were the first stem cells to be prospectively identified. Since then, an ever increasing number of new types of stem cells, including embryonic stem cell cells, tissue resident stem cells and cancer stem cells, have been identified and characterized. Currently, the derivation of induced pluripotent stem cells (iPS cells) from differentiated, post-mitotic cells that behave similar to ESC is further extending this exciting field. iPS cells may eventually combine the advantages of ESCs and autologous cell transplantation, allowing for a generation of patient specific derived stem cells for unrestricted tissue regeneration and without ethical issues.

Both general types of stem cells, those possessing pluripotent differentiation potential like ESC or iPS cells, as well as those with multipotent differentiation potential like tissue stem cells, are of equal interest. They can be useful in understanding general cellular processes in, e.g., embryogenesis, organogenesis, cancer or ageing, but also as a vehicle for the generation of transgenic mice for functional gene analysis and disease models. Further applications are in industrial research as cell based screenings for drug target discovery, drug discovery or predictive toxicology, and in clinical research as a potential source for tissue regeneration.

The reproducibility of culturing cells at a defined stage, as well as differentiating them to a certain endpoint, is a prerequisite for each of the listed applications. Therefore, a number of different protocols have been published for the isolation and enrichment of stem cells including selective culturing, immunopanning, flow cytometric sorting, or magnetic sorting.

In this chapter we give an overview of existing sorting technologies and protocols, outline the phenotypic characteristics of a number of different stem cells and summarize their potential for clinical applications.

2 Methods and Technologies

2.1 *Stem Cell Enrichment Using Flow Cytometry*

2.1.1 Flow Sorting

Flow cytometric cell sorting utilizes optical differences between target cells and nontarget cells. Light scattering and fluorescent properties are the optical parameters and are either intrinsic to the cell population (size and granularity for forward and sideward scatter) or generated by differential binding or incorporation of fluorescent dyes into cell populations.

Single cell suspensions in a flowing stream are embedded in a second fluid stream (sheath fluid) and are subject to a hydrodynamic focussing process that allows for passing an illumination and sensing unit within a defined distance. Cells in the center of the sample stream are illuminated, for example, by a laser beam, interact with the light and respond with light emission at different angles, intensities, and wavelengths. The optical signal of individual cells sequentially passing the sensor are compared with previously defined criteria for target and nontarget cells, and the fluid stream containing the cell suspension is split to direct different portions of the stream into different collection containers [1].

Different technologies are used to split the stream. Droplet sorters are the most widely used flow sorting technology. The fluid stream is broken into droplets – for example, by a vibrating nozzle. Some droplets contain cells, and droplets containing the desired target cells or unwanted nontarget cells can be identified through prior optical analysis and directed into collection containers.

Droplet frequencies of 2,000–100,000 per second can be achieved, limiting the sorting frequency to 50,000 (presort) cells per second.

Enclosed sorters are significantly slower ($<1000\ s^{-1}$) than droplet sorters at significant lower costs and can be realized by different technology: catcher tube sorters move a collection tube into the liquid in air stream when a target cells has been detected; fluidic-switching sorters actuate valves in a branched fluid path, switching between different paths the cells can pass; and destructive sorters destroy nontarget cells, for example, by an intense laser beam.

The unique property of flow sorting is that a combination of multiple optical parameters can be used to identify the cell subset of choice.

2.1.2 Surface Staining

Stem cells differ from other cell populations by specific proteins expressed at the cell surface (cell surface markers). Monoclonal antibodies can selectively bind to cell surface proteins, and fluorescent dyes conjugated to the antibody thus tag the cell of interest. CD34 and CD133 cell surface molecules are frequently used to identify and sort human hematopoietic stem cells. The presence or absence of additional markers can be used to further define the target cell population (e.g., CD38-negative; see Sect. 4 for details).

2.1.3 Side Population Sorting

Stem cells are frequently described as being a "side population", which is by definition a rare cell population distinguished from most other cells by specific characteristics. Stem cells differ from nonstem cells in their ability to transport Hoechst stains (Hoechst 33342) out of the cell. Hoechst 33342 is a DNA-binding fluorescent dye, excitable by ultraviolet light at 350 nm and emitting at 461 nm. A multidrug-like transporter in stem cells causes an increased efflux of Hoechst 33342 by an active biological process. Figure 1 shows a typical flow cytometric characterization of side population cells.

2.1.4 Aldefluor®

Stem and progenitor cells possess a different aldehyde dehydrogenase (ALDH) activity compared to nonstem cells. This enzyme converts a nonfluorescent substrate (an aminoacetaldehyde) into a fluorescent product (an aminoacetate) that is retained within living cells with an intact membrane. Cells with different ALDH enzyme activity can thus be differentially stained with the fluorescent product, and stem cells can be isolated based on their enzyme activity [2, 3]

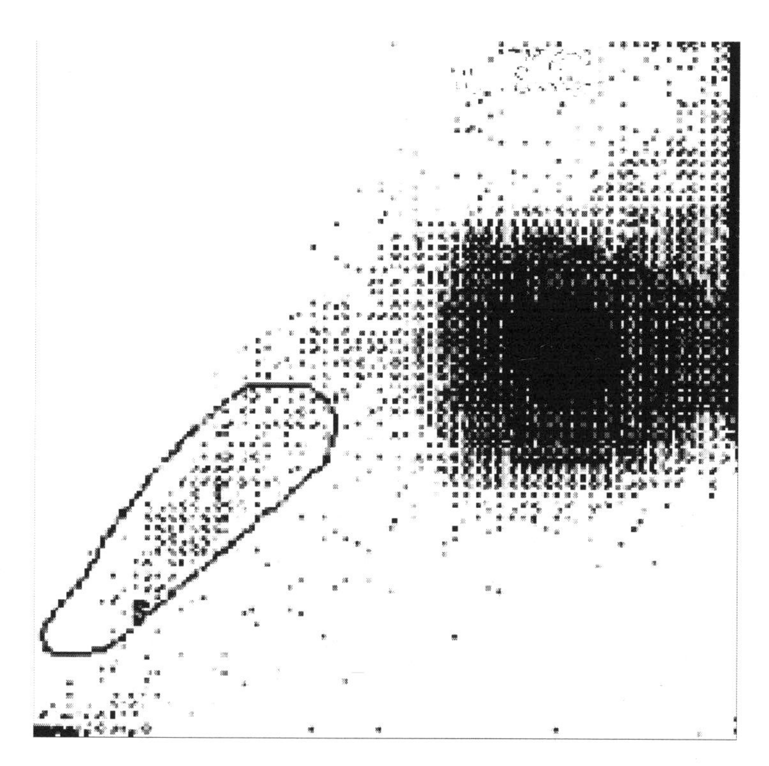

Fig. 1 Side population cells (*encircled*) are characterized by a low staining intensity for Hoechst dyes, and represent a low proportion of measured cells

2.2 *Magnetic Cell Sorting*

2.2.1 Introduction

Magnetic cell sorting has become a standard method for cell separation in many different fields. Numerous publications have demonstrated its use, from lab bench to the clinic; small to large scale; from abundant cells to rare cells with complex phenotypes; from human and mouse cells to many other species. Isolation of almost any cell type is possible from complex cell mixtures, such as peripheral blood, hematopoietic tissue (spleen, lymph nodes, thymus, bone marrow), nonhematopoietic tissue (e.g., solid tumors, epidermis, dermis, liver, thyroid gland, muscle, connective tissue) or cultured cells [4–11]. There are various magnetic cell separation systems currently available. They differ principally in two features: the composition and size of the magnetic particles used for cell labeling [5, 12] and the mode (i.e., "positive isolation"/ enrichment or "negative isolation"/depletion) of magnetic separation.

2.2.2 Technology

The MACS® System is characterized by the use of nano-sized superparamagnetic particles (approx. 50 nm in diameter), unique separation columns, and MACS Separators providing the required strong magnetic field [5, 8, 10].

Magnetic cell separation using MACS Technology is performed in three steps as outlined in Fig. 2. The entire procedure can be performed in less than 30 min, and both cell fractions, magnetically labeled and untouched cells, are immediately ready for further use, such as flow cytometry, molecular analysis, cell culture, transfer into animals, or clinical cell therapy applications.

Dynabeads® represent an example of larger, e.g., cell-sized, magnetic beads, to be used in a tube-based system. They are super-paramagnetic and are made from a synthetic polymer [13, 14]. The starting sample is incubated with the beads, and the test tube is then placed in the field of a strong permanent magnet. Complexes of cells and beads are attracted to the wall of the tube, and the supernatant can thus be removed.

Both cell fractions can be used – bead-captured cells and untouched cells. Should captured cells be subjected to functional studies, the beads need be removed [15], e.g., by enzymatic cleavage or binding competition with affinity molecules (peptides, antibodies, biotin) disrupting the binding of antibodies to the target molecules.

2.2.3 Magnetic Separation Strategies

Magnetic cell separation is a very simple but flexible technique, with two basic strategies ("modes"): positive selection or negative selection ("depletion"). The optimal separation strategy depends on the abundance of target cells in the cell sample, their phenotype compared with other cells in the sample, the availability of reagents, and a full consideration of how the target cells are to be used, including any restrictions with respect to purity, yield, and activation status.

Positive selection means that the desired target cells are magnetically labeled and isolated directly, representing the positive cell fraction (see Fig. 2). It is the most direct and specific way to isolate the target cells from a heterogenous cell suspension and requires a cell surface marker specific for the target cells. Positive selection is particularly well suited for the isolation of rare cells, such as hematopoietic stem cells, from complex cell mixtures, such as blood cells (for an example see Fig. 3).

Both fractions – labeled and unlabeled – can be recovered and used. Due to their composition of iron oxide and polysaccharide, MicroBeads are biodegradable and typically degrade and disappear rapidly when the cells are cultured. MicroBeads attached to receptors that are internalized and recycled to the cell surface may even be degraded much faster.

Depending on the cell type, on the target surface molecules used for magnetic labeling, and on the labeling moiety of the MicroBeads (mAb or ligand), the functional status of the cells can be influenced. This is inherent to labeling with Ab or ligands that recognize and crosslink cell surface receptors and thus may induce or suppress signal transduction. Labeling with antibody-conjugated MicroBeads has no additive effect compared to labeling with an unconjugated crosslinking Ab.

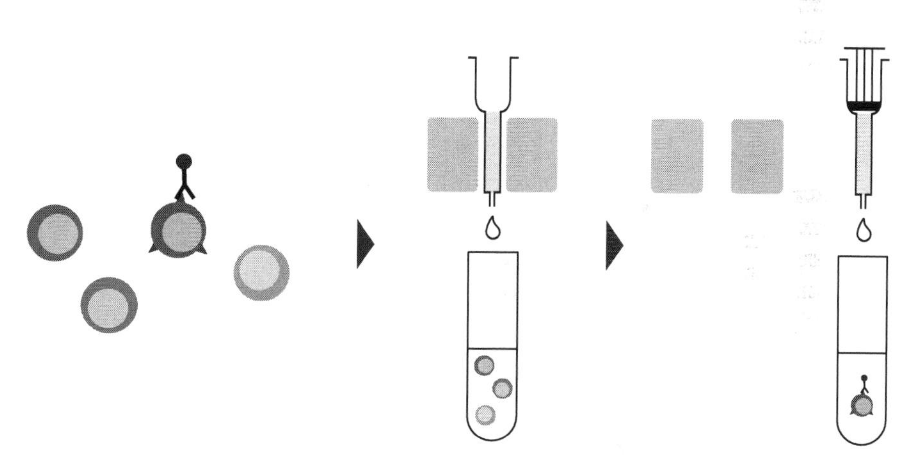

Fig. 2 Principle of high-gradient magnetic cell sorting. The procedure comprises three steps. Magnetic labeling (*left*): The cell preparation and labeling methods are similar to those used in flow cytometry. Individual cells of a cell suspension are immunomagnetically labeled using MACS MicroBeads, which typically are covalently conjugated to a monoclonal antibody (mAb) or to a ligand specific for a certain cell type. Magnetic separation (*middle*): The cell suspension is passed through the separation column that contains a ferromagnetic matrix and is placed in a MACS Separator. The separator contains a strong permanent magnet creating a high-gradient magnetic field in the magnetizable column matrix. Labeled target cells are retained in the column via magnetic force, whereas unlabeled cells flow through. By simply rinsing the column with buffer, the entire untouched cell fraction can be eluted. Elution of the labeled cell fraction (*right*): After removing the column from the magnetic field of the MACS Separator, the retained labeled cells can easily be eluted with buffer

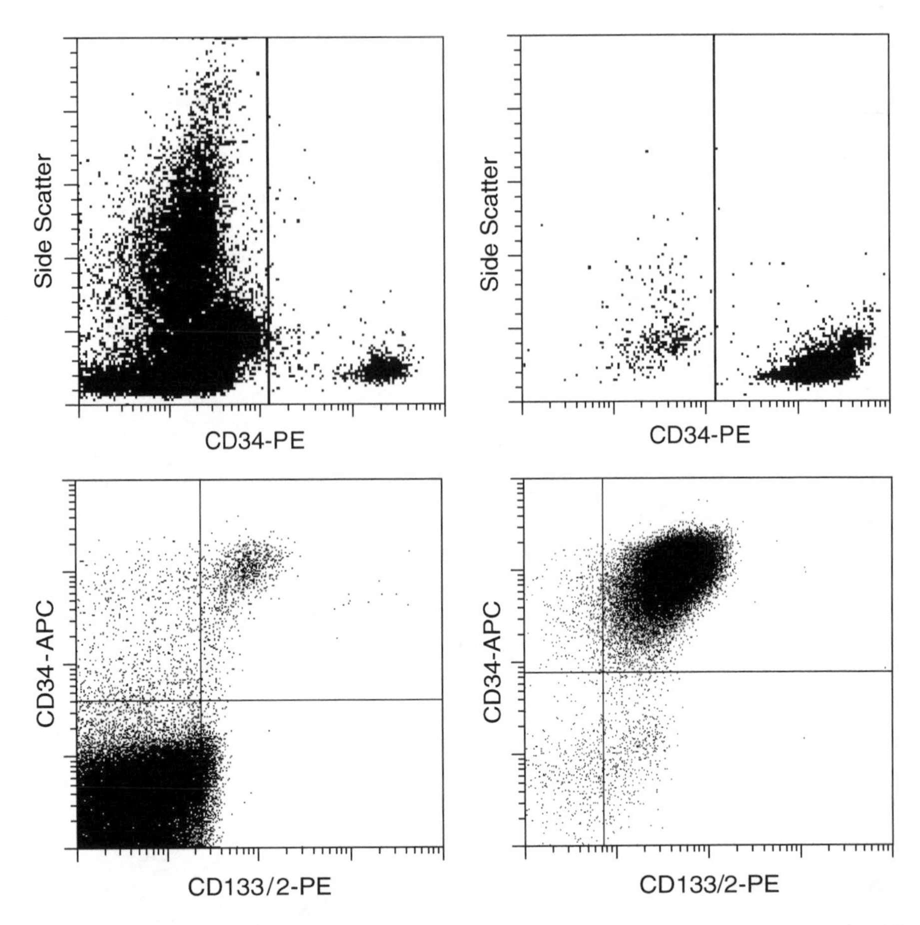

Fig. 3 FACS diagrams showing isolation of stem cells by positive selection with CD34 and CD133 directly conjugated antibodies and the CliniMACS Plus Instrument. CD34 cells were enriched from mobilized leukapheresis product (*upper row*), CD133 cells from bone marrow aspirate (*lower row*). Diagrams show the cellular composition before (*left column*) and after (*right column*) enrichment. Mononuclear cells from peripheral blood (PBMC), cord blood, bone marrow, fetal liver or leukapheresis harvest are obtained by density gradient centrifugation using Ficoll Paque®. For CliniMACS separation, hematopoietic stem and progenitor cells are directly magnetically labeled using MACS MicroBeads specific for CD34 and CD133, respectively. After enrichment, 99.2 or 96.7% pure stem cell fractions are obtained starting from frequencies of 0.92 and 3.1%

In summary, positive selection should be considered for (1) excellent purity, especially for enrichment of rare cells, (2) excellent recovery, and (3) fast procedures.

Depletion or negative isolation, on the other hand, means that the unwanted cells are magnetically labeled to eliminate them from the cell mixture, whereas the nonmagnetic, untouched fraction contains the cells of interest (Fig. 4). Potential effects on the functional status of cells can thus be minimized. A single depletion procedure can remove up to 99.99% of the magnetically labeled cells, leaving a highly pure fraction of unlabeled cells.

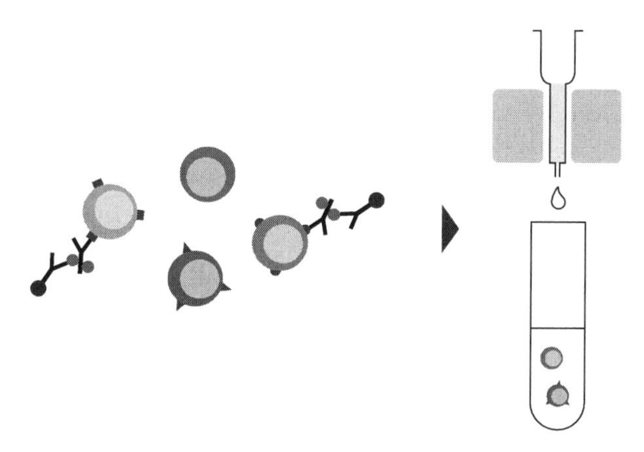

Fig. 4 Depletion strategy. The unwanted cells are labeled with immunomagnetic MicroBeads or a cocktail thereof and applied to the column. Labeled cells are eliminated on the column, and the untouched fraction with the cells of interest is collected in the flow-through

In particular, this strategy may be advantageous if functional studies have to be performed with the target cells, such as T cell activation studies or gene expression profiling. If the desired target cells are heterogeneous or do not have a well-defined phenotype, removing well-characterized cells by depletion is an efficient way to isolate the target cell population. Commonly used examples of depletion approaches include the depletion of cancer cells from autologous stem cell grafts and the depletion of T cells and B cells from allogeneic stem cell grafts.

In summary, a depletion strategy should be considered (1) for the removal of unwanted cells, (2) if no specific antibody is available for target cells, (3) if binding of antibody to target cells is not desired, and (4) for the subsequent isolation of a cell subset by means of positive selection (see below).

Multiparameter magnetic cell sorting is the strategy for isolating target cells that cannot be defined by a single cell surface marker, but by multiple cell surface antigens. Using only magnetic separation, sequential isolation of even complex targets cells can be achieved, combining both depletion and positive selection steps. There are several different routes for multiparameter magnetic sorting.

Commonly, a first step is "debulking" of the start population by using a panel of reagents directed against multiple cell surface antigens to deplete for several markers simultaneously.

Second, depletion may be followed by positive selection. The nonretained cells from the first separation are again magnetically labeled and enriched on a second column. In order to obtain highest purity, different stringencies may be used for the two separations. The depletion step can be performed on a steel-wool column with the highest retention rates for labeled cells and the enrichment step is performed on an iron-sphere matrix column with the lowest unspecific retention rates for unlabeled cells. This reduces the probability that labeled cells will be carried over from the first separation step into the second.

A third option is sequential positive selection. This can be accomplished by using colloidal superparamagnetic particles, which can be rapidly released from the cell (MultiSort MicroBeads) using an enzyme. Since the specificity of the enzyme is unique to the magnetic particles, cell surface molecules are not modified. MultiSort MicroBeads are typically used for a first positive selection. After this first step, release of the MultiSort MicroBeads takes less than 10 min. The cells are then ready for further labeling and another separation cycle.

The concept of positive selection followed by depletion is very attractive for the depletion of contaminating tumor cells or alloreactive T cells from purified $CD34^+$ hematopoietic progenitor cells for therapeutic autologous or allogeneic stem cell grafting. This concept requires either the combination of MultiSort MicroBeads and MicroBeads or the use of MicroBeads followed by larger magnetic beads.

2.2.4 Magnetic Labeling Strategies and Reagents

Direct labeling is the fastest way of magnetic labeling. Only one labeling step is required if a monoclonal antibody specific for a certain cell surface antigen can be directly coupled to the MicroBeads (Fig. 5, left).

Direct labeling minimizes the number of washing steps and thereby prevents cell loss. For many human, mouse, rat, and nonhuman primate cell surface markers, antibody-conjugated MicroBeads are available as one-step reagents.

Indirect labeling (Fig. 5, right) is performed if no direct MicroBeads are available, if a panel of antibodies directed against multiple cell surface antigens is used, or if two-step magnetic labeling is significantly more efficient compared to one-step labeling, for example, with weakly expressed antigens or antibodies of low affinity.

Cells are labeled with a primary antibody that is unconjugated, biotinylated, or fluorochrome-conjugated. In a second step, three different indirect magnetic labeling methods can be used:

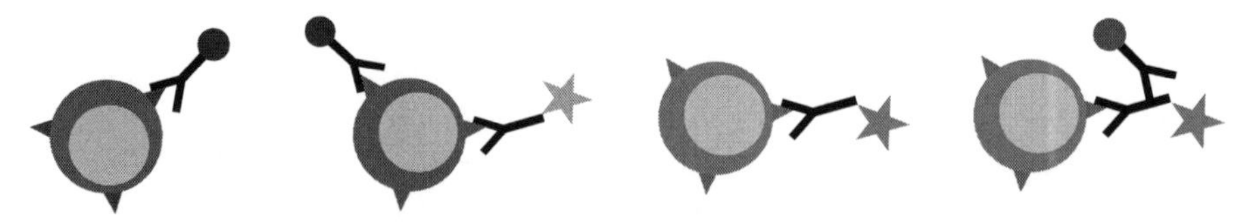

Fig. 5 Principles of magnetic labeling with superparamagnetic MACS MicroBeads. Direct labeling (*left*): One-step magnetic labeling, where a cell surface-antigen specific mAb is directly conjugated to the MicroBeads. Indirect labeling (anti Ig) (*right*): Two-step magnetic labeling with a primary cell surface-antigen specific Ab and anti immunoglobulin Ab-conjugated MicroBeads. Like any staining reagent, each magnetic bead reagent must be titrated for optimal cell separation, using different concentrations of MicroBeads for one otherwise standardized separation and determining the concentration with the best performance with respect to purity and yield of the cells of interest

1. MicroBeads conjugated with antiimmunoglobulin antibody to detect unlabeled primary antibody
2. MicroBeads conjugated with streptavidin or antibiotin antibody to detect biotinylated primary antibody
3. MicroBeads conjugated with antifluorochrome antibody (e.g., antiFITC) to detect fluorochrome-labeled primary antibody

A cocktail of antibodies can also be used for isolating or depleting a number of cell types concurrently. This amplifies the magnetic labeling and thus, indirect labeling may be the method of choice if dimly expressed markers are targeted for magnetic separation.

2.2.5 Superparamagnetic MicroBeads

MACS MicroBeads are superparamagnetic particles made of an iron oxide core and a dextran coating. They are nano-sized, ranging between 20 and 150 nm in diameter (see Fig. 6), and form colloidal solutions, i.e., they remain dispersed [5, 8]. Superparamagnetism means that in a magnetic field the iron oxide cores magnetize strongly like ferromagnetic material, but when removed from the magnetic field the particles do not retain any residual magnetism. The dextran coating of the MicroBeads permits chemical conjugation of biomolecules. Numerous highly specific mAb, fluorochromes, oligonucleotides and various other moieties have all been covalently linked to MicroBeads, thereby transferring additional biochemical and physical properties to them [5, 6].

The nano-sized iron-dextran particles confer several unique features on MACS Technology. MACS MicroBeads are biodegradable and do not alter cell function. Effects on the functional status of cells by magnetic labeling with MicroBeads are primarily dependent on the target cell surface antigen and on the degree of cross-linking by mAb or ligands conjugated to the MicroBeads, but not on the MicroBeads themselves. Cells labeled with MicroBeads have been used for numerous functional in vitro assays, experimental transfers into animals, and therapeutic transplantations in humans.

2.2.6 Column Technology and Separators

MACS MicroBeads are extremely small, and the amount of magnetizable material bound to cells is very low. Specific devices are required to generate a high-gradient magnetic field powerful enough to retain the labeled cells. MACS Technology uses high gradient magnetic cell separation units consisting of a strong permanent magnet of 0.4–1 Tesla and a separation column with a matrix of iron spheres.

When the columns are placed between the poles of the magnet of a MACS Separator, high magnetic gradients up to some 10^4 T/m are generated in the vicinity of the ferromagnetic matrix. The magnetic force is then sufficient to retain the target cells labeled with a very small number of MicroBeads. Once the column is removed

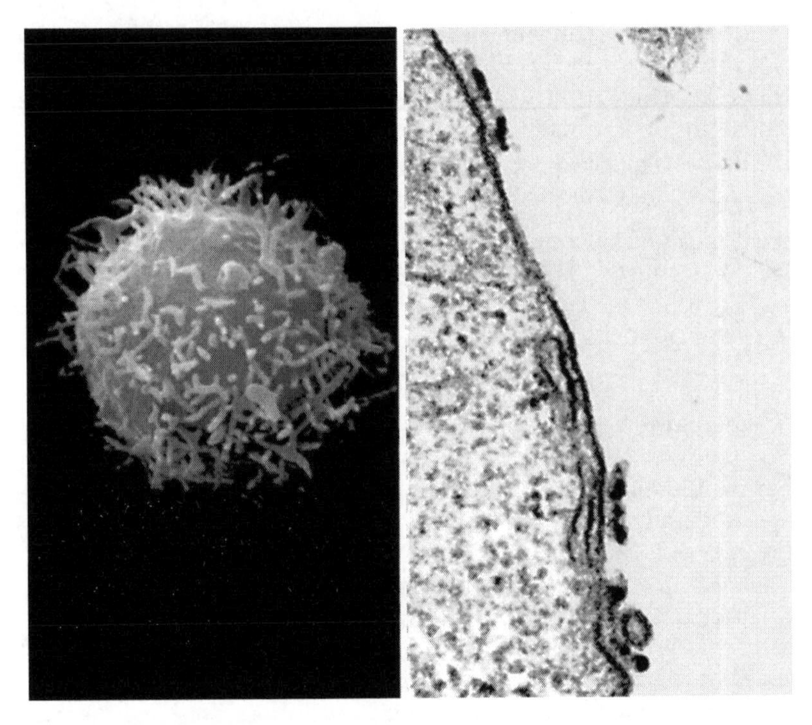

Fig. 6 Scanning (*left*) and transmission (*right*) electron micrograph of a CD8$^+$ T cell. The cell was isolated with MACS Technology using CD8 Ab-conjugated superparamagnetic MicroBeads (EM courtesy of Prof. Groscurth, Zürich, Switzerland). Some superparamagnetic MicroBeads attached to the membrane are visible on the micrograph image. They are about 50 nm in diameter, form colloidal solutions, and are biodegradable. Their small size enables high kinetics of the MicroBead-cell reaction and minimizes unspecific binding. Thus, cell enrichment of more than a 10,000-fold is possible from frequencies below 10^{-8}

from the magnet, the column matrix rapidly demagnetizes, and retained cells can be easily and completely eluted simply by rinsing the column with buffer.

MACS Columns for research use are available in various sizes (Fig. 7) for fast (5–30 min) processing of different amounts of cells. Up to 2×10^{10} cells, containing up to 10^9 target cells can be routinely handled. This is in striking contrast to fluorescence-activated cell sorting (FACS, see Sect. 2.1.1), where cells are sorted one after the other, limiting the sorting speed to about 50,000 cells per second, that is, 10^8 cells in 33 min, or a leukapheresis pack with 10^{10} cells in 56 h.

With the autoMACS and autoMACS Pro Separators, column-based magnetic cell separation can also be automated in order to standardize frequent cell separations.

2.2.7 Clinical-scale Cell Separation

Magnetic cell separation technologies have provided novel tools to use specified cell populations for treatment of patients. Desired effects such as reconstitution of

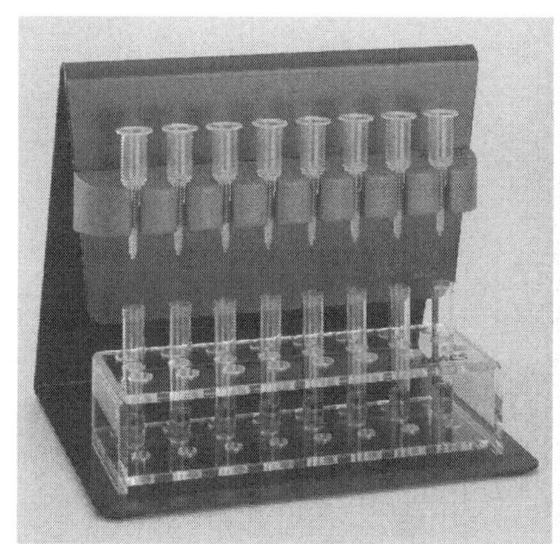

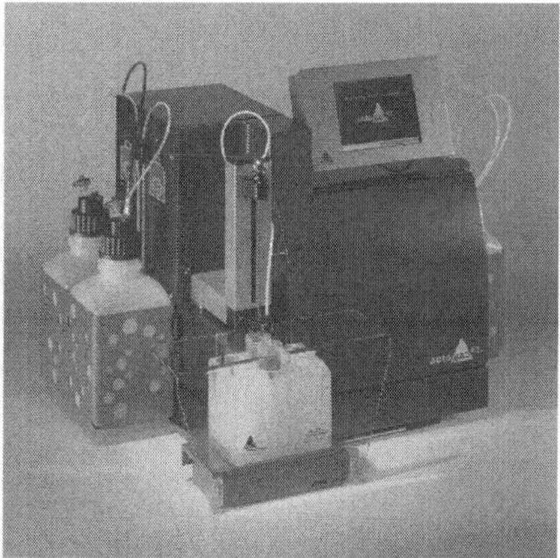

Fig. 7 Hardware and instruments. A variety of different MACS Separators and Columns is available, each individually designed for specific applications. The *OctoMACS* Separator (*left*), for example, is a device for separations of up to 10^8 labeled cells and up to 2×10^9 total cells in combination with LS columns. The *autoMACS Pro Separator* (*right*) is an automated benchtop magnetic cell sorter for high cell numbers or multiple samples. It is capable of sorting up to 10 million cells per second from samples of up to 4×10^9 cells

the immune system can be utilized while sparing unwanted effects of nontarget cells such as immune reactions vs patient tissue [16]. Two devices for isolation of stem cells by magnetic cell separation technologies are available. They differ in the size of magnetic particles used (see Sect. 2.2.2).

CliniMACS® Plus Instrument

The CliniMACS Plus Instrument is an automated cell separation device based on MACS Technology. It enables the operator to perform large-scale magnetic cell separation in a closed and sterile system (Fig. 8).

The use of clinical-grade isolation or depletion of cells has grown dramatically over the past few years, and is now a standard technique established in many cellular therapy centers. The CliniMACS Plus Instrument is a flexible system for separating cells labeled with clinical-grade MicroBeads. Cells are processed and labeled in a closed bag system using standard clean-room techniques. The processed cells are then attached to a tubing set and processed using the preset programs of the CliniMACS Plus Instrument. Target cells are recovered in a transfer pack or cell culture bag ready for downstream processing, again using a closed system.

Stem cells isolated with the CliniMACS Plus Instrument are used for stem cell grafts ("graft engineering") to reconstitute the immune system in the context of tumor therapies (chemotherapy, whole body irradiation) and for regeneration of patient tissues (regenerative medicine, tissue engineering; see Chap. 5 for details). Graft engineering procedures can be performed by both positive isolation (CD34 or CD133 enrichment) and negative isolation (CD3/CD19 depletion); see Sect. 2.2.3.

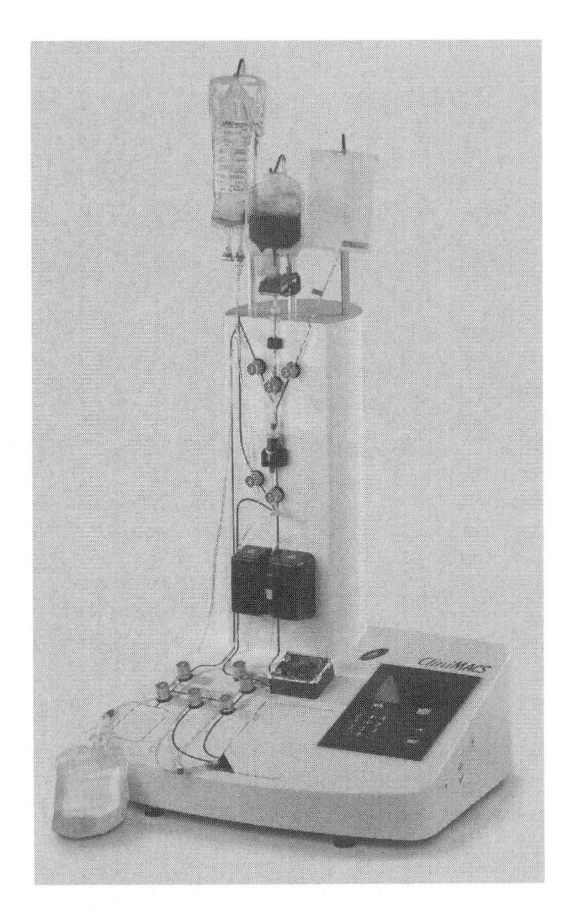

Fig. 8 CliniMACS® Plus Instrument. The CliniMACS System is an automated cell separation system for clinical-scale magnetic enrichment of target cells or depletion of unwanted cells in a closed and sterile system. For separation, a single-use tubing set, including a separation column, is attached to the CliniMACS Plus Instrument. Then the cell preparation bag containing the labeled cells is connected to the tubing set. After starting the separation program, the system automatically applies the cell sample to the separation column, performs a series of washing steps, and finally elutes the purified target cells. The CliniMACS® System components (Reagents, Tubing Sets, Instruments and PBS/EDTA Buffer) are manufactured and controlled under an ISO 13485 certified quality system. In Europe, the CliniMACS System components are available as CE-marked medical devices. In the USA, the CliniMACS System components including the CliniMACS Reagents are available for use only under an approved Investigational New Drug (IND) application or Investigational Device Exemption (IDE). CliniMACS® MicroBeads are for research use only and not for use in humans

Isolex® 300i

Baxter Healthcare Corporation has adapted Dynabead-based stem cell isolation to an automated process in a clinical scale. The Isolex 300i Magnetic Cell Selection System allows for separation of CD34-positive cells. Magnetic beads are removed from the isolated stem cells using a competing peptide [15, 17, 18].

2.2.8 Evaluation of Separation Performance

Different technologies for isolation and enrichment of stem cells are available, and thus, cell separation performance parameters are useful to compare those methods.

The most evident performance parameter for isolation of stem cells is the purity of target cells, i.e., the frequency of stem cells within a given processed target cell population:

$$\text{purity} = \frac{\#\,\text{stem cells}}{\#\,\text{all cells}} 100\%.$$

Nevertheless, purity of stem cells alone is not a sufficient performance parameter, as one always needs a specific number of stem cells for either basic research or clinical applications. Thus recovery of almost all of the stem cells contained in the initial cell product is desirable:

$$\text{yield} = \frac{\#\,\text{stem cells\_in\_processed\_sample}}{\#\,\text{stem cells\_in\_unprocessed\_sample}} 100\%.$$

It is obvious that 100% purity of target cells with 100% yield during processing would be optimal, at best combined with a low processing time. In practice, and for a given technology, optimizing one parameter can only be done at the expense of another, moving a coordinate within the area of a triangle (see Fig. 9).

Purity and yield characterize the processed cell product. Both may significantly depend on the input product, e.g., abundance of target cells before processing. Additional parameters have been defined that characterize a relative separation performance:

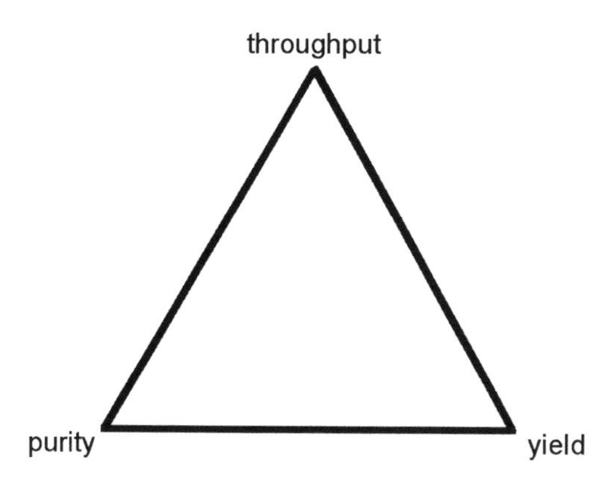

Fig. 9 The separation triangle. Each point within the triangle represents a possible parameter set in three dimensions (purity, yield, throughput). For a given process performance, parameters can only be optimized by compromises in other parameters, i.e., 100% purity and 100% yield of target cells cannot be combined with maximum throughput

$$\text{Enrichmentrate } f_E = \frac{\%\text{pos}_{\text{pos}} / \%\text{neg}_{\text{pos}}}{\%\text{pos}_{\text{ori}} / \%\text{neg}_{\text{ori}}}$$

$$\text{Depletion rate } f_D = \frac{\%pos_{\text{ori}} / \%neg_{\text{ori}}}{\%pos_{\text{neg}} / \%neg_{\text{neg}}}.$$

In these equations, %pos means the frequency of cells "positive" for a specific marker, e.g., CD34, and %neg means the frequency of cells "negative" for the same marker (100%–%pos).

Stem and progenitor cells are usually very rare in cell samples being used for isolation. Thus, high enrichment rates are required to obtain optimal purity. For a given technology the final purity will depend on the input frequency of target cells (see Fig. 10). Using MACS Technology, enrichment rates of up to 5,000 can be achieved.

Typical depletion rates are 5–200, and for a positive isolation strategy they assess how many labeled target cells are lost into the flow-through fraction. For a negative isolation strategy the depletion rates measure how effective labeled nontarget cells are removed from the sample.

Both enrichment rate and depletion rate use frequencies of cell populations for calculation and do not take into account possible bulk cell loss during processing. Graft engineering procedures thus typically use different parameters for evaluation of separation performance, based on absolute cell numbers. The probability P defines the fraction of nontarget cells (e.g., CD34-negative cells) that are still contained in the final cell product:

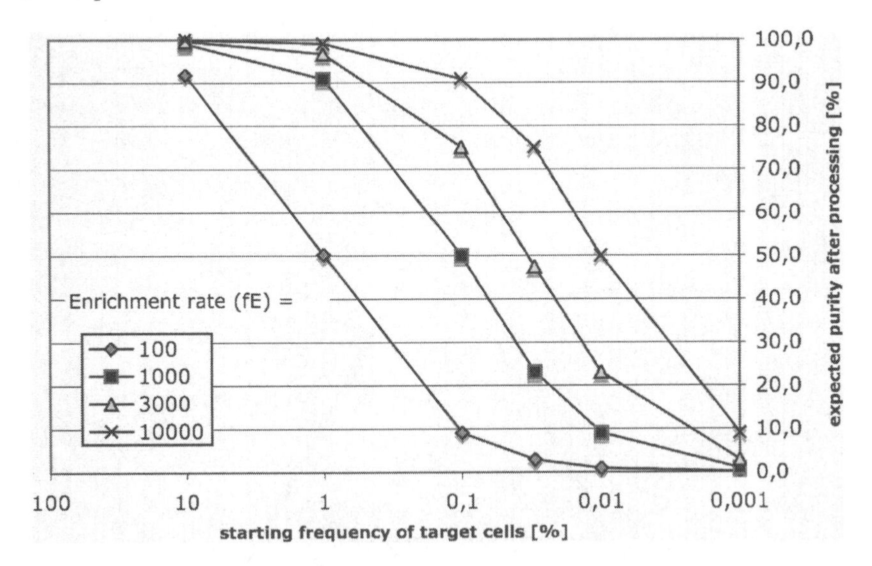

Fig. 10 Dependence of purity on enrichment rate and starting frequency. The final purity of a cell separation procedure depends on the frequency of target in the unprocessed sample and on the enrichment rate of the respective separation technology used. With MACS Technology, enrichment rates of up to 5,000 can be achieved. In conclusion, high purity is only achievable with a high enrichment rate and moderate starting cell frequency

$$P = \frac{\#\,\text{neg}_{\text{pos}}}{\#\,\text{neg}_{\text{ori}}},$$

where “#neg” is the number of negative (i.e., nontarget) cells. A typical probability of a CliniMACS separation procedure using CD34 as a target molecule to carry over nontarget cells to the final cell product is below 0.4×10^{-4}, i.e., > 99.96% of CD34-negative cells are removed.

P is usually very small for high performance cell separation systems. Therefore, the logarithmic scale is used:

$$-\log P = -\log 10 \frac{\#\,\text{neg}_{\text{pos}}}{\#\,\text{neg}_{\text{ori}}}.$$

CliniMACS CD34 procedures typically achieve *a* >3.5 log depletion of CD34-negative cells.

When stem cell isolation is used clinically for graft engineering of hematopoietic stem cell grafts for allogeneic transplantation, the removal of T cells is of utmost importance for patient safety. T cells in the graft may cause life-threatening immune reactions versus patient tissue (graft vs host disease, GVHD). Therefore, graft engineering performance is frequently characterized by the efficiency of T cell depletion rather depletion of all CD34-negative nontarget cells.

When a stem cell isolation system, such as the CliniMACS Plus Instrument, is characterized with regard to nontarget cell carry-over (e.g., –log P of 3.5), the stem cell purity of the final product mainly depends on the starting frequency, and Fig. 11 may be used to predict stem cell purity for samples with different stem cell content.

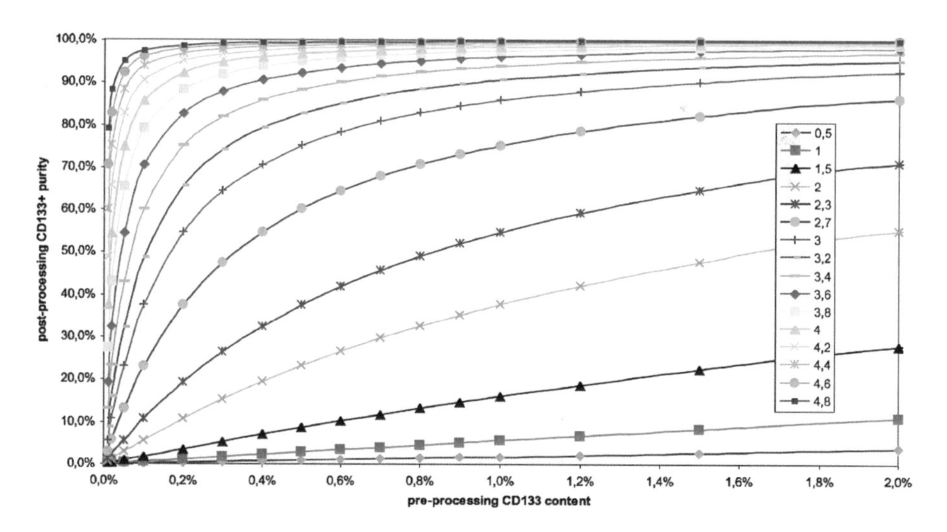

Fig. 11 Dependence of purity on depletion efficiency of nontarget cells and preprocessing stem cell content. The final purity of a cell separation procedure depends on the frequency of target cells in the *pre-processing stem cell* sample and on the depletion efficiency of nontarget cells of the respective separation technology used. With MACS Technology, depletion efficiencies of up to 4.5 orders of magnitude can be achieved

3 Isolation and Enrichment of Embryonic Stem Cells

3.1 *Introduction*

Embryonic stem cells (ESC) are not continuously present in an organism but can be derived during a very limited period of time from the inner cell mass of blastocysts. The indefinite in vitro self renewal of mouse ESCs (mESC) and moreover their pluripotency, that is, the capacity to differentiate into every cell type in the body, was described for the first time more than 25 years ago [19, 20]. Later on, ESCs were derived from a number of different species and finally also from human preimplantation embryos [21]. Due to their unique properties, ESCs have been used in a variety of different fields: (1) in basic research to understand general cellular processes in, for example, embryogenesis, organogenesis, cancer, or ageing, and as a vehicle for the generation of transgenic mice for functional gene analysis and disease models; (2) in industrial research as cell-based screenings for drug target discovery, drug discovery, or predictive toxicology; (3) in clinical research as a potential source for tissue regeneration. The recently described derivation of induced pluripotent stem cells (iPS cells) from differentiated, postmitotic cells that behave similar to ESCs have sparkled the whole field even more [22]. With iPS cells combining the advantages of ESCs and autologous cell transplantation, the generation of patient specific derived stem cells for unrestricted tissue regeneration and without ethical issues can be envisaged.

The broad application of ESCs, but also the fact that they are kept in culture for a prolonged time, has led to a number of different protocols for their derivation, isolation, and enrichment at a pluripotent stage or after differentiation into a certain cell type.

The techniques involved for the isolation and enrichment are generally the same as described above, including selective culturing, immunopanning, flow cytometric sorting, or magnetic sorting.

3.2 *Selective Culturing of Embryonic Stem Cells*

For historical reasons, the most eminent protocol for enrichment of pluripotent mouse and human ESCs is based on selective culturing. A detailed description for the derivation of ESCs can also be found elsewhere in this book (see Itskovitz-Eldor). In brief, mouse ESCs are derived from embryonic day 3.5 blastocysts by letting them attach and expand on mitotically inactivated murine embryonic fibroblast layer in a medium containing leukemia inhibitory factor (LIF) [23]. Expanded blastocysts are repeatedly trypsinized and single clones derived. As the original protocol was quite inefficient, with a success rate of up to 30% and strong dependency on the mouse strain, many improvements have been introduced. Such improvements include use of specifically conditioned medium [24], genetically modified blastocysts [25], microdissection of the blastocyst [26], treatment with pharmacological drugs [27], and use of serum replacement (SR) [28].

Human ESCs have been derived using similar protocols as originally described for mouse ESCs. However, at least the first hESC lines had a higher tendency for spontaneous differentiation and a lower proliferation rate which made the handling much more difficult – up to the point that individual colonies need to be selected by a micropipette according to their undifferentiated morphology and then mechanically dissociated into clumps in order to proliferate them at an undifferentiated stage [21].

Despite almost 10 years of research, currently available hESCs show heterogeneous phenotypes, and the consistency of culture is still a challenge for many labs. No generally applicable culture protocol has evolved [29]. Different laboratories culture hESCs either feeder-free ("matrix culture") [30], with mouse embryonic fibroblasts (MEF) [21] or with different kinds of human fibroblasts (HEF) as feeder cells [30]. Also, the propagation of hESCs is either done by mechanical ("cut and paste") or enzymatical dissociation of cell colonies using serum-containing or serum-free/xeno-free media. The main difficulties still arise from the observation that singularized hESCs tend to differentiate spontaneously if culturing conditions are not tightly controlled.

To address these problems, a study (ISCI II) has been started which is coordinated by the International Stem Cell Forum (http://www.stemcellforum.org/) and follow the ISCI I ring study which originally aimed to characterize 59 human embryonic stem cell lines [31]. The ISCI II study is carried out in four reference laboratories and seeks to clarify if certain media are able to support pluripotent growth of hESC for 40 passages (1 year) while maintaining a stable karyotype.

3.3 Isolation and Enrichment of ESCs Based on Surface Markers

As one way to standardize culturing of ESCs and to synchronize undifferentiated but also differentiated ESCs, populations can be envisaged by using cell sorting techniques which are based on the expression of stage-specific surface markers.

With regard to sorting of pluripotent embryonic stem cells, different monoclonal antibodies reacting with surface markers of undifferentiated (pluripotent) ESCs have been described. These markers differ partly between mouse and human ESCs. For mouse ESCs, these are mainly E-cadherin (CD324) and SSEA-1 (CD15).

For human ESCs, CD90, GCTM2, GCTM343, SSEA-3, SSEA4, CD9, TRA-1-60, TRA-1-81 and HLA A/B/C have been suggested [31]. The enrichment of pluripotent ESCs has been used for different purposes. For example, a synchronization of mESC cultures by sorting with SSEA-1 (CD15) MicroBeads has been described by Cui et al. [32].

In another report, immunomagnetic sorting has been used to separate pluripotent mESC from mouse embryonic feeder cells with a primary SSEA-1 antibody. In a slightly different approach, Annexin V MicroBeads have been used to remove apoptotic cells from mESC during normal cultivation or differentiation [33].

Similarly, SSEA-3 has been used for flow separation of undifferentiated human ESC [34]. Based on SSEA-3 expression, the authors propose a cellular differentiation hierarchy for maintenance cultures of hESC. While SSEA-3^{+} cells represent pluripotent stem cells, normal SSEA-3^{-} cells have exited this compartment, but retained multilineage differentiation potential. However, adapted SSEA-3^{+} and SSEA-3^{-} cells cosegregate within the stem cell territory, implying that adaptation reflects an alteration in the balance between self-renewal and differentiation.

SSEA-3 and SSEA4 have not been classified as ultimate markers of pluripotency, due to their slow kinetics upon differentiation. Search for those markers is still ongoing and several groups claim to have identified such fast downregulated markers [35].

Besides the selection of pluripotent stem cells to ease and standardize the propagation of undifferentiated ESC, the capability of undifferentiated hESCs to form teratomas is a risk factor worth considering when applying hESC-derivatives to cellular therapy. Again, cell sorting techniques might help to enrich target cell types and to deplete unwanted cell types or undifferentiated hESCs. Lastly, the removal of residual pluripotent mESC from differentiated cells can also be used to purify ESC-derived cell populations. Hedlund et al., for example, have reported the selection of murine dopaminergic neurons by sorting of TH-EGFP positive and SSEA-1 negative cells before transplantation [36].

3.4 Sorting of Cell Types Derived from Embryonic Stem Cells

A number of protocols have been reported for the targeted differentiation of ESCs to progenitor or postmitotic cell types. By exposure of pluripotent ESCs to growth factors, such as basic fibroblast growth factor (bFGF), transforming growth factor beta1 (TGF-beta1), activin-A, bone morphogenic protein 4 (BMP-4), hepatocyte growth factor (HGF), epidermal growth factor (EGF), beta nerve growth factor (betaNGF), or retinoic acid, almost every somatic mouse and human cell type has been generated [37]. This includes neurons, glia, skin, muscle, bone, and many others [37, 38].

However, the characterization of the derived cell types is often limited to surface marker description which is obviously not an unambiguous proof for a given cell type. Also, most protocols do not direct the differentiation exclusively to one cell type, but to multiple routes of differentiation and a mixture of different stages of differentiation. This again makes it desirable to enrich specific cell types of interest or to deplete unwanted cell types. A great number of surface differentiation markers – essentially all those which are also used for the characterization or isolation of somatic cells – have been described for mouse or human ESC-derived progenitors or differentiated cell types, amongst others: A2B5, PSA-NCAM, CD56 (NCAM), O1, O4, CD309 (VEGFR-2/KDR/Flk-1), Sca-1, CD117 (c-kit), CD34, CD133 (Prominin), CXCR4, CD324 (E-cadherin). Enrichment of mESC-derived hematopoietic/endothelial (hemangio) precursor cells has been achieved by indirect immunomagnetic sorting [39] and in another report, direct labeling with Sca-1

MicroBead-conjugated antibodies has been used for mESC-derived vascular progenitors [40]. Recently, it was also shown that CD56-positive neural cells derived from hESCs can be sorted magnetically with good survival rates [41].

Notably, epitopes like CD324 (E-cadherin) might also be used as markers for particular differentiation stages. Considering this, a general feature of surface marker-based cell sorting becomes apparent. It does not essentially have to be a marker exclusively expressed on a certain cell type at a certain differentiation stage. A unique expression in relation to the other cell types present in a given organ or cell culture can be sufficient for cell sorting.

The success of efficient enrichment of undifferentiated cells depends – besides other factors – on the turnover rate of these markers, especially when differentiation of ESCs starts, on the number of marker protein per cell, and on the specificity and avidity of the monoclonal antibodies.

As already mentioned above, by using a negative sorting strategy, early markers of differentiation can also be used to enrich untouched undifferentiated cells by depletion protocols [36].

3.5 *Sorting Based on Genetically Modified Embryonic Stem Cells*

Despite the obvious advantages of marker-based cell sorting, so far, magnetic cell separation has just started to be used for the enrichment or depletion of pluripotent ESCs and ESC-derivatives. Cell separation by flow cytometry is already used more routinely, especially with the help of genetically modified mESCs which express EGFP under control of a given cell-type specific promoter [42, 43]. Interestingly, the approach of using genetically modified ESCs to enrich differentiated derivatives can also be used for magnetic cell sorting (Fig. 12). For example, David et al. [44] reported the labeling of stably transfected ES cells expressing a human CD4 molecule lacking its intracellular domain (DeltaCD4) under control of the phosphoglycerate kinase promoter for magnetic cell sorting. The membrane-bound protein allowed for immunomagnetic sorting with purities greater than 97%. The viability of selected cells was demonstrated by reaggregation and de novo formation of embryoid bodies developing all three germ layers.

It was concluded that expression of DeltaCD4 in differentiated ES cells can be used for a rapid high-yield purification of a desired cell type for tissue engineering and transplantation studies.

Combined selections of GFP-expressing and surface marker-positive cells have recently been described for the enrichment of mESC-derived cardiomyocyte precursors. Here, GFP-expression was controlled by promoters of mesodermal- or cardiomyocyte-specific transcription factors and coselection performed with antibodies against the surface markers CD309 (VEGFR-2/KDR/Flk-1) or CD117 (c-kit) [45–47].

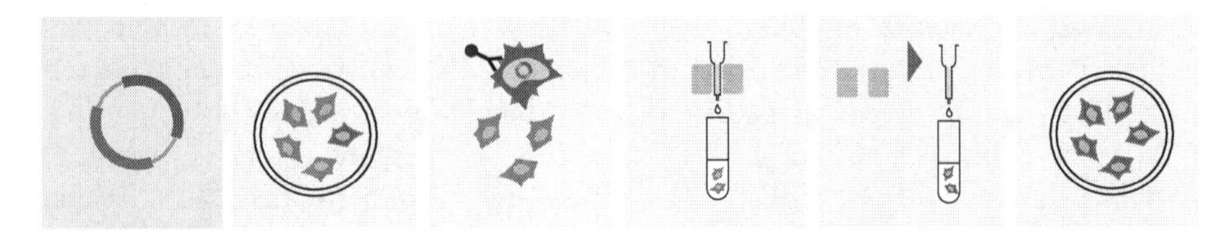

Fig. 12 Immunomagnetic enrichment of ESCs or derivatives thereof using genetically modified embryonic stem cells. ES cells are stably transfected with a vector carrying a certain cell type–specific promoter, which drives the expression of a vector-coded surface resident protein. This surface marker can then be used for immunomagnetic labeling and separation by MACS Technology

3.6 Concluding Remarks

In the past, isolation and enrichment of ESCs and their derivatives was mainly achieved by selective culturing. With the advent of genetically tagged ESCs, and supported by an increasing availability of antibodies reacting with cell surface markers expressed only on specific cell types or at certain differentiation stages, flow and magnetic cell sorting has become more popular. The surface marker-based sorting of cells offers great potential to optimize further the routine culturing but also the differentiation protocols of ESCs. Both the starting population as well as intermediate and postmitotically differentiated ESCs can be enriched to high purity. Especially magnetic cell sorting with its advantage of swift processing of high cell numbers, also in a closed setting, will help translate ESC research to clinical applications. The recently generated induced pluripotent stem cells (iPS cells), which essentially behave like ESCs and are thought to pave the way for autologous tissue regeneration approaches, will greatly profit from the knowhow currently generated with ESCs.

4 Adult Stem Cells

4.1 Stem Cells from the Hematopoietic System with Hematopoietic Differentiation Potential

For many applications that are under development for future clinical applications, mice are used as model organisms, facilitating the translation from basic in vitro research to the in vivo environment. Cell populations include cells with hematopoietic and nonhematopoietic differentiation potential, as well as pluripotent stem cells and differentiated progenitors.

Blood contains a complex mixture of cells, such as erythrocytes, the oxygen-transporting cells, the white blood cells comprising the cells of immune response, such as lymphocytes (T cells, B cells, dendritic cells, etc.) and macrophages, as well as the platelets that trigger blood clotting in case of tissue damage. Hematopoietic stem cells (HSCs) generate all these cells and can thus be considered as being multipotent and capable of regenerating the complex hematopoietic system. HSCs give rise to more specialized progenitor cells with more limited differentiation potential, which are the progenitors of red blood cells, platelets, and the two main categories of white blood cells, the lymphoid and the myeloid progenitors.

4.1.1 Phenotype and Isolation of Mouse Hematopoietic Stem Cells

Several marker combinations have been identified that describe murine HSCs, including negative or low expression of lineage commitment markers such as CD5, CD45R (B220), CD11b, Gr-1 (Ly-6G/C), 7–4, and Ter-119, and high expression of markers such as stem cell factor receptor CD117 (c-kit/SCFR) and Sca-1 [48, 49]. This cell population is then called KSL. Additional markers have been defined to be not or only weakly expressed on the KSL population, such as CD90.1 and CD34.

Another strategy for defining hematopoietic stem and progenitor cells is the use of SLAM markers. A specific set of these markers, the "slam code," is supposed to characterize hematopoietic stem cells and more committed progenitors for their potential [50]. SLAM cell surface markers delineate differentiation steps in early hematopoiesis. Originating with multipotent hematopoietic stem cells (HSCs), differentiation steps include multipotent progenitor cells (MPPs) and lineage-restricted progenitor cells (LRPs). Each is characterized by a different complement of SLAM markers: HSCs are $CD150^{+}$ $CD48^{-}$ $CD244^{-}$; MPPs are $CD150^{-}$ $CD48^{-}$ $CD244^{+}$; LRPs are $CD150^{-}$ $CD48^{+}$ $CD244^{+}$. It should be noted that CD48 is a ligand for CD244, and thus $CD150^{+}$ $CD48^{-}$ is sufficient to distinguish HSCs from MPPs and LRPs.

Other ways to define these cells apart from by surface marker expression is the use of fluorescent mitochondrial and DNA-binding dyes, such as rhodamine-123 and Hoechst 33342. Primitive hematopoietic cells are able to transport the dye outward, resulting in a $Hoechst^{low}$ phenotype.

Because of its characteristic flow cytometric profile, the $Hoechst^{low}$ stem cell population has been designated as the "side population" (SP) phenotype [2]. SP cells are lineage-negative, which means that negative preselection approaches can be used to deplete mature cells from the sample, thus reducing the flow cytometric sorting time required to isolate SP cells. The SP phenotype has been attributed to high expression of membrane transporters. Although several multidrug transporter molecules are expressed in primitive cells, one transporter molecule, ABCG2 (or BCRP1), has been shown to be necessary and sufficient to mediate the Hoechst dye efflux ability of SP cells. Since ABCG2 expression is highest in primitive cells and gets downregulated during differentiation, this molecule might also be a potentially useful marker to identify and isolate primitive HSCs.

Other approaches have been made to define and isolate better the population of long-term repopulating HSCs (LTR–HSCs) – the most primitive HSCs in mouse bone marrow. Chen and colleagues isolated a population of LTR–HSCs based on the expression of Sca-1 and CD105 in combination with Rhodamine 123 staining [51–53].

In addition to the hematopoietic potential described for stem cell populations KSL and SP, these populations also show a certain nonhematopoietic differentiation potential, although this issue is still controversial. Highly purified HSCs from mouse bone marrow have been reported to contribute to hematopoietic regeneration and also to hepatic regeneration with functional differentiation producing serum transaminases and bilirubin, as well as certain amino acids, such as phenylalanine [54]. Furthermore, these cells have been used to regenerate cardiac [55] and muscle [56] tissue and have been shown to contribute to neovascularization [57] as well as regeneration of the neural system [58]. However, the mechanism of their contribution has not yet been fully elucidated.

4.1.2 Phenotype and Isolation of Human Hematopoietic Stem Cells

Human CD34 was the first differentiation marker recognized on hematopoietic stem and progenitor cells from hematopoietic sources, such as fetal liver, cord blood, peripheral blood, and bone marrow. It is therefore the classical marker used to obtain enriched populations of human hematopoietic stem and progenitor cells (HSCs/HPCs) for research and clinical use. CD34 is expressed on approximately 1–3% of the nucleated cells in normal human bone marrow (BM) and on 0.1–0.5% of the nucleated cells in human peripheral blood. The majority of human cells capable of producing multilineage hematopoietic engraftment in myeloablated recipients express CD34. The engraftment potential of enriched populations of human $CD34^+$ cells has also been demonstrated clinically in numerous autologous and allogeneic transplantation trials (see Chap. 5). The $CD34^+$ subset also includes hematopoietic stem cells and more committed progenitor cells, such as lymphocyte progenitor cells, but is not expressed on the majority of terminally differentiated cells. Cytokine treatment and/or cytotoxic therapy increase the level of $CD34^+$ cells in the blood to more than 1%. $CD34^+$ cell mobilization regimens have become well-established methods to collect by leukapheresis sufficient amounts of HSCs for clinical transplantation (see Chap. 5 and references therein). Human $CD34^+$ cells can be isolated by FACS or by immunomagnetic methods using monoclonal antibodies against CD34 coupled to superparamagnetic MicroBeads.

For the immunomagnetic depletion of mature cells from stem cell-enriched fractions, cells expressing lineage commitment markers can be depleted in a single negative selection step by using combinations of lineage-specific antibodies, such as CD2, CD3, CD11b, CD14, CD15, CD16, CD19, CD56, CD123, and CD235a (Glycophorin A). Furthermore, CD34-enriched but CD38-depleted populations have been used to enrich for early hematopoietic progenitor cells [59, 60].

The usefulness of CD34 as a hematopoietic stem and progenitor cell marker for human cells is well established. There is evidence, however, of the existence of a very primitive population of $CD34^+$ cells with HSC and lymphopoietic potential in human cord blood and adult hematopoietic sources. Thus far, the phenotype of primitive CD34+ HSCs has been characterized by the concurrent absence of CD38, and the positive expression of CD133 [61]. CD133 has been described as a marker of more primitive hematopoietic stem and progenitor cells. It was originally found on HSCs and HPCs deriving from human fetal liver, bone marrow, and peripheral blood [62]. Phenotypical analysis of CD133-expressing cells (CD133+ cells) revealed a high expression on primitive hematopoietic and myeloid progenitor cells [63].

Functional studies showed that CD133 is lightly or not at all expressed on late progenitors, such as pre-B cells, CFU-E (colony forming units-erythrocytes), CFU-G (colony forming unit-granulocytes). Long-term culture–initiating cells (LTC–ICs), the most primitive human hematopoietic cells that can be assayed in vitro, are highly enriched among CD133+ cells [64, 65]. Thus, CD133+ cells in the hematopoietic system appear to be ancestral to CD34+ cells, especially as the latter can be generated in vitro from CD133+ CD34− cells [66]. Furthermore, CD133+ cells from cord blood display a higher proliferative activity [66, 67] and a more primitive gene expression profile [68] than CD34+ cells.

Thus far, the phenotype of CD34-negative HSCs has been characterized by the concurrent absence of CD38, lack of lineage-specific cell surface antigens, as well as by expression of CD133 [69]. In contrast, CD133− CD34+ cells were shown to mostly consist of B cell progenitors, late erythroid progenitors [61], and other more committed hematopoietic progenitors [64]. CD34, although well established, might therefore not be the best choice as a marker for the isolation of primitive human hematopoietic stem cells, due to its variable expression on late hematopoietic progenitors (see Fig. 13).

Enumeration of hematopoietic stem cells by phenotyping, although useful, does not always predict the abundance, viability, and hematopoietic potential of the cells that support hematopoiesis after transplantation, in particular after cryopreservation, expansion in culture or other ex vivo manipulations. Analysis of the functional properties of HSCs can be done by diverse in vivo and in vitro assays, e.g., repopulation assays in mouse, by which the transplanted cell (population) is tested for its ability to regenerate the complete hematopoietic system. In vitro assays are commonly used to investigate the differentiation potential of HSCs and their progenitors in the myeloid lineage, e.g., by the HSC–CFU assay.

In addition to the hematopoietic potential of stem cells isolated from hematopoietic sources, such as bone marrow, cord or peripheral blood, a nonhematopoietic differentiation potential has been described for this population. Therefore, these cells are also of great interest for tissue engineering and regenerative research applications.

4.2 Stem Cells from the Hematopoietic System with Nonhematopoietic Differentiation Potential

Ongoing investigations have led to the proposal that HSCs, as well as other stem cells from the hematopoietic system (bone marrow, peripheral blood, cord blood), have the capacity to differentiate into a wide range of nonhematopoietic tissues.

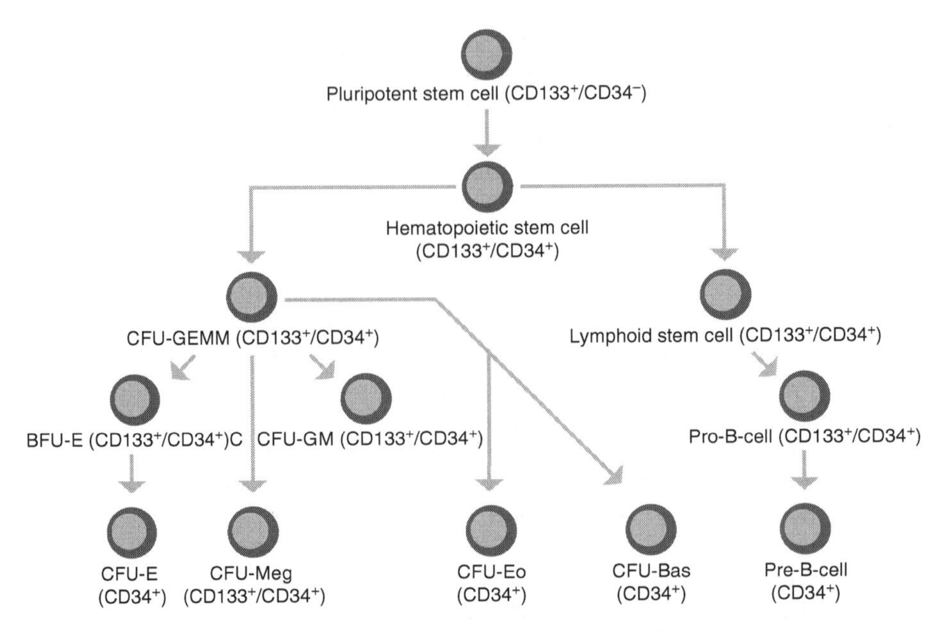

Fig. 13 In contrast to CD34, no expression of surface marker CD133 can be found on late progenitors, such as pre-B cells, colony forming unit erythrocytes (CFU-E), and colony forming unit granulocytes (CFU-G). CD133 and CD34 are coexpressed on early hematopoietic progenitors with multipotent differentiation potential, such as colonies consisting of granulocytes, erythrocytes, macrophages and megacaryocytes (CFU-GEMM), granulocytes and macrophages (CFU-GM), as well as the early burst forming unit erythrocytes (BFU-E)

One example is the hemangioblast, the common progenitor of HSCs and endothelial progenitor cells (EPCs), which can differentiate not only into blood cells but also into endothelial cells [70, 71].

4.3 *Vascular Tissue*

Vascularization of tissues is a major challenge of tissue engineering. In the last decade, a number of experimental data and clinical observations have suggested that bone marrow represents a reservoir of immature cells that permanently reconstitute the hematopoietic system and also participate in regeneration and repair of many peripheral tissues. These stem or progenitor cells are activated and mobilized to the blood stream by environmental stimuli for physiological and pathological tissue regeneration. Asahara first described the isolation of human progenitor cells from peripheral blood, their ability to differentiate to endothelial cells in vitro and to form new blood vessels and thus contribute to vascular repair. This cell population was termed endothelial progenitor cells (EPCs) [72] and has been defined by the expression of the markers CD34 and CD309 (VEGFR-2/KDR/Flk-1), as well as in combination with CD133 to distinguish between early and (matured) EPCs in human [73]. In mouse, the phenotype for EPCs is described as Lin^- $Sca\text{-}1^+$ $c\text{-}kit^+$ CD309 $(VEGFR\text{-}2/KDR/Flk\text{-}1)^+$ [74].

Considering the importance of blood vessel development for organogenesis, vasculogenesis by EPCs may be an essential cascade for tissue and organ regeneration following pathological damage in various critical diseases [75].

Regeneration of vascular tissue is also an important topic in therapeutic research, especially for the potential treatment of atherosclerosis and the revascularization of ischemic tissues, for example, in the heart or peripheral vascular disease. Due to the role EPCs play in postnatal neoangiogenesis and neovascularization, they have come into focus for tissue engineering applications and for the potential treatment of ischemic or injured tissue [74, 76], as well as after myocardial infarction [77, 78].

In mice, it has been shown in serial studies that EPCs can be mobilized from bone marrow in response to endogeneous and exogeneous stimuli and can therefore be isolated from populations of $Sca\text{-}1^+$ cells from mouse blood and can "home" and incorporate into foci of neovascularization [57]. Bone marrow-derived $Sca\text{-}1^+$ CD309 $(VEGFR\text{-}2/KDR/Flk\text{-}1)^+$ progenitor cells isolated from mouse peripheral blood showed the potential to differentiate into endothelial and epithelial cells in vivo after induced lung injury [79]. EPCs from mouse bone marrow have been enriched by their expression of CD117 (c-kit, SCFR) and play a key role in therapeutic angiogenesis. After transplantation of $CD117^+$ cells into the ischemic hindlimbs of mice, the cells survived and were incorporated in microvessels within 14 days in contrast to the $CD117^-$ cells [80].

In humans, $CD133^+$ cells isolated from bone marrow [81], cord blood [76, 82], mobilized [71, 83] and unmobilized peripheral blood [84] are capable of giving rise to endothelial cells in vitro.

Vascular progenitor cells isolated from embroid bodies by CD34 expression showed in vitro differentiation potential to endothelial cells and smooth muscle cells. Implantation studies in nude mice showed that both cell types contribute to the formation of human microvasculature in vivo [85]. Isolated from cord blood, $CD133^+$ cells incorporated into capillary networks, augmented neovascularization, and improved ischemic limb salvage after transplantation into nude mice suffering from ischemic hind limb [76].

$CD133^+$ cells have been used in studies that show significantly improved vascular network restoration in an ischemic hind limb rat model [86]. Biodegradable scaffolds are also being employed for the three-dimensional tissue engineering of microvessels, also using $CD133^+$ cells [84].

Tissue engineering may offer patients new options if replacement or repair of an organ is needed. However, most tissues will require a microvascular network to supply oxygen and nutrients. One strategy for creating a microvascular network would be promotion of vasculogenesis in situ by seeding vascular progenitor cells within a three-dimensional biodegradable construct. Isolated $CD34^+$ $CD133^+$ endothelial progenitor cells (EPC) from human umbilical cord blood were expanded ex vivo as EPC-derived endothelial cells (EC). EPC-derived EC formed capillary-like structures and microvessels when seeded on scaffolds in combination with human smooth muscle cells, indicating that EPCs may be well suited for creating microvascular networks within tissue-engineered constructs [82].

The work of Suuronen and colleagues demonstrates a novel approach for the expansion and delivery of blood CD133$^+$ cells resulting in improved implantation and vasculogenic capacity. Adult human CD133$^+$ progenitor cells from peripheral blood were expanded and delivered within an injectable collagen-based matrix into the ischemic hindlimb of athymic rats. Controls received injections of phosphate-buffered saline, matrix, or CD133-negative cells alone. Immunohistochemistry of hindlimb muscle 2 weeks after treatment revealed that the number of CD133-positive cells retained within the target site was more than twice as great when delivered by matrix than when delivered alone ($P < 0.01$). The transplanted CD133$^+$ cells incorporated into vascular structures, and the matrix itself was also vascularized. Rats that received matrix and CD133-positive cells demonstrated greater intramuscular arteriole and capillary density than other treatment groups ($P < 0.05$ and $P < 0.01$, respectively). Compared with other experimental approaches, treatment of ischemic muscle tissue with generated CD133-positive progenitor cells delivered in an injectable collagen-based matrix significantly improved the restoration of a vascular network [84].

CD133-positive vascular progenitor cells (hVPCs) from the human fetal aorta were able to differentiate into mixed populations of mature endothelial cells, smooth muscle cells, and pericytes after stimulation of progenitor cells. When embedded in a three-dimensional collagen gel, hVPCs reorganized into cohesive cellular cords that resembled mature vascular structures. Transplantation of such cells into the ischemic limb muscle of immunodeficient mice indicated the therapeutic efficacy of a small number of transplanted hVPCs that markedly improved neovascularisation and inhibited the loss of endogenous endothelial cells and myocytes, thus ameliorating the clinical outcome from ischemia [87].

4.4 *Multipotent Mesenchymal Stem Cells (MSCs)*

A brief review of the history of MSC research is given in the article by Geraerts and Verfaillie (see p. #). For tissue engineering applications, it is crucial to start with a defined cell population to develop standardizes protocols and obtain reliable results. Therefore, a broad range of approaches for the isolation of defined stem cell populations from different tissues have been developed (Table 1).

Several cell surface antigens have been used for the isolation of MSCs, such as antifibroblast antigen [88], CD117 [89], CD105 [90, 91], Stro-1 and CD146 [92], CD133 [93], CD271 [61] and MSCA-1 (W8B2) [61]. Clone W8B2 recognizes the mesenchymal stem cell antigen 1 (MSCA-1), a so far unknown antigen. MSCA-1 was shown to be restricted to mesenchymal stem cells (MSCs) in the CD271^{bright} population in bone marrow. These CD271^{bright}CD45^{dim} MSCs have a much higher clonogenic capacity compared with the CD271$^+$ CD45$^+$ fraction in bone marrow [61]. MSCA-1 is therefore suited to identify MSCs with a high proliferative potential. Remarkably, CD105$^+$ cells, isolated from bone marrow, also showed the capacity to form bone in vivo without prior cultivation or differentiation [91].

Table 1 Strategy of isolation for human and mouse MSCs

Strategy for the isolation of fresh MSCs	Cell source	Reference
Human primary cells		
Positive selection of CD271 (LNGFR/p75NTR)	Bone marrow	[61, 99]
Positive selection of CD117$^+$ cells	Bone marrow	[89]
	Amniotic fluid/amniocentesis cultures	[97]
Positive selection of CD133$^+$ cells	Peripheral blood, bone marrow, cord blood	[93, 100]
Depletion of CD45$^+$CD31$^+$ cells	Lipoaspirate/stromal vascular cells (SVF)	[101, 102]
Positive selection of CD34$^+$ cells	Lipoaspirate, stromal vascular fraction	[103]
Isolation of CD34$^+$CD31$^-$ cells	Lipoaspirate/stromal vascular fraction	[104]
Positive selection of Stro-1$^+$ cells	Bone marrow	[92, 105–110]
Positive selection of Stro-1$^+$ cells	Bone marrow, fetal liver, fetal brain	[109]
STRO-1		
Positive selection of Stro-1 + CD146 + cells	Bone marrow and dental pulp	[92]
Stro-1/CD106 (VCAM)+	Bone marrow	[110]
Positive selection of CD63 (HOP-26) + cells	Bone marrow	[108, 111]
Positive selection of CD49a (α1-integrin subunit) + cells	Bone marrow	[100, 108, 112]
Positive selection of CD166 (SB-10) + cells		[108]
SSEA-4	Bone marrow	[113]
Positive selection of GD2 (neural ganglioside) + cells	Bone marrow	[114]
Depletion of GlyA + CD45 + cells	Bone marrow	[99, 116]
Depletion of GlyA + CD45 + cells	Maternal blood	[117]
Mouse primary cells		
Lineage depletion	Bone marrow	[56, 118]
Depletion of CD45 + cells	Bone marrow	[119]
Cultured MSCs	*Cell source*	
Positive selection of CD117 + cells	Amniotic fluid/amniocentesis cultures	[97]
Positive selection of Sca-1 + cells	Bone marrow	[100]
Positive selection of CD49a (α1-integrin subunit) + cells	Bone marrow	[100]
Positive selection of CD271 (LNGFR/p75NTR)	Adipose tissue	[120]
Depletion of CD11b + cells	Bone marrow–derived MSCs after culture	[95]
Depletion of CD45 + CD34 + cells	Bone marrow–derived MSCs	[94]

Mouse MSCs are often heterogeneous populations that are contaminated by lymphohematopoietic (CD34$^+$, CD45$^+$) cells [94], hematopoietic stem cells and macrophages [95], until late passages. Contaminating cells have been depleted from MSC cultures by their expression of CD11b [95] or by their expression of CD34 and CD45 [94] as well as by depletion using a combination of Anti-Ter119 and CD45 MicroBeads [96]. Multipotent plastic-adherent fetal stem cells have been positively selected from amniocentesis cultures by their expression of CD117 [97] and showed broad differentiation potential. MSCs expanded from mouse bone marrow culture are also described to be positive for Sca-1, CD117 (c-kit), and CD105 – among other markers [98].

4.5 Multipotent Adult Progenitor Cells (MAPCs)

Unique cells in human and rodent postnatal marrow are the extremely rare (1 in 10^7 to 1 in 10^8 marrow cells) multipotent adult progenitor cells (MAPCs). MAPCs were selected by depletion from adult bone marrow of hematopoietic cells expressing CD45 (human and mouse) and glycophorin-A (human) or Ter-119 (mouse), followed by long-term culture on fibronectin with EGF, PDGF and low-serum condition. The emerging cell population did not undergo proliferative senescence, due to telomerase expression and maintenance of long telomeres that showed no shortening over 80 doublings. For more details on MAPCs see the article by Geraerts and Verfaillie in this book (see p. ###).

4.6 Tissue Resident Stem Cells and Cancer Stem Cells

Several varieties of tissue resident stem cells and progenitor cells have been identified and also partly isolated in vivo and in vitro. All are characterized by their dual ability to both self-renew and to reconstitute and differentiate into a given number of different somatic or postmitotic cell lineages, depending on their potency. Included are stem cells for oocyte, intestine, breast, kidney, skin, pancreas, hair, lung, ovary, teeth, or stomach formation. In the following, only the most prominent tissue stem cells – neural, cardiac, spermatogonia, and liver (hepatic) stem cells – are described in more detail.

In general, when considering the isolation of tissue resident (stem) cells, an appropriate processing of the tissue prior to cell sorting is crucial. The dissociation might influence the relative composition of cell types or even lead to complete loss of certain cell types, e.g., of large or fragile cells by shear stress or vulnerable cells by high concentrations of proteases. Further, rare cell types might be lost because of incomplete dissociation, making a careful perfusion of the respective organ mandatory. Strong aggregation and adhesion of certain cell types to each other might lead to false interpretation of markers as well. Finally, dead cells and cell debris can influence the purity and recovery of sorted cell types. But not only cells can be harmed by tissue dissociation. Already the protease sensitivity of certain epitopes will influence the sorting outcome when approaches based on surface markers are used. Loss of antigen epitopes can either decrease the yield of target cells, or the outcome might change when using a separation strategy combining several markers. It is therefore important to choose the appropriate protease for each experiment according to the antigen epitope used for isolation. We have carefully analyzed the influence of different concentrations of papain and trypsin on cell viability, recovery and epitope integrity. Papain is often viewed as a mild protease, while trypsin treatment is regarded as harsh and causing detrimental effects on epitopes. We could show that this perception does not apply to a number of antigen epitopes and that even the opposite can be the case [121].

To address the problems concerned with enzymatic dissociation of brain tissue, for example, an enzyme mix has been developed for whole mouse brain tissue or of specific regions, such as the subventricular zone (NTDK, Miltenyi Biotec). In addition, to facilitate and standardize mechanical tissue dissociation, the process can be performed with semiautomated mechanical dissociation systems, such as the gentleMACS? Dissociator (Miltenyi Biotec). Thereby, fluctuations in the yield of viable cells caused by different mincing of the tissue can be avoided.

4.6.1 Neural Stem Cells

The existence of neural stem cells in the rodent brain is widely accepted, as it has been shown that there are restricted regions in the postnatal and adult brain where developmental processes such as neuronal generation and migration continue [122, 123]. Prominent neuronal migration is evident in the cerebellum, hippocampus and rostral migratory stream (RMS) [124]. For humans, a comparable migratory system was not found until 2007 when Curtis et al. presented data demonstrating the presence of a human RMS, which is organized around a lateral ventricular extension reaching the olfactory bulb, and illustrating the respective neuroblasts [125].

CD133, which has emerged as an important surface marker for many stem cell types, was originally described as "Prominin" in murine embryonic neural stem cells (later termed Prominin-1) [126, 127] and as an antigen on human fetal and adult hematopoietic stem and progenitor cells [128]. CD133 antibodies were used to isolate human neural stem cells from fetal brain but not from later developmental stages [129–131]. Likewise, Prominin antibody stained murine neural stem cells in very early (E11.5 and E12) embryos [126, 132]. Lee et al. [133] reported the isolation of a CD133-positive cell population with neural stem cell properties from postnatal murine cerebellum. Whether or not it persists in the adult cerebellum remains elusive. Finally, Pfenninger et al. showed that CD133 is present on neural stem cells in the embryonic brain, on an intermediate radial glial/ependymal cell type in the early postnatal stage, and on ependymal cells in the adult brain [134].

Markers for neural and glial (astrocytes and oligodendrocytes) precursor cells like PSA-NCAM and A2B5 have also been identified and used in many studies [135, 136].

4.6.2 Cardiac Stem Cells

With respect to mouse cardiac stem cells, the report by Hierlihy and colleagues in 2002 [137] was the first identifying a stem cell-like population in adult hearts. Their findings were based on the specific ability of stem cells to efflux Hoechst dye, as shown for many different types of stem cells, also known as side population (SP) [2]. In 2003, Beltrami et al. thoroughly described a population of rat cardiac stem cells (CD117 (c-kit)$^+$ cells) found in clusters and residing among cardiomyocytes in adult hearts [138]. In vitro, cardiac c-kit$^+$ cells were able to undergo self-renewal and differentiation into cardiac cell lineages, i.e., cardiomyocytes, endothelial, and smooth muscle cells. These c-kit$^+$ cells, when implanted in mouse hearts following myocard infarct, retained the capacity for differentiation into cardiomyocytes in vivo. Oh et al. employed a different stem cell marker, Sca-1, to identify yet another population of resident cardiac progenitor cells in adult hearts [139, 140]. Similarly, these Sca1$^+$ cells were found to be capable of differentiation into cardiomyocytes in vitro and in vivo. Then, Pfister et al. demonstrated that, among mouse cardiac SP cells, cardiomyogeneic differentiation is restricted to cells negative for CD31 expression and positive for Sca-1 expression (CD31$^-$/Sca-1$^+$ SP cells) [141]. Besides the described stem cell types, a fourth population of cardiac stem cells is characterized by its expression of the transcription factor Isl1 in rat, mouse, and human myocardium [142]. Isl1-expressing cells are also present in the adult mammalian heart, but they are limited to the right atrium, are found in smaller numbers than in embryonic hearts, and have an unknown physiological role [143]. The most important part of the heart with respect to obstructive heart failure, however, is the left ventricle.

In addition to the Isl1-expressing human cardiac stem cells, Bearzi et al. described their isolation and expansion from human myocardial samples obtained by a minimally invasive biopsy procedure. Following their findings, human cardiac stem cells are positive for the stem cell antigen c-kit, but negative for the hematopoietic and endothelial antigens CD45, CD34, CD31, and KDR. CD45 and KDR are typically expressed in a subset of bone marrow c-kit$^+$ cells that have the ability to migrate to the heart after injury; CD31 (PECAM-1) on mature endothelial cells, platelets, and on some white blood cells, such as monocytes, NK cells, granulocytes, B cells, and T cell subsets [144]. Smith et al. use a simple explant outgrowth and cardiosphere expansion method [145]. While the description of mouse cardiac stem cells is quite accepted, these first findings on human cardiac stem cells have to be fully confirmed.

4.6.3 Spermatogonial Stem Cells

Germ cells are defined by their innate potential to transmit genetic information to the next generation through fertilization. Males produce numerous sperm for long periods to maximize chances of fertilization. Key to the continuous production of large numbers of sperm are germline stem cells and their immediate daughter cells, functioning as transit amplifying cells [146]. Several possible options for preservation and reestablishment of the reproductive potential have been described. Apart from fertility preservation, SSC studies are useful for other applications as well, such as gene targeting [22, 147], transgenerational gene therapy, and cell-based organ regeneration therapy [148, 149]. A marker for mouse spermatogonial stem cells has recently been described by Seandel et al. [150]. The authors show that highly proliferative adult spermatogonial progenitor cells (SPCs) can be efficiently obtained by cultivation on mitotically inactivated testicular feeders containing $CD34^+$ stromal cells. SPCs exhibit testicular repopulating activity in vivo and maintain the ability in long-term culture to give rise to multipotent adult spermatogonia-derived stem cells (MASCs). Furthermore, both SPCs and MASCs express GPR125, an orphan adhesion-type G-protein-coupled receptor.

4.6.4 Hepatic Stem Cells (HpSC)

Widespread use of liver transplantation in the treatment of hepatic diseases is restricted by the limited availability of donated organs [151]. Stem cells are a promising source for liver repopulation after cell transplantation. However, it is still not clear whether or not the adult mammalian liver contains hepatic stem cells [152].

According to Schmelzer et al., human hepatic stem cells (hHpSCs), which are pluripotent precursors of hepatoblasts and thence of hepatocytic and biliary epithelia, are located in ductal plates in fetal livers and in Canals of Hering in adult livers [153, 154] and can be isolated by positive immunoselection for the epithelial cell adhesion molecule CD326 ($EpCAM^+$). The hHpSCs are approximately 9 mm in diameter, and express cytokeratins 8, 18, and 19, CD133/1, telomerase, CD44H, claudin 3, and albumin (weakly). They are negative for a-fetoprotein (AFP), intercellular adhesion molecule 1(ICAM-1), and for markers of adult liver cells (cytochrome P450s), hematopoietic (progenitor) cells (CD45, CD34, CD14, CD38, CD90 (Thy1), CD235a (Glycophorin A), and mesenchymal cells (vascular endothelial growth factor receptor and desmin) [153, 154].

As for rodent HpSCs, Yovchev et al. studied progenitor/oval cell surface markers in the liver of rats subjected to 2-acetylaminofluorene treatment, followed by partial hepatectomy (2-AAF/PH). Further, they compared hepatic cells isolated by two surface markers, epithelial cell adhesion molecule (EpCAM) and thymus cell antigen 1 (Thy-1). They found that CD326 $(EpCAM)^+$ and CD90 $(Thy\text{-}1)^+$ cells represent two different populations of cells in the oval cell niche. $EpCAM^+$ cells express the classical oval cell markers (alpha-fetoprotein, cytokeratin-19, OV-1 antigen, a6 integrin, and connexin 43), as well as cell surface markers identified previously by the same researchers (CD44, CD24, CD326 (EpCAM), aquaporin 5, claudin-4, secretin receptor, claudin-7, v-ros sarcoma virus oncogene homolog 1, cadherin 22, mucin-1, and CD133). Oval cells do not express previously reported hematopoietic stem cell markers Thy-1, c-kit, CD34, or CD56, the neuroepithelial marker neural cell adhesion molecule 1 (NCAM-1). It was shown that $Thy\text{-}1^+$ cells are mesenchymal cells with characteristics of myofibroblasts/activated stellate cells. Transplantation experiments reveal that $EpCAM^+$ cells are true progenitors capable of repopulating injured rat liver [155, 156].

4.6.5 Cancer Stem Cells (CSCs)

Although the concept that cancers arise from "stem cells" or "germ cells" was first proposed about 150 years ago, it is only recently that advances in stem cell biology have given new impetus to the cancer stem cell hypothesis. What has become clear in the past 10 years is that tumor cells are functionally heterogeneous. They are organized in a hierarchy of heterogeneous cell populations with different biological properties. Specifically, only a minority of tumor cells has the capacity to regenerate the tumor and sustain its growth when injected into an immune-compromised mouse model [157].

In the last 5 years, investigation of solid-tumor stem cells has gained momentum. Using similar approaches and principles as for ALM of serial dilution and serial transplantation, solid-tumor stem cells have been prospectively identified in several tissues, such as blood, brain, colon, liver, lung, pancreas, prostate, skin, and breast cancers. The experimental strategy most often combines sorting of tumor cell subpopulations, identified on the basis of the different expression of surface markers, with functional transplantation into appropriate animal models.

Blood or Hematopoietic CSCs

John Dick and colleagues isolated and identified $CD34^+$ $CD38^-$ leukemic stem cells (LSCs) from human AML by FACS and demonstrated that these cells initiated leukemia in NOD-SCID mice compared with the $CD34^+$ $CD38^+$ and $CD34^-$ fractions [158]. An engrafted leukemia could be serially transplanted into secondary recipients, providing functional evidence for self-renewal. Xenotransplantation, followed by serial transplantation, is now regarded as an essential criterion in defining cancer stem cells. The ability to recapture tumor pathophysiology is an important defining functional criterion of cancer stem cells prospectively isolated [157].

Breast CSCs

A minor, phenotypically distinct tumor cell population has been isolated that is able to form mammary tumors in NOD–SCID mice, whereas cells with alternative phenotypes are nontumorigenic even when implanted at significantly higher cell numbers, thereby demonstrating enrichment of tumor-initiating cells in selected fractions. The tumorigenic cells can be serially passaged, demonstrating self-renewal capacity, and are able to generate tumor heterogeneity, producing differentiated, nontumorigenic progeny. Thus, like AML, breast cancer growth appears to be driven by a rare population of tumor-initiating cells [157]. Breast cancer stem cells have been reported to be ESA^{+} CD44^{+} CD24$^{-/low}$Lineage^{-} [159–161]. ESA stands for "epithelial specific antigen," also known as EpCAM (epithelial cell adhesion molecule) or CD326. Usage of those markers allows for a more than 50-fold enrichment to form tumors (0.6% of cancer cells).

Brain CSCs

Singh et al. [161, 162] reported the identification and purification of cancer stem cells from human brain tumors of different phenotypes that possess a marked capacity for proliferation, self-renewal, and differentiation. The increased self-renewal capacity of the brain tumor stem cell (BTSC) was highest among the most aggressive clinical samples of medulloblastoma compared with low-grade gliomas. The BTSC was exclusively isolated with the cell fraction expressing the neural stem cell surface marker CD133. The CD133^{+} fraction among highly aggressive glioblastomas (GBMs) ranged from 19 to 29%, and among medulloblastomas ranged from 6 to 21%, and correlated closely with an in vitro primary sphere formation assay (which was used to quantify stem cell frequency).

Lung CSCs

Lung cancer stem cells were first identified by Kim et al. [163] who describe a niche in the bronchioalveolar duct junction of adult mouse lung that harbors stem cells from which adenocarcinomas are likely to arise. More importantly, these double-positive cells appear enriched in FACS-sorted Sca-1^{+}/CD34^{+} cell populations and show enhanced capacity for both self-renewal and differentiation. Subsequently it was shown that the human lung cancer-derived A549 cell line also harbors CSCs with a side population (SP) phenotype revealing several stem cell properties [164]. Very recently, Eramo et al. [165] found that the tumorigenic cells in small-cell and nonsmall-cell lung cancer are a rare population of undifferentiated cells expressing CD133.

Liver CSCs

Although liver CSCs have been identified in hepatocellular carcinoma (HCC) cell lines, no data have shown the presence of these cells in human settings until very recently. Now, Yang et al. [166] have delineated CSCs serially from HCC cell lines, human liver cancer specimens, and blood samples, using CD90 as a potential marker. The number of CD90^{+} cells increased with the tumorigenicity of HCC cell lines. CD45^{-} CD90^{+} cells were detected in all the tumor specimens, but not in the normal, cirrhotic, and parallel nontumorous livers. In addition, CD45^{-} CD90^{+} cells were detectable in 90% of blood samples from liver cancer patients, but none in normal subjects or patients with cirrhosis.

Prostate CSCs

According to Richardson et al. [167], prostatic stem cells are alpha2beta1^{+}/CD133^{+}. Collins et al. [168] have shown that cancer stem cells in prostate have been identified with a CD44^{+}/integrin alpha2beta1^{high}/CD133^{+} phenotype. Approximately 0.1% of cells in any tumor expressed this phenotype, and there was no correlation between the number of CD44^{+}/a2b1^{high}/CD133^{+} cells and tumor grade. In addition, Miki et al. have shown that expression of CXCR4 was also detected in CD133^{+} cancer cells.

Colon CSCs

O'Brien et al. [169] and Ricci-Vitiani et al. [170] showed that the tumorigenic population in colon cancer is restricted to CD133^{+} cells, which are able to reproduce the original tumor in permissive recipients. Ricci-Vitiani et al. state that the vast majority of the samples analyzed showed the presence of rare cells (2.5±1.4%) clearly positive for CD133 while CD133 expression in normal colon tissues was extremely rare (barely detectable upon extensive analysis of histological sections using CD133/1 and CD133/2 antibodies. O'Brien et al. state that purification experiments established that all colon cancer–initiating cells (CC–ICs) were CD133-positive; the CD133-negative cells comprising the majority of the tumor were unable to initiate tumor growth.

Melanoma CSCs

According to Fang et al. [171] and Kamstrup et al. [172], a small subpopulation of CD20^{+} melanoma cells harbors multipotent stem cells. Most interestingly, CD20 has been identified by gene expression profiling as one of the top 22 genes that define aggressive melanomas. In metastatic melanomas, they have identified individual CD20^{+} tumor cells. Monoclonal antibodies against CD20 have become a standard treatment for nonHodgkin's lymphoma. CD20 seems to be a potential target for melanoma as well, although a correlation between differentiation ability and tumorigenicity is still under investigation by comparing CD20^{+} with CD20^{-} fractions.

Pancreas CSCs

Li et al. [173] identified a highly tumorigenic subpopulation of pancreatic cancer cells expressing the cell surface markers CD44, CD24, and epithelial- specific antigen (ESA; EpCAM; CD326). Pancreatic cancer cells with the CD44^{+} CD24^{+} ESA^{+} phenotype (0.2–0.8% of pancreatic cancer cells) had a 100-fold increased tumorigenic potential compared with nontumorigenic cancer cells, with 50% of animals injected with as few as 100 CD44^{+} CD24^{+} ESA^{+} cells forming tumors that were histologically indistinguishable from the human tumors from which they originated.

As a conclusion, the field of tissue stem cells (apart from hematopoietic stem cells) is still at the very beginning. This is partly due to the fact that potential markers for these tissue stem cells have not or have only recently been defined and are still intensively debated. Also, standardized processes for appropriate dissociation of tissues are currently hardly available, resulting in poor comparability of sorting results. It can be estimated that sorting of tissue resident (stem) cells will play a dramatically increasing role in the future, because this will offer the option for a detailed analysis and understanding of malignant and disease-causing cells, as well as of cell types urgently needed for tissue regeneration and engineering approaches.

5 Clinical Applications of Stem Cells

5.1 Allogeneic Hematopoietic Stem Cell Transplantation

To date, the major application of enriched and purified stem cells is the transplantation of hematopoietic stem cells derived from bone marrow or peripheral blood. For a number of patients suffering from malignant and nonmalignant diseases, allogeneic stem cell transplantation, i.e., when the stem cells originate from a healthy individual, is the only curative treatment option [174].

For many years, the clinical outcome of an allogeneic transplantation has been determined to a major degree by the matching between donor and recipient of the genes encoding histocompatibility antigens – the human leukocyte antigen (HLA) system in human beings. However, the "ideal" donor, an HLA-matched sibling, can only be found for about 30% of the patients. With the help of worldwide registries of unrelated donors, the probability of identifying a matched unrelated donor (MUD) depends on the diversity of HLA antigens within a population and on the race, and ranges from about 75% for Caucasians to less than 50% for ethnic minorities [174]. A MUD donor search is time-consuming, and may take more than 3 months in most cases or even longer, so that not every patient might benefit from a potentially life-saving allogeneic transplantation. For those patients who do not have a matched related or unrelated donor or who are at high risk for disease progression during the donor search, an alternative approach is the use of mismatched related family donors (MMFD). Most of these donors share only one HLA haplotype with the patient and are referred to as haploidentical donors. In fact, virtually every patient has a potentially suited haploidentical donor among parents or children. However, in the past haploidentical transplantations have been hampered by clinical complications linked to the high degree of donor–recipient HLA disparity, such as graft failure, prolonged and profound immunodeficiency, and severe acute or chronic graft vs host disease (GVHD). Acute GVHD is a major cause of death after allogeneic transplantation (Fig. 14). Additionally, chronic graft vs host disease is

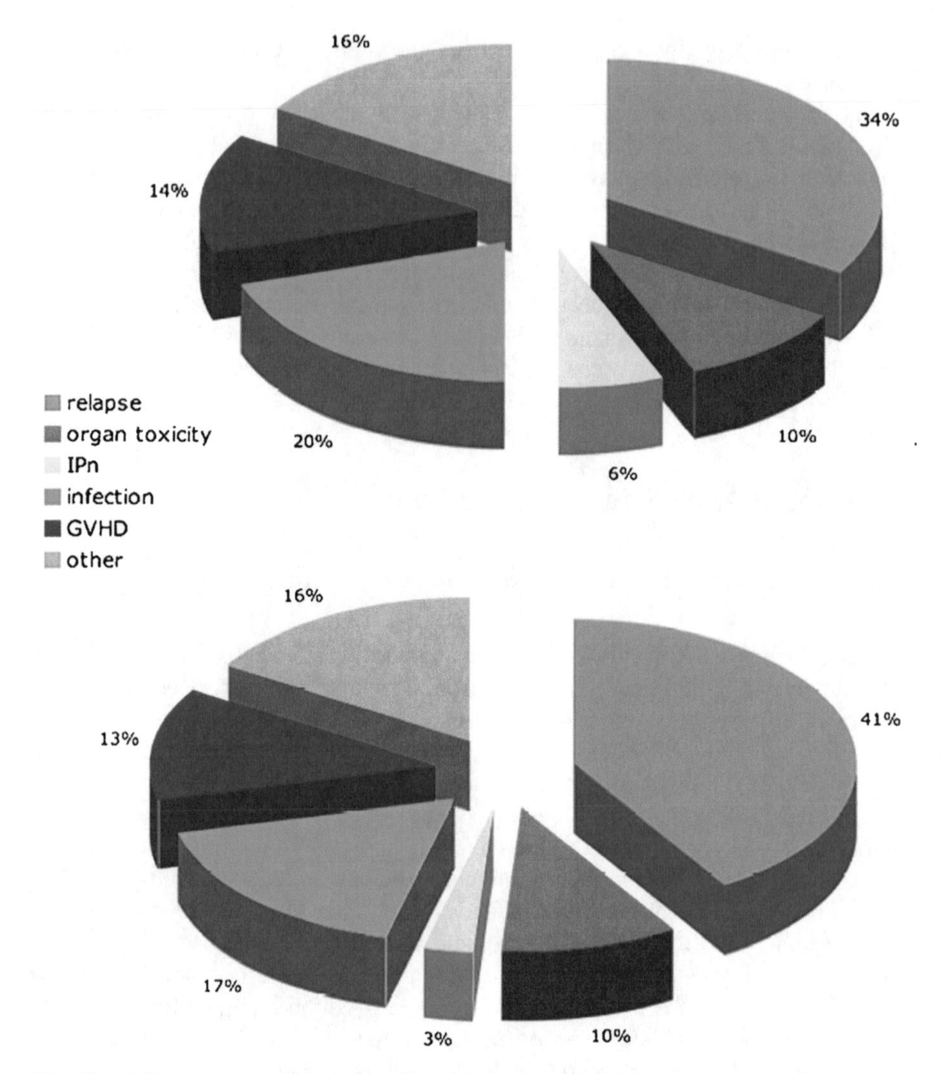

Fig. 14 a,b Causes of death after allogeneic stem cell transplantation (2001–2006). Graft vs host disease (GVHD) is a major complication after allogeneic stem cell transplantation. A total of 14% and 13% of patients die of GVHD after unrelated and HLA-identical sibling donor transplantation, respectively. **a** Unrelated donor transplantation. **b** HLA-identical sibling donor transplantation (Source: CIBMTR Newsletter, Dec 2007)

the major predictor of long-term outcome and overall survival after allogeneic stem cell transplantation [175]. Since the severity of GVHD, among other factors, depends on the degree of histocompatibility between donor and recipient and mediated by alloreactive donor T cells, a thorough T cell depletion is mandatory in haploidentical transplantation, where only one HLA haplotype is shared between patient and mismatched family donor.

In recent years both experimental concepts and technical developments have eventually paved the way for the clinical application of haploidentical stem cell transplantation.

In 2001, Martelli and Reisner [176] published their pioneering work about the "megadose" concept, demonstrating that transplantation of high doses of purified stem cells can overcome the HLA barrier between donor and recipient and can lead to tolerance.

At the same time, the finding that growth factors can release high amounts of stem cells from bone marrow into the peripheral blood, and the development of effective in vitro T cell depletion systems have made haploidentical transplantation technically feasible. In vitro CD34$^+$ selection is one of the most potent technologies for an effective T cell depletion, resulting in a ten- to hundred-thousand-fold depletion of T cells from the graft [177]. The use of mobilized peripheral stem cells in addition to bone marrow, or as a sole source of stem cells in combination with effective CD34$^+$ selection strategies, allowed for the transplantation of megadoses of highly purified stem cells. This resulted in high engraftment rates and complete prevention of GVHD without the need for a post-transplant immunosuppressive treatment [174].

The CliniMACS Plus Instrument is an automated cell selection system based on MACS® Technology (see Chap. Methods and Technologies). It enables the operator to perform clinical-scale magnetic cell selection of target cells or the depletion of unwanted cells in a functionally closed and sterile system. A number of publications and reports substantiate the reliable and excellent performance (Table 2) of CD34$^+$ cell separation with the CliniMACS Plus Instrument, which provides a high recovery and purity of target cells as well as efficient T cell depletion rates.

The Laboratory of Hematology at the University of Perugia is one of the most experienced transplant centers for haploidentical transplantation. Recently, Aversa and coworkers gave an update of the clinical outcome after haploidentical transplantation for adult patients suffering from acute leukemia [181]. The event-free survival (EFS) after haploidentical transplantation is closely related to disease and disease status at transplant. In the Perugia experience, 62 patients with acute lymphoid leukemia (ALL) transplanted in clinical remission (CR) have about 25% probability of surviving event-free. Results are better for the 83 patients with acute

Table 2 CliniMACS CD34 separation results from PBSCs of allogeneic donors [177–180]

	$n > 500$	$n = 293$	$n = 136$	$n = 30$	$n = 335$	$n = 73$
Purity CD34 (%)	95–99	97.5	90	96	93	92
Recovery CD34 (%)	70	77	81	64	81	71
CD3 log depletion	>5	4.6	4.6	5	4.8	5.1
CD19 log depletion	nd	nd	3.2	3.2	3.3	3.7

nd = not determined

myeloid leukemia (AML) transplanted in CR: the 3-year EFS ranges between 40%, for those who were transplanted in second or later CR, to 50% for patients transplanted in first CR. With no chronic GVHD, all these long-term survivors enjoy an excellent quality of life. It has been shown by the Perugia group [182, 183] that natural killer (NK) cell alloreactivity can exert an antileukemic effect in the absence of killer inhibitory receptors (KIR). Donor vs recipient NK-cell alloreactivity impacts favorably upon survival in AML patients: Thirty patients in any CR who received a transplant from an NK cell alloreactive donor enjoy 67% EFS vs 18% in 31 patients transplanted from nonNK alloreactive donors [181].

5.2 Autologous Hematopoietic Stem Cell Transplantation in Autoimmune Diseases

Recently, the transplantation of highly enriched stem cells has emerged as a novel treatment option for some therapy-refractive autoimmune diseases. In a recently published article [184], Vonk et al. report on promising data of the outcome after high dose immunosuppressive therapy and autologous $CD34^+$-selected cell transplantation in severe systemic sclerosis (SSc), also referred to as scleroderma. This disease is a generalized autoimmune disease causing morbidity and reduced life expectancy, particularly in patients with rapidly progressive diffuse cutaneous SSc. Since no proven treatment exists, autologous hematopoietic stem cell transplantation (the stem cells are derived from the patient) is considered as a new therapeutic strategy in patients with poor prognosis. $CD34^+$ cell enrichment is performed in order to remove any autoreactive T and B cells. Vonk and colleagues [184] report that after a median follow-up of 5.3 (1–7.5) years, 21 out of 26 patients (81%) demonstrated a clinically beneficial response. Event-free survival, defined as survival without mortality, relapse or progression of disease, was 64.3% at 5 years and 57.1% at 7 years. Alexander and coworkers [185] report on the long-term outcome (median follow-up period: 60 months) of seven patients suffering from refractory systemic lupus erythematosus (SLE) who had been treated by transplantation of $CD34^+$-enriched stem cells after a profound immunoablation. The presented data show that the long-term therapy-free clinical remissions observed in SLE patients after complete immunoablation and ASCT are accompanied by a loss of immunological memory and a fundamental reset of the immune system.

5.3 Stem Cells for Tissue Regeneration in Cardiac Diseases

Current pharmacological, interventional, or surgical approaches fail to regenerate nonviable myocardium. As a consequence, restoration of functional myocardium following cardiac infarction remains an ambitious challenge for clinicians. The heart has the ability to elicit a regenerative response designed to restore cardiac function through replacement of damaged cells. There is rising evidence that this is accomplished by the activation of resident cardiac stem cells [144, 186] or through the recruitment of a stem cell population from other tissues such as bone marrow [187, 188].

Hence, stem cell therapy is a promising new strategy for myocardial repair. Since the availability of autologous cardiac stem cells in large numbers is poor and requires ex vivo expansion, highly proliferative, totipotent embryonic stem cells (ESCs) seemed to be a promising alternative. However, there are persisting ethical and legal issues as well as concerns about the tumorigenic and infectious potential of allogeneic ESCs. Furthermore, the early deaths of ESCs as a result of ischemia currently impede their use in clinical studies [189].

The discovery of adult tissue-specific stem cells (ASCs), which have the ability to transdifferentiate into other tissues, led to extensive use of these cells for hematological, cancer and myocardial infarct-related administration [190].

In skeletal muscle the so-called satellite cells function as progenitor cells and are responsible for normal muscle growth and regeneration [191]. Transplantation of skeletal muscle into mouse myocardium alone [192, 193] and in combination with CABG into human heart [194–197] proved to be feasible, safe, and efficient in restoring functional myocardium. However, although improved symptoms and LVEF were achieved [198, 199], the lack of gap junction formation of the graft may lead to failure in electromechanical coupling and, thus, to a higher arrhythmic risk. There are several strategies aiming at an enhancement of electromechanical coupling, e.g., by modifying the expression of connexin-43 [200]. Further investigation may provide long-term improvement.

Bone marrow (BM), among several other organs, possesses multipotent adult stem cells with high plasticity, as demonstrated convincingly in the mouse model of infarcted myocardium [201]. Only recently it has been demonstrated in a meta-analysis including 18 randomized controlled trials and cohort studies that BMC transplantation seems safe and is associated with modest improvements in physiological and anatomical parameters in patients with both acute myocardial infarction and chronic ischemic heart disease, above and beyond conventional therapy [202].

Mesenchymal stem cells (MSCs) are bone marrow–populating cells (stromal cells), which possess an extensive proliferative potential and the ability to differentiate into various cell types, including osteocytes, adipocytes, chondrocytes, myocytes, cardiomyocytes and neurons [203] (see above). Apart from bone marrow, MSCs are located in other tissues, such as adipose tissue, peripheral blood, cord blood, liver, and fetal tissues. Their multilineage potential and their ability to elude detection by the host's immune system, as well as their relative ease of expansion in culture, make MSCs a promising source of stem cells for transplantation [204]. However, it was recently shown in mice that the developmental fate of bone marrow-derived MSCs is not restricted by the surrounding tissue after myocardial infarction, but by induced calcification and/or ossification [205].

Probably the simplest approach to use stem cells for myocardial cell therapy is to harvest mononuclear cells either from bone marrow or mobilized peripheral blood [206–210]. However, the functionality of the different cells has not yet been clearly identified. When delivered intracoronarily, improvement in cardiac function was sparse [211–213]. Furthermore, concerns exist regarding limited efficiency due to the small numbers of progenitor cells in nonenriched peripheral blood and bone marrow that are delivered intramyocardially. The risk of foreign tissue differentiation following local stroma cell injections was also found with mononuclear cells [205]. Thus, neither the preferred source and type of stem cell nor the optimal method of delivery of stem cells to the target area have been defined so far. Positively selected human $CD34^+/CD133^+$ cells from mobilized peripheral blood for intravenous injection in nude rats resulted not only in a substantial increase in left ventricular ejection fraction but also in a fivefold increase in the number of capillaries compared with the control [214].

Hematopoietic stem cells (HSCs) expressing CD133, a marker for more 'primitive' multipotent stem cells, are considered to be particularly important in the context of myocardial repair [186, 201, 215–217]. The cell surface antigen CD133 is expressed on primitive HSCs and endothelial progenitor cells (EPCs), which collaborate to promote vascularization of ischemic tissues [218]. $CD133^+$ cells can integrate into sites of neovascularization and differentiate into mature endothelial cells. For enriched $CD34^+$ cells, homing into the border zone of infarcted myocardium has been demonstrated [219].

With respect to functionality, a direct side-by-side comparison of human $CD133^+$ bone marrow cells and human skeletal myoblasts in a myocardial ischemia model in immuno-incompetent rats demonstrated similar functional improvement in both groups, although only the myoblasts reached robust engraftment [220].

Since intracoronary administration of mononuclear cells did not yield the expected functional benefit [213], and the heart is easily accessible during CABG procedures, several groups started to inject mononuclear [221], purified selected cord blood-derived [222, 223], bone marrow-derived, or blood-derived $CD133^+$ stem cells [78, 224–226] intramyocardially during chirurgical intervention.

In 2003, Stamm et al. published first clinical results of a phase I study on patients suffering from chronic ischemic heart disease, treated with $CD133^+$ stem cells in conjunction with Coronary Artery Bypass Grafting (CABG) confirming safety and feasibility [77]. Functional benefit could also been demonstrated in a subsequent controlled randomized trial [227]. The therapeutic potential of $CD133^+$ cells was further confirmed by the positive response of 10 patients with end-stage chronic ischemic cardiomyopathy, only treated with cell injection [228].

In summary, results from current trials support conducting large randomized trials to evaluate the impact of cell therapy on patient-related outcomes.

References

1. Hoffman R, Houck D (1998) In: Recktenwald D, Radbruch A (eds) Cell separation methods and applications. Marcel Dekker, New York, p 237
2. Jones R, Barber J, Vala M (1995) Blood 85:2742
3. Storms R, Trujillo A, Springer J (1999) Proc Natl Acad Sci U S A 96:9118

4. Molday R, MacKenzie D (1982) J Immunol Methods 52:353
5. Miltenyi S, Muller W, Weichel W, Radbruch A (1990) Cytometry 11:231
6. Kato K, Radbruch A (1993) Cytometry 14:384
7. Radbruch A, Mechtold B, Thiel A, Miltenyi S, Pfluger E (1994) Methods Cell Biol 42 Pt B:387
8. Kantor A, Gibbons I, Miltenyi S, Schmitz J (1997) In: Recktenwald D, Radbruch A (eds) Cell separation methods and applications. Marcel Dekker, New York
9. McNiece I, Briddell R, Stoney G, Kern B, Zilm K, Recktenwald D, Miltenyi S (1997) J Hematother 6:5
10. Miltenyi S, Schmitz J (1999) In: Radbruch A (ed) Flow cytometry and cell sorting. Springer, Berlin Heidelberg New York, p 218
11. Apel M, Heinlein U, Miltenyi S, Schmitz J, Campbell J (2007) In: Andrä W, Nowak H (eds) Magnetism in medicine – a handbook. Wiley, Weinheim, p 571
12. Ugelstad J, Prestvik W, Stenstad P, Kilaas L, Kvalheim G (1998) In: Andrä W, Nowak H (eds) Magnetism in medicine – a handbook. Wiley, Weinheim, p 473
13. Ugelstad J, Mørk P, Herder Kaggerud K, Ellingsen T, Berge A (1980) Adv Colloid Interface Sci 13:101
14. Funderud S, Nustad K, Lea T, Vartdal F, Gaudernack G, Stenstad P, Ugelstad J (1987) In: Klaus G (ed) Lymphocytes: a practical approach. IRL, Oxford, p 55
15. Neurauter A, Bonyhadi M, Lien E, Nokleby L, Ruud E, Camacho S, Aarvak T (2007) Adv Biochem Eng Biotechnol 106:41
16. Dreger P, Ritgen M, Ho A (2006) In: Ho A, Hoffman R, Zanjani E (eds) Stem cell transplantation – biology, processes, therapy. Wiley, New Jersey, p 247
17. Dreger P, Viehmann K, von Neuhoff N, Glaubitz T, Petzoldt O, Glass B, Uharek L, Rautenberg P, Suttorp M, Mills B, Mitsky P, Schmitz N (1999) Bone Marrow Transplant 24:153
18. Rowley SD, Loken M, Radich J, Kunkle LA, Mills BJ, Gooley T, Holmberg L, McSweeney P, Beach K, MacLeod B, Appelbaum F, Bensinger WI (1998) Bone Marrow Transplant 21:1253
19. Bradley A, Evans M, Kaufman MH, Robertson E (1984) Nature 309:255
20. Evans MJ, Kaufman MH (1981) Nature 292:154
21. Thomson JA, Itskovitz-Eldor J, Shapiro SS, Waknitz MA, Swiergiel JJ, Marshall VS, Jones JM (1998) Science 282:1145
22. Takahashi K, Yamanaka S (2006) Cell 126:663
23. Hogan BL (1994) Manipulating the mouse embryo: a laboratory manual. Cold Spring Harbor Laboratory, Plainview, NY
24. Schoonjans L, Kreemers V, Danloy S, Moreadith RW, Laroche Y, Collen D (2003) Stem Cells 21:90
25. McWhir J, Schnieke AE, Ansell R, Wallace H, Colman A, Scott AR, Kind AJ (1996) Nat Genet 14:223
26. Brook FA, Gardner RL (1997) Proc Natl Acad Sci U S A 94:5709
27. Buehr M, Smith A (2003) Philos Trans R Soc Lond B Biol Sci 358:1397
28. Cheng J, Dutra A, Takesono A, Garrett-Beal L, Schwartzberg PL (2004) Genesis 39:100
29. Hoffman LM, Carpenter MK (2005) Stem Cell Rev 1:139
30. Amit M, Itskovitz-Eldor J (2006) Methods Enzymol 420:37
31. Adewumi O, Aflatoonian B, Ahrlund-Richter L, Amit M, Andrews PW, Beighton G, Bello PA, Benvenisty N, Berry LS, Bevan S, Blum B, Brooking J, Chen KG, Choo AB, Churchill GA, Corbel M, Damjanov I, Draper JS, Dvorak P, Emanuelsson K, Fleck RA, Ford A, Gertow K, Gertsenstein M, Gokhale PJ, Hamilton RS, Hampl A, Healy LE, Hovatta O, Hyllner J, Imreh MP, Itskovitz-Eldor J, Jackson J, Johnson JL, Jones M, Kee K, King BL, Knowles BB, Lako M, Lebrin F, Mallon BS, Manning D, Mayshar Y, McKay RD, Michalska AE, Mikkola M, Mileikovsky M, Minger SL, Moore HD, Mummery CL, Nagy A, Nakatsuji N, O'Brien CM, Oh SK, Olsson C, Otonkoski T, Park KY, Passier R, Patel H, Patel M, Pedersen R, Pera MF, Piekarczyk MS, Pera RA, Reubinoff BE, Robins AJ, Rossant J, Rugg-Gunn P, Schulz TC, Semb H, Sherrer ES, Siemen H, Stacey GN, Stojkovic M, Suemori H, Szatkiewicz J, Turetsky T, Tuuri T, van den Brink S, Vintersten K, Vuoristo S, Ward D, Weaver TA, Young LA, Zhang W (2007) Nat Biotechnol 25:803

32. Cui L, Johkura K, Yue F, Ogiwara N, Okouchi Y, Asanuma K, Sasaki K (2004) J Histochem Cytochem 52:1447
33. Bashamboo A, Taylor AH, Samuel K, Panthier JJ, Whetton AD, Forrester LM (2006) J Cell Sci 119:3039
34. Enver T, Soneji S, Joshi C, Brown J, Iborra F, Orntoft T, Thykjaer T, Maltby E, Smith K, Dawud RA, Jones M, Matin M, Gokhale P, Draper J, Andrews PW (2005) Hum Mol Genet 14:3129
35. Choo AB, Tan HL, Ang SN, Fong WJ, Chin A, Lo J, Zheng L, Hentze H, Philp RJ, Oh SK, Yap M (2008) Stem Cells 26:1454
36. Hedlund E, Pruszak J, Ferree A, Vinuela A, Hong S, Isacson O, Kim KS (2007) Stem Cells 25:1126
37. Schuldiner M, Yanuka O, Itskovitz-Eldor J, Melton DA, Benvenisty N (2000) Proc Natl Acad Sci U S A 97:11307
38. Carpenter MK, Inokuma MS, Denham J, Mujtaba T, Chiu CP, Rao MS (2001) Exp Neurol 172:383
39. Rolny C, Nilsson I, Magnusson P, Armulik A, Jakobsson L, Wentzel P, Lindblom P, Norlin J, Betsholtz C, Heuchel R, Welsh M, Claesson-Welsh L (2006) Blood 108:1877
40. Xiao Q, Zeng L, Zhang Z, Margariti A, Ali ZA, Channon KM, Xu Q, Hu Y (2006) Arterioscler Thromb Vasc Biol 26:2244
41. Pruszak J, Sonntag KC, Aung MH, Sanchez-Pernaute R, Isacson O (2007) Stem Cells 25:2257
42. Perez-Campo FM, Spencer HL, Elder RH, Stern PL, Ward CM (2007) Exp Cell Res 313:3604
43. Tomishima MJ, Hadjantonakis AK, Gong S, Studer L (2007) Stem Cells 25:39
44. David R, Groebner M, Franz WM (2005) Stem Cells 23:477
45. Kattman SJ, Huber TL, Keller GM (2006) Dev Cell 11:723
46. Moretti A, Caron L, Nakano A, Lam JT, Bernshausen A, Chen Y, Qyang Y, Bu L, Sasaki M, Martin-Puig S, Sun Y, Evans SM, Laugwitz KL, Chien KR (2006) Cell 127:1151
47. Wu SM, Fujiwara Y, Cibulsky SM, Clapham DE, Lien CL, Schultheiss TM, Orkin SH (2006) Cell 127:1137
48. Hubin F, Humblet C, Belaid Z, Lambert C, Boniver J, Thiry A, Defresne MP (2005) Stem Cells 23:1626
49. Schiedlmeier B, Santos AC, Ribeiro A, Moncaut N, Lesinski D, Auer H, Kornacker K, Ostertag W, Baum C, Mallo M, Klump H (2007) Proc Natl Acad Sci U S A 104:16952
50. Kiel MJ, Yilmaz OH, Iwashita T, Terhorst C, Morrison SJ (2005) Cell 121:1109
51. Chen CZ, Li M, de Graaf D, Monti S, Gottgens B, Sanchez MJ, Lander ES, Golub TR, Green AR, Lodish HF (2002) Proc Natl Acad Sci U S A 99:15468
52. Chen CZ, Li L, Li M, Lodish HF (2003) Immunity 19:525
53. Zhang CC, Lodish HF (2005) Blood 105:4314
54. Lagasse E, Connors H, Al-Dhalimy M, Reitsma M, Dohse M, Osborne L, Wang X, Finegold M, Weissman IL, Grompe M (2000) Nat Med 6:1229
55. Orlic D, Kajstura J, Chimenti S, Limana F, Jakoniuk I, Quaini F, Nadal-Ginard B, Bodine DM, Leri A, Anversa P (2001) Proc Natl Acad Sci U S A 98:10344
56. Wong SH, Lowes KN, Bertoncello I, Quigley AF, Simmons PJ, Cook MJ, Kornberg AJ, Kapsa RM (2007) Stem Cells 25:1364
57. Takahashi T, Kalka C, Masuda H, Chen D, Silver M, Kearney M, Magner M, Isner JM, Asahara T (1999) Nat Med 5:434
58. Sanchez-Ramos J, Song S, Cardozo-Pelaez F, Hazzi C, Stedeford T, Willing A, Freeman TB, Saporta S, Janssen W, Patel N, Cooper DR, Sanberg PR (2000) Exp Neurol 164:247
59. Giebel B, Zhang T, Beckmann J, Spanholtz J, Wernet P, Ho AD, Punzel M (2006) Blood 107:2146
60. Copland M, Hamilton A, Elrick LJ, Baird JW, Allan EK, Jordanides N, Barow M, Mountford JC, Holyoake TL (2006) Blood 107:4532
61. Buhring HJ, Seiffert M, Bock TA, Scheding S, Thiel A, Scheffold A, Kanz L, Brugger W (1999) Ann N Y Acad Sci 872:25

62. Yin AH, Miraglia S, Zanjani ED, Almeida-Porada G, Ogawa M, Leary AG, Olweus J, Kearney J, Buck DW (1997) Blood 90:5002
63. Freund D, Oswald J, Feldmann S, Ehninger G, Corbeil D, Bornhauser M (2006) Cell Prolif 39:325
64. de Wynter E, Buck D, Hart C, Heywood R, Coutinho L, Clayton A, Rafferty J, Burt D, Gueneche G, Bueren J, Gagen D, Fairbairn L, Lord B, Testa N (1998) Stem Cells 16:387
65. Matsumoto K, Yasui K, Yamashita N, Horie Y, Yamada T, Tani Y, Shibata H, Nakano T (2000) Stem Cells 18:196
66. Summers Y, Heyworth C, de Wynter E, Hart C, Chang J, Testa N (2004) Stem Cells 22:704
67. Goussetis E, Theodosaki M, Paterakis G, Peristeri J, Petropoulos D, Kitra V, Papassarandis C, Graphakos S (2000) J Hematother Stem Cell Res 9:827
68. Hemmoranta H, Hautaniemi S, Niemi J, Nicorici D, Laine J, Yli-Harja O, Partanen J, Jaatinen T (2006) Stem Cells Dev 15:839
69. Gallacher L, Murdoch B, Wu DM, Karanu FN, Keeney M, Bhatia M (2000) Blood 95:2813
70. Cogle CR, Wainman DA, Jorgensen ML, Guthrie SM, Mames RN, Scott EW (2004) Blood 103:133
71. Loges S, Fehse B, Brockmann MA, Lamszus K, Butzal M, Guckenbiehl M, Schuch G, Ergun S, Fischer U, Zander AR, Hossfeld DK, Fiedler W, Gehling UM (2004) Stem Cells Dev 13:229
72. Asahara T, Masuda H, Takahashi T, Kalka C, Pastore C, Silver M, Kearne M, Magner M, Isner JM (1999) Circ Res 85:221
73. Peichev M, Naiyer AJ, Pereira D, Zhu Z, Lane WJ, Williams M, Oz MC, Hicklin DJ, Witte L, Moore MA, Rafii S (2000) Blood 95:952
74. Rafii S, Lyden D (2003) Nat Med 9:702
75. Asahara T, Kawamoto A (2004) Am J Physiol Cell Physiol 287:C572
76. Yang C, Zhang ZH, Li ZJ, Yang RC, Qian GQ, Han ZC (2004) Thromb Haemost 91:1202
77. Stamm C, Westphal B, Kleine HD, Petzsch M, Kittner C, Klinge H, Schumichen C, Nienaber CA, Freund M, Steinhoff G (2003) Lancet 361:45
78. Stamm C, Kleine HD, Westphal B, Petzsch M, Kittner C, Nienaber CA, Freund M, Steinhoff G (2004) Thorac Cardiovasc Surg 52:152
79. Yamada M, Kubo H, Kobayashi S, Ishizawa K, Numasaki M, Ueda S, Suzuki T, Sasaki H (2004) J Immunol 172:1266
80. Li TS, Hamano K, Nishida M, Hayashi M, Ito H, Mikamo A, Matsuzaki M (2003) Am J Physiol Heart Circ Physiol 285:H931
81. Quirici N, Soligo D, Caneva L, Servida F, Bossolasco P, Deliliers GL (2001) Br J Haematol 115:186
82. Wu X, Rabkin-Aikawa E, Guleserian KJ, Perry TE, Masuda Y, Sutherland FW, Schoen FJ, Mayer JE Jr, Bischoff J (2004) Am J Physiol Heart Circ Physiol 287:H480
83. Adams V, Lenk K, Linke A, Lenz D, Erbs S, Sandri M, Tarnok A, Gielen S, Emmrich F, Schuler G, Hambrecht R (2004) Arterioscler Thromb Vasc Biol 24:684
84. Suuronen EJ, Veinot JP, Wong S, Kapila V, Price J, Griffith M, Mesana TG, Ruel M (2006) Circulation 114:I138
85. Ferreira LS, Gerecht S, Shieh HF, Watson N, Rupnick MA, Dallabrida SM, Vunjak-Novakovic G, Langer R (2007) Circ Res 101:286
86. Tamaki S, Eckert K, He D, Sutton R, Doshe M, Jain G, Tushinski R, Reitsma M, Harris B, Tsukamoto A, Gage F, Weissman I, Uchida N (2002) J Neurosci Res 69:976
87. Invernici G, Emanueli C, Madeddu P, Cristini S, Gadau S, Benetti A, Ciusani E, Stassi G, Siragusa M, Nicosia R, Peschle C, Fascio U, Colombo A, Rizzuti T, Parati E, Alessandri G (2007) Am J Pathol 170:1879
88. Jones EA, Kinsey SE, English A, Jones RA, Straszynski L, Meredith DM, Markham AF, Jack A, Emery P, McGonagle D (2002) Arthritis Rheum 46:3349
89. Huss R, Moosmann S (2002) Br J Haematol 118:305
90. Majumdar MK, Keane-Moore M, Buyaner D, Hardy WB, Moorman MA, McIntosh KR, Mosca JD (2003) J Biomed Sci 10:228
91. Aslan H, Zilberman Y, Kandel L, Liebergall M, Oskouian RJ, Gazit D, Gazit Z (2006) Stem Cells 24:1728

92. Shi S, Gronthos S (2003) J Bone Miner Res 18:696
93. Tondreau T, Meuleman N, Delforge A, Dejeneffe M, Leroy R, Massy M, Mortier C, Bron D, Lagneaux L (2005) Stem Cells 23:1105
94. Kinnaird T, Stabile E, Burnett MS, Lee CW, Barr S, Fuchs S, Epstein SE (2004) Circ Res 94:678
95. Schrepfer S, Deuse T, Lange C, Katzenberg R, Reichenspurner H, Robbins RC, Pelletier MP (2007) Stem Cells Dev 16:105
96. Jiang Y, Jahagirdar BN, Reinhardt RL, Schwartz RE, Keene CD, Ortiz-Gonzalez XR, Reyes M, Lenvik T, Lund T, Blackstad M, Du J, Aldrich S, Lisberg A, Low WC, Largaespada DA, Verfaillie CM (2002) Nature 418:41
97. De Coppi P, Bartsch G Jr, Siddiqui MM, Xu T, Santos CC, Perin L, Mostoslavsky G, Serre AC, Snyder EY, Yoo JJ, Furth ME, Soker S, Atala A (2007) Nat Biotechnol 25:100
98. Sun S, Guo Z, Xiao X, Liu B, Liu X, Tang PH, Mao N (2003) Stem Cells 21:527
99. Quirici N, Soligo D, Bossolasco P, Servida F, Lumini C, Deliliers GL (2002) Exp Hematol 30:783
100. Gindraux F, Selmani Z, Obert L, Davani S, Tiberghien P, Herve P, Deschaseaux F (2007) Cell Tissue Res 327:471
101. Boquest AC, Shahdadfar A, Fronsdal K, Sigurjonsson O, Tunheim SH, Collas P, Brinchmann JE (2005) Mol Biol Cell 16:1131
102. Noer A, Sorensen AL, Boquest AC, Collas P (2006) Mol Biol Cell 17:3543
103. Astori G, Vignati F, Bardelli S, Tubio M, Gola M, Albertini V, Bambi F, Scali G, Castelli D, Rasini V, Soldati G, Moccetti T (2007) J Transl Med 5:55
104. Miranville A, Heeschen C, Sengenes C, Curat CA, Busse R, Bouloumie A (2004) Circulation 110:349
105. Simmons PJ, Torok-Storb B (1991) Blood 78:55
106. Gronthos S, Graves SE, Ohta S, Simmons PJ (1994) Blood 84:4164
107. Gronthos S, Simmons PJ (1995) Blood 85:929
108. Stewart K, Monk P, Walsh S, Jefferiss CM, Letchford J, Beresford JN (2003) Cell Tissue Res 313:281
109. Airey JA, Almeida-Porada G, Colletti EJ, Porada CD, Chamberlain J, Movsesian M, Sutko JL, Zanjani ED (2004) Circulation 109:1401
110. Gronthos S, Zannettino AC, Hay SJ, Shi S, Graves SE, Kortesidis A, Simmons PJ (2003) J Cell Sci 116:1827
111. Zannettino AC, Harrison K, Joyner CJ, Triffitt JT, Simmons PJ (2003) J Cell Biochem 89:56
112. Deschaseaux F, Gindraux F, Saadi R, Obert L, Chalmers D, Herve P (2003) Br J Haematol 122:506
113. Gang EJ, Bosnakovski D, Figueiredo CA, Visser JW, Perlingeiro RC (2007) Blood 109:1743
114. Martinez C, Hofmann TJ, Marino R, Dominici M, Horwitz EM (2007) Blood 109:4245
115. Reyes M, Lund T, Lenvik T, Aguiar D, Koodie L, Verfaillie CM (2001) Blood 98:2615
116. Niemeyer P, Kasten P, Simank HG, Fellenberg J, Seckinger A, Kreuz PC, Mehlhorn A, Sudkamp NP, Krause U (2006) Cytotherapy 8:354
117. O'Donoghue K, Choolani M, Chan J, de la Fuente J, Kumar S, Campagnoli C, Bennett PR, Roberts IA, Fisk NM (2003) Mol Hum Reprod 9:497
118. Wu Y, Chen L, Scott PG, Tredget EE (2007) Stem Cells 25:2648
119. Rombouts WJ, Ploemacher RE (2003) Leukemia 17:160
120. Yamamoto N, Akamatsu H, Hasegawa S, Yamada T, Nakata S, Ohkuma M, Miyachi E, Marunouchi T, Matsunaga K (2007) J Dermatol Sci 48:43
121. Reiß S, Herzig I, Schmitz J, Bosio A, Pennartz S (2008) Society for Neuroscience Meeting 2007: Abstract 235.7
122. Lois C, Alvarez-Buylla A (1993) Proc Natl Acad Sci U S A 90:2074
123. Luskin MB (1993) Neuron 11:173
124. Ghashghaei HT, Lai C, Anton ES (2007) Nat Rev Neurosci 8:141

125. Curtis MA, Kam M, Nannmark U, Anderson MF, Axell MZ, Wikkelso C, Holtas S, van Roon-Mom WM, Bjork-Eriksson T, Nordborg C, Frisen J, Dragunow M, Faull RL, Eriksson PS (2007) Science 315:1243
126. Weigmann A, Corbeil D, Hellwig A, Huttner WB (1997) Proc Natl Acad Sci U S A 94:12425
127. Fargeas CA, Corbeil D, Huttner WB (2003) Stem Cells 21:506
128. Miraglia S, Godfrey W, Yin AH, Atkins K, Warnke R, Holden JT, Bray RA, Waller EK, Buck DW (1997) Blood 90:5013
129. Hall PE, Lathia JD, Miller NG, Caldwell MA, ffrench-Constant C (2006) Stem Cells 24:2078
130. Uchida N, Buck DW, He D, Reitsma MJ, Masek M, Phan TV, Tsukamoto AS, Gage FH, Weissman IL (2000) Proc Natl Acad Sci U S A 97:14720
131. Yu S, Zhang JZ, Zhao CL, Zhang HY, Xu Q (2004) Biotechnol Lett 26:1131
132. Sawamoto K, Nakao N, Kakishita K, Ogawa Y, Toyama Y, Yamamoto A, Yamaguchi M, Mori K, Goldman SA, Itakura T, Okano H (2001) J Neurosci 21:3895
133. Lee A, Kessler JD, Read TA, Kaiser C, Corbeil D, Huttner WB, Johnson JE, Wechsler-Reya RJ (2005) Nat Neurosci 8:723
134. Pfenninger CV, Roschupkina T, Hertwig F, Kottwitz D, Englund E, Bengzon J, Jacobsen SE, Nuber UA (2007) Cancer Res 67:5727
135. Pennartz S, Belvindrah R, Tomiuk S, Zimmer C, Hofmann K, Conradt M, Bosio A, Cremer H (2004) Mol Cell Neurosci 25:692
136. Seidenfaden R, Desoeuvre A, Bosio A, Virard I, Cremer H (2006) Mol Cell Neurosci 32:187
137. Hierlihy AM, Seale P, Lobe CG, Rudnicki MA, Megeney LA (2002) FEBS Lett 530:239
138. Beltrami AP, Barlucchi L, Torella D, Baker M, Limana F, Chimenti S, Kasahara H, Rota M, Musso E, Urbanek K, Leri A, Kajstura J, Nadal-Ginard B, Anversa P (2003) Cell 114:763
139. Oh H, Bradfute SB, Gallardo TD, Nakamura T, Gaussin V, Mishina Y, Pocius J, Michael LH, Behringer RR, Garry DJ, Entman ML, Schneider MD (2003) Proc Natl Acad Sci U S A 100:12313
140. Oh H, Chi X, Bradfute SB, Mishina Y, Pocius J, Michael LH, Behringer RR, Schwartz RJ, Entman ML, Schneider MD (2004) Ann N Y Acad Sci 1015:182
141. Pfister O, Mouquet F, Jain M, Summer R, Helmes M, Fine A, Colucci WS, Liao R (2005) Circ Res 97:52
142. Laugwitz KL, Moretti A, Lam J, Gruber P, Chen Y, Woodard S, Lin LZ, Cai CL, Lu MM, Reth M, Platoshyn O, Yuan JX, Evans S, Chien KR (2005) Nature 433:647
143. Liao R, Pfister O, Jain M, Mouquet F (2007) Prog Cardiovasc Dis 50:18
144. Bearzi C, Rota M, Hosoda T, Tillmanns J, Nascimbene A, De Angelis A, Yasuzawa-Amano S, Trofimova I, Siggins RW, Lecapitaine N, Cascapera S, Beltrami AP, D'Alessandro DA, Zias E, Quaini F, Urbanek K, Michler RE, Bolli R, Kajstura J, Leri A, Anversa P (2007) Proc Natl Acad Sci U S A 104:14068
145. Smith RR, Barile L, Cho HC, Leppo MK, Hare JM, Messina E, Giacomello A, Abraham MR, Marban E (2007) Circulation 115:896
146. Ogawa T (2008) Int J Urol 15:121
147. Takehashi M, Kanatsu-Shinohara M, Miki H, Lee J, Kazuki Y, Inoue K, Ogonuki N, Toyokuni S, Oshimura M, Ogura A, Shinohara T (2007) Dev Biol 312:344
148. Ellen G, Herman T (2007) Curr Stem Cell Res Ther 2:189
149. Yamanaka S (2007) Cell Stem Cell 1:39
150. Seandel M, James D, Shmelkov SV, Falciatori I, Kim J, Chavala S, Scherr DS, Zhang F, Torres R, Gale NW, Yancopoulos GD, Murphy A, Valenzuela DM, Hobbs RM, Pandolfi PP, Rafii S (2007) Nature 449:346
151. Sahin MB, Schwartz RE, Buckley SM, Heremans Y, Chase L, Hu WS, Verfaillie CM (2008) Liver Transplant 14:333
152. Oertel M, Shafritz DA (2008) Biochim Biophys Acta 1782:61
153. Schmelzer E, Zhang L, Bruce A, Wauthier E, Ludlow J, Yao HL, Moss N, Melhem A, McClelland R, Turner W, Kulik M, Sherwood S, Tallheden T, Cheng N, Furth ME, Reid LM (2007) J Exp Med 204:1973

154. Schmelzer E, Reid LM (2008) Front Biosci 13:3096
155. Yovchev MI, Grozdanov PN, Joseph B, Gupta S, Dabeva MD (2007) Hepatology 45:139
156. Yovchev MI, Grozdanov PN, Zhou H, Racherla H, Guha C, Dabeva MD (2008) Hepatology 47:636
157. Tang C, Ang BT, Pervaiz S (2007) FASEB J 21:3777
158. Bonnet D, Dick JE (1997) Nat Med 3:730
159. Al-Hajj M, Wicha MS, Benito-Hernandez A, Morrison SJ, Clarke MF (2003) Proc Natl Acad Sci U S A 100:3983
160. Ponti D, Costa A, Zaffaroni N, Pratesi G, Petrangolini G, Coradini D, Pilotti S, Pierotti MA, Daidone MG (2005) Cancer Res 65:5506
161. Singh SK, Clarke ID, Terasaki M, Bonn VE, Hawkins C, Squire J, Dirks PB (2003) Cancer Res 63:5821
162. Singh SK, Hawkins C, Clarke ID, Squire JA, Bayani J, Hide T, Henkelman RM, Cusimano MD, Dirks PB (2004) Nature 432:396
163. Kim CF, Jackson EL, Woolfenden AE, Lawrence S, Babar I, Vogel S, Crowley D, Bronson RT, Jacks T (2005) Cell 121:823
164. Seo DC, Sung JM, Cho HJ, Yi H, Seo KH, Choi IS, Kim DK, Kim JS, El-Aty AA, Shin HC (2007) Mol Cancer 6:75
165. Eramo A, Lotti F, Sette G, Pilozzi E, Biffoni M, Di Virgilio A, Conticello C, Ruco L, Peschle C, De Maria R (2008) Cell Death Differ 15:504
166. Yang ZF, Ngai P, Ho DW, Yu WC, Ng MN, Lau CK, Li ML, Tam KH, Lam CT, Poon RT, Fan ST (2008) Hepatology 47:919
167. Richardson GD, Robson CN, Lang SH, Neal DE, Maitland NJ, Collins AT (2004) J Cell Sci 117:3539
168. Collins AT, Berry PA, Hyde C, Stower MJ, Maitland NJ (2005) Cancer Res 65:10946
169. O'Brien CA, Pollett A, Gallinger S, Dick JE (2007) Nature 445:106
170. Ricci-Vitiani L, Lombardi DG, Pilozzi E, Biffoni M, Todaro M, Peschle C, De Maria R (2007) Nature 445:111
171. Fang D, Nguyen TK, Leishear K, Finko R, Kulp AN, Hotz S, Van Belle PA, Xu X, Elder DE, Herlyn M (2005) Cancer Res 65:9328
172. Kamstrup M, Gniadecki R, Skovgard G (2007) Exp Dermatol 16: 1030
173. Li C, Heidt DG, Dalerba P, Burant CF, Zhang L, Adsay V, Wicha M, Clarke MF, Simeone DM (2007) Cancer Res 67:1030
174. Handgretinger R, Chen X, Pfeiffer M, Mueller I, Feuchtinger T, Hale GA, Lang P (2007) Ann N Y Acad Sci 1106:279
175. Lee SJ, Kim HT, Ho VT, Cutler C, Alyea EP, Soiffer RJ, Antin JH (2006) Bone Marrow Transplant 38:305
176. Reisner Y, Martelli MF (2000) Curr Opin Immunol 12:536
177. Handgretinger R, Klingebiel T, Lang P, Gordon P, Niethammer D (2003) Pediatr Transplant 7(Suppl 3):51
178. Lang P, Bader P, Schumm M, Feuchtinger T, Einsele H, Fuhrer M, Weinstock C, Handgretinger R, Kuci S, Martin D, Niethammer D, Greil J (2004) Br J Haematol 124:72
179. Elmaagacli AH, Peceny R, Steckel N, Trenschel R, Ottinger H, Grosse-Wilde H, Schaefer UW, Beelen DW (2003) Blood 101:446
180. Falzetti F (2002) Bone Marrow Transplant 29(Suppl 2):199
181. Aversa F, Reisner Y, Martelli MF (2008) Blood Cells Mol Dis 40:8
182. Ruggeri L, Capanni M, Urbani E, Perruccio K, Shlomchik WD, Tosti A, Posati S, Rogaia D, Frassoni F, Aversa F, Martelli MF, Velardi A (2002) Science 295:2097
183. Ruggeri L, Mancusi A, Burchielli E, Aversa F, Martelli MF, Velardi A (2006) Cytotherapy 8:554
184. Vonk MC, Marjanovic Z, van den Hoogen FH, Zohar S, Schattenberg AV, Fibbe WE, Larghero J, Gluckman E, Preijers FW, van Dijk AP, Bax JJ, Roblot P, van Riel PL, van Laar JM, Farge D (2008) Ann Rheum Dis 67:98
185. Alexander T, Thiel A, Massenkeil G, Sattler A, Kohler S, Mei H, Radtke H, Burmester G, Radbruch A, Arnold R, Hiepe F (2008) 34th Annual Meeting of the European Group for Blood and Marrow Transplantation Abstract, p O103

186. Beltrami A, Urbanek K, Kajstura J, Yan S, Finato N, Bussani R, Nadal-Ginard B, Silvestri F, Leri A, Beltrami C, Anversa P (2001) N Engl J Med 344:1750
187. Gill M, Dias S, Hattori K, Rivera M, Hicklin D, Witte L, Girardi L, Yurt R, Himel H, Rafii S (2001) Circ Res 88:167
188. Shintani S, Murohara T, Ikeda H, Ueno T, Honma T, Katoh A, Sasaki K, Shimada T, Oike Y, Imaizumi T (2001) Circulation 103:2776
189. Gepstein L (2002) Circ Res 91:866
190. Kuehnle I, Goodell M (2002) Br Med J 325:372
191. Schultz E, McCormick K (1994) Rev Physiol Biochem Pharmacol 123:213
192. Rubart M, Soonpaa M, Nakajima H, Field L (2004) J Clin Invest 114:775
193. Murtuza B, Suzuki K, Bou-Gharios G, Beauchamp J, Smolenski R, Partridge T, Yacoub M (2004) Proc Natl Acad Sci U S A 101:4216
194. Menasché P, Hagège A, Scorsin M, Pouzet B, Desnos M, Duboc D, Schwartz K, Vilquin J, Marolleau J (2001) Lancet 357:279
195. Herreros J, Prósper F, Perez A, Gavira J, Garcia-Velloso M, Barba J, Sánchez P, Cañizo C, Rábago G, Martí-Climent J, Hernández M, López-Holgado N, González-Santos J, Martín-Luengo C, Alegria E (2003) Eur Heart J 24:2012
196. Dib N, McCarthy P, Campbell A, Yeager M, Pagani F, Wright S, MacLellan W, Fonarow G, Eisen H, Michler R, Binkley P, Buchele D, Korn R, Ghazoul M, Dinsmore J, Opie S, Diethrich E (2005) Cell Transplant 14:11
197. Gavira J, Herreros J, Perez A, Garcia-Velloso M, Barba J, Martin-Herrero F, Cañizo C, Martin-Arnau A, Martí-Climent J, Hernández M, López-Holgado N, González-Santos J, Martín-Luengo C, Alegria E, Prósper F (2006) J Thorac Cardiovasc Surg 31:799
198. Menasché P, Hagège A, Vilquin J, Desnos M, Abergel E, Pouzet B, Bel A, Sarateanu S, Scorsin M, Schwartz K, Bruneval P, Benbunan M, Marolleau J, Duboc D (2003) J Am Coll Cardiol 114:108
199. Hagège A, Marolleau J, Vilquin J, Alhéritière A, Peyrard S, Duboc D, Abergel E, Messas E, Mousseaux E, Schwartz K, Desnos M, Menasché P (2006) Circulation 114:108
200. Tolmachov O, Ma Y, Themis M, Patel P, Spohr H, Macleod K, Ullrich N, Kienast Y, Coutelle C, Peters N (2006) BMC Cardiovasc Disord 6:25
201. Orlic D, Kajstura J, Chimenti S, Jakoniuk I, Anderson SM, Li B, Pickel J, McKay R, Nadal-Ginard B, Bodine DM, Leri A, Anversa P (2001) Nature 410:701
202. Abdel-Latif A, Bolli R, Tleyjeh I, Montori V, Perin E, Hornung C, Zuba-Surma E, Al-Mallah M, Dawn B (2007) Arch Intern Med 167:989
203. Pittenger M, Mackay A, Beck S, Jaiswal R, Douglas R, Mosca J, Moorman M, Simonetti D, Craig S, Marshak D (1999) Science 284:143
204. Silva GV, Litovsky S, Assad JA, Sousa AL, Martin BJ, Vela D, Coulter SC, Lin J, Ober J, Vaughn WK, Branco RV, Oliveira EM, He R, Geng YJ, Willerson JT, Perin EC (2005) Circulation 111:150
205. Breitbach M, Bostani T, Roell W, Xia Y, Dewald O, Nygren JM, Fries JW, Tiemann K, Bohlen H, Hescheler J, Welz A, Bloch W, Jacobsen SE, Fleischmann BK (2007) Blood 110:1362
206. Strauer BE, Brehm M, Zeus T, Kostering M, Hernandez A, Sorg RV, Kogler G, Wernet P (2002) Circulation 106:1913
207. Perin EC, Dohmann HF, Borojevic R, Silva SA, Sousa AL, Mesquita CT, Rossi MI, Carvalho AC, Dutra HS, Dohmann HJ, Silva GV, Belem L, Vivacqua R, Rangel FO, Esporcatte R, Geng YJ, Vaughn WK, Assad JA, Mesquita ET, Willerson JT (2003) Circulation 107:2294
208. Wollert KC, Meyer GP, Lotz J, Ringes-Lichtenberg S, Lippolt P, Breidenbach C, Fichtner S, Korte T, Hornig B, Messinger D, Arseniev L, Hertenstein B, Ganser A, Drexler H (2004) Lancet 364:141
209. Fernandez-Aviles F, San Roman JA, Garcia-Frade J, Fernandez ME, Penarrubia MJ, de la Fuente L, Gomez-Bueno M, Cantalapiedra A, Fernandez J, Gutierrez O, Sanchez PL, Hernandez C, Sanz R, Garcia-Sancho J, Sanchez A (2004) Circ Res 95:742
210. Galinanes M, Loubani M, Davies J, Chin D, Pasi J, Bell PR (2004) Cell Transplant 13:7

211. Assmus B, Schachinger V, Teupe C, Britten M, Lehmann R, Dobert N, Grunwald F, Aicher A, Urbich C, Martin H, Hoelzer D, Dimmeler S, Zeiher AM (2002) Circulation 106:3009
212. Schachinger V, Assmus B, Britten MB, Honold J, Lehmann R, Teupe C, Abolmaali ND, Vogl TJ, Hofmann WK, Martin H, Dimmeler S, Zeiher AM (2004) J Am Coll Cardiol 44:1690
213. Lunde K, Solheim S, Aakhus S, Arnesen H, Abdelnoor M, Egeland T, Endresen K, Ilebekk A, Mangschau A, Fjeld JG, Smith HJ, Taraldsrud E, Grogaard HK, Bjornerheim R, Brekke M, Muller C, Hopp E, Ragnarsson A, Brinchmann JE, Forfang K (2006) N Engl J Med 355:1199
214. Kocher AA, Schuster MD, Szabolcs MJ, Takuma S, Burkhoff D, Wang J, Homma S, Edwards NM, Itescu S (2001) Nat Med 7:430
215. Orlic D, Kajstura J, Chimenti S, Bodine DM, Leri A, Anversa P (2001) Ann N Y Acad Sci 938:221
216. Quaini F, Urbanek K, Beltrami AP, Finato N, Beltrami CA, Nadal-Ginard B, Kajstura J, Leri A, Anversa P (2002) N Engl J Med 346:5
217. Goodell MA (2003) Curr Opin Hematol 10:208
218. Camargo FD, Chambers SM, Goodell MA (2004) Cell Prolif 37:55
219. Hofmann M, Wollert KC, Meyer GP, Menke A, Arseniev L, Hertenstein B, Ganser A, Knapp WH, Drexler H (2005) Circulation 111:2198
220. Agbulut O, Vandervelde S, Al Attar N, Larghero J, Ghostine S, Leobon B, Robidel E, Borsani P, Le Lorc'h M, Bissery A, Chomienne C, Bruneval P, Marolleau JP, Vilquin JT, Hagege A, Samuel JL, Menasche P (2004) J Am Coll Cardiol 44:458
221. Hamano K, Nishida M, Hirata K, Mikamo A, Li TS, Harada M, Miura T, Matsuzaki M, Esato K (2001) Jpn Circ J 65:845
222. Leor J, Guetta E, Feinberg MS, Galski H, Bar I, Holbova R, Miller L, Zarin P, Castel D, Barbash IM, Nagler A (2006) Stem Cells 24:772
223. Ma N, Ladilov Y, Moebius JM, Ong L, Piechaczek C, David A, Kaminski A, Choi YH, Li W, Egger D, Stamm C, Steinhoff G (2006) Cardiovasc Res 71:158
224. Klein HM, Ghodsizad A, Borowski A, Saleh A, Draganov J, Poll L, Stoldt V, Feifel N, Piecharczek C, Burchardt ER, Stockschlader M, Gams E (2004) Heart Surg Forum 7:E416
225. Ghodsizad A, Klein HM, Borowski A, Stoldt V, Feifel N, Voelkel T, Piechaczek C, Burchardt E, Stockschlader M, Gams E (2004) Cytotherapy 6:523
226. Pompilio G, Cannata A, Peccatori F, Bertolini F, Nascimbene A, Capogrossi MC, Biglioli P (2004) Ann Thorac Surg 78:1808
227. Stamm C, Kleine HD, Choi YH, Dunkelmann S, Lauffs JA, Lorenzen B, David A, Liebold A, Nienaber C, Zurakowski D, Freund M, Steinhoff G (2007) J Thorac Cardiovasc Surg 133:717
228. Klein HM, Ghodsizad A, Marktanner R, Poll L, Voelkel T, Mohammad Hasani MR, Piechaczek C, Feifel N, Stockschlaeder M, Burchardt ER, Kar BJ, Gregoric I, Gams E (2007) Heart Surg Forum 10:E66

Adv Biochem Engin/Biotechnol (2009) 114: 73-106
DOI: 10.1007/10_2008_49

Published online: 28 March 2009

Transdifferentiation of Stem Cells: A Critical View

Ina Gruh and Ulrich Martin

Abstract Recently a large amount of new data on the plasticity of stem cells of various lineages have emerged, providing new perspectives especially for the therapeutic application of adult stem cells. Previously unknown possibilities of cell differentiation beyond the known commitment of a given stem cell have been described using keywords such as "blood to liver," or "bone to brain." Controversies on the likelihood, as well as the biological significance, of these conversions almost immediately arose within this young field of stem cell biology. This chapter will concentrate on these controversies and focus on selected examples demonstrating the technical aspects of stem cell transdifferentiation and the evaluation of the tools used to analyze these events.

Keywords Adult stem cells, Transdifferentiation

Contents

I. Gruh and U. Martin (✉)
Leibniz Research Laboratories for Biotechnology and Artificial Organs LEBAO, Hannover Medical School, Carl-Neuberg-Str. 1, 30625, Hannover, Germany
e-mail: gruh.ina@mh-hannover.de, martin.ulrich@mh-hannover.de

Abbreviations

BM	Bone marrow
BMC	Bone marrow cells
CAC	Circulating angiogenic cells
eGFP	Enhanced green fluorescent protein
EPC	Endothelial progenitor cells
FISH	Fluorescence in situ hybridization
GFP	Green fluorescent protein
HNF	Hepatocyte nuclear factor
HSC	Hematopoietic stem cells
HUVEC	Human umbilical cord vein cells
MAPC	Multipotent adult progenitor cells
MHC	Myosin heavy chain
MNC	Mononuclear cells
MSC	Mesenchymal stem cells
NOD-SCID	Nonobese diabetic severe combined immunodeficient
NRCM	Neonatal rat cardiomyocytes
USSC	Unrestricted somatic stem cells

1 Introduction

During the past decade, stem cell research has become a rapidly evolving field providing new insights into developmental biology, as well as new hope for therapeutic applications. The most versatile stem cells to date are pluripotent embryonic stem cells (ESC) with the capability of differentiating into the whole panel of somatic cell types derived from all three germ layers, i.e., endoderm, mesoderm and ectoderm. Some cell types which can be generated from adult tissue have now been described to have similar characteristics; these cells include the so called "induced pluripotent" stem (iPS) cells [1, 2] or germ-line derived stem cells [3]. Notably, it is not clear at present whether the adult testis contains rare pluripotent stem cells in vivo. It is considered more likely that isolated unipotent spermatogonial stem cells can be reprogrammed into pluripotent stem cells under certain culture conditions. In contrast to ESC, the natural potential of stem and progenitor cells found in various organs of the adult body appears to be limited and was initially considered restricted to cells related to the respective organs, or at least derived from the same germ layer. This concept was challenged by reports on the plasticity of stem cells of various lineages going beyond these boundaries, an event which is often referred to as "transdifferentiation."

However, a critical view on the "transdifferentiation of stem cells" should start with a critical view on the term itself. The observation that one cell type can change its phenotype and become another cell type in vivo was described in 1922 by Maccarty et al. to occur in ovarian tumors [4]. This phenomenon was termed "metaplasia" and believed to be mainly a response to physiological or pathological stress.

A classical example of metaplasia is the epithelial–mesenchymal transition (EMT), a highly conserved and fundamental process, mediated by transforming growth factor β (TGF-β) signaling, that governs morphogenesis in embryonic development and may also contribute to cancer metastasis [5]. The most prominent feature of EMT is the complete loss of epithelial traits, such as E-cadherin expression, by the former epithelial cells and the acquisition of mesenchymal characteristics, such as vimentin and fibronectin expression, gaining invasive motility and others [6, 7].

In the adult organism, examples of metaplasia can be found in the eye, with reports dating back as early as 1934 [8]. More recent reports include the conversion of limbal basal epithelial cells into corneal epithelial cells [9], retinal pigmental epithelial (RPE) cells into neural epithelium [10], conjunctival epithelial cells into corneal epithelium [11] and neural retina into lens epithelium [12]. Another form of metaplasia in the eye, the conversion of lens epithelial cells into myofibroblasts [13], reflects a common mechanism of the body in response to injury, i.e., the replacement of functional tissue specific cells by myofibroblasts, e.g., in scar formation. This process is mediated by increased levels of tumor necrosis factor α (TNF-α) and/or TGF-β, and has been described for a large variety of cell types including, but not limited to, fat storing cells in the liver [14], tubular epithelial cells in the kidney [15], keratocytes in the skin [16], fibroblasts in the lung [17], the heart [18], and the prostate [19], as well as Schwann cells in the brain [20].

"Transdifferentiation is a subclass of metaplasia and by definition an irreversible switch of one already differentiated cell to another, resulting in the loss of one phenotype and the gain of another" [21].

Like other sources, this statement by Eberhard and Tosh explicitly defines "transdifferentiation" as a "nonstem cell" transformation. Therefore, under a critical view, the expression "transdifferentiation of adult stem cells" seems to be contradictory in itself. However, in recent years this classical definition has been broadened when it became evident that adult stem cells with a presumed commitment not only underwent differentiation into anticipated progenies, but differentiation also resulted in phenotypes beyond the expected lineage of the respective stem cells. This "plasticity," which has been defined as the ability to undergo transdifferentiation, can be seen, for example, in the differentiation of hematopoietic stem cells into nonblood cells. Subsequently, we will use this broadened definition to investigate the alleged transdifferentiation of stem cells into, or from various tissues, reviewing conflicting reports in this relatively new field of stem cell research with a focus on technical aspects of the given data, the methods used, and their power to prove differentiation events unequivocally.

2 Mechanism of Stem Cell Transdifferentiation

Stem cells were thought to differentiate usually into one or more typical cell types of the very tissue from which the respective stem cell originated. In addition to this lineage-restricted multipotentiality, stem cells, under certain circumstances seem to be able to cross lineage boundaries and differentiate into atypical cell types, or, as Rota et al. expressed it, to "break the law of tissue fidelity" [22]. Theoretically, this transdifferentiation can occur directly, or via the generation of an intermediate cell type. In this case, a *de*-differentiation of the stem cell would be followed by a subsequent differentiation into another cell type [23] (Fig. 1).

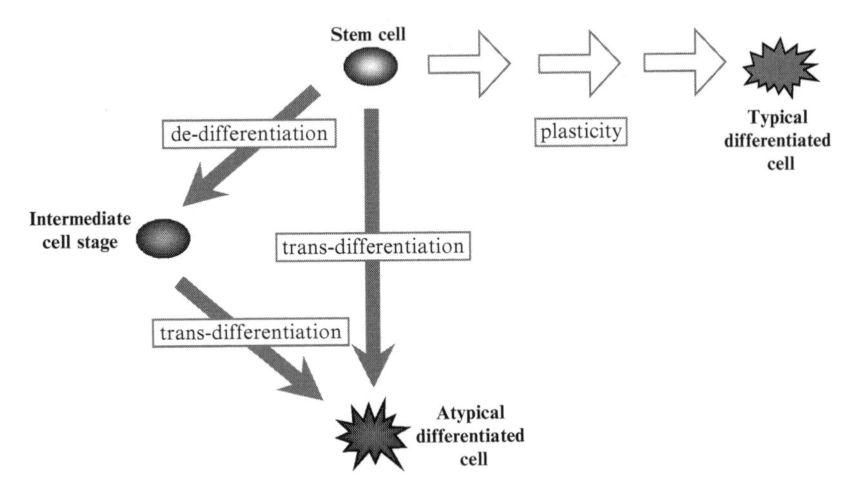

Fig. 1 Mechanism of stem cell transdifferentiation. Modified after Koestenbauer et al. [23]

Early reports on this previously unobserved form of differentiation were fairly surprising: "… But just because we scientists were surprised, it does not mean that the cells themselves were surprised by their broad potential! …" Eisenberg and Eisenberg [24].

3 Examples of Adult Stem Cell Transdifferentiation

3.1 Transdifferentiation into Hepatocytes

3.1.1 Hematopoietic Stem Cells

In vivo, liver progenitor/oval cells differentiate into hepatocytes and biliary epithelial cells, repopulating the liver when the regenerative capacity of hepatocytes is impaired. Bone marrow (BM) derived hematopoietic stem cells (HSC), which, apart from their putative main function in the body, i.e., replenishing blood cells, have also been associated with organ repair. Transdifferentiation has been proposed as one underlying mechanism. After Petersen et al. identified BM as source of hepatic cells in 1999 [25], Lagasse et al. confirmed a therapeutic effect of HSC transplantation in mice with an inherited liver disease [26] and Theise et al. demonstrated similar effects in humans [27]. This idea was then challenged by contradictory reports by Wagers et al. [28], Dahlke et al. [29] and others [30, 31], introducing cell fusion as an alternative mechanism to transdifferentiation. Closer investigation of the methods used to analyze stem cell transdifferentiation in these respective studies provides insight into some of the contradicting results.

Petersen et al. recognized the bone marrow as a potential source of hepatic oval cells using cross-sex or cross-strain BM and whole liver transplantation in rats to trace the origin of the repopulating liver cells [25]. Following liver injury a proportion of the regenerated hepatic cells were shown to be donor-derived as identified by markers for Y-chromosome, dipeptidyl peptidase IV (DPP IV) enzyme, and L21-6 antigen. Immunohistochemical staining of hepatocyte-specific cytokeratins and fluorescence in situ hybridization (FISH) for X- and Y-chromosomes identified hepatocyte engraftment. This was observed both in human females receiving male BM transplants and in male recipients of orthotopic female liver transplants [27]. In this study, peak values were observed in one of the liver transplant recipients with recurrent hepatitis C. Therefore, this setting resembled an injury approach in an animal model like that of Lagasse et al. who demonstrated that mice with an inherited liver disease (corresponding to human tyrosinaemia type 1) could be cured by HSC transplantation leading to the reconstitution of functioning mature hepatocytes [26].

In follow-up studies, more sophisticated methods have been used to investigate the controversial fate of hematopoietic stem cells in the liver. Using chimeric animals, as well as green fluorescent protein (GFP)-positive:GFP-negative parabiotic mice, Wagers et al. showed that single HSC robustly reconstituted the BM, as well as peripheral blood leukocytes in these animals, but did not contribute appreciably to nonhematopoietic tissues, including brain, kidney, gut, liver, and muscle. It was concluded that transdifferentiation of circulating HSC and/or their progeny is an extremely rare event, if it occurs at all [28]. Wang et al. performed serial transplantation of BM-derived hepatocytes [31]. Southern blot analysis and cytogenetic analysis of hepatocytes transplanted from female donor mice into male recipients provided evidence of fusion between donor and host cells rather than liver-specific (trans-) differentiation of hematopoietic stem cells. Reviewing the role of various stem cell populations, including hematopoietic stem cells in liver regeneration, Dahlke et al. claimed that closer scrutiny of the data published by Lagasse et al. [26] also reveals that cell fusion rather than transdifferentiation appears to be responsible for liver regeneration in their model [29].

Further studies investigating whether BM-derived liver progenitor/oval cells can repopulate the liver were unable to confirm the early data by Theise et al. and Lagasse et al. One possible explanation for this discrepancy might be the time point of analysis. Menthena et al. transplanted lethally irradiated female DPP IV-negative mutant F344 rats with wild-type male F344 BM cells [30]. Initially, donor-derived cells were detected in all liver sections of recipient rats after the application of different liver injury protocols. However, most of the donor-derived clusters disappeared over time and very few oval cells (less than 1%) and none of the small hepatocytic clusters showed double labeling for the donor-derived DPP IV and hepatocyte markers. Consequently, the authors conclude that the sources of oval cells and small hepatocytes in the injured liver are endogenous liver progenitors which do not arise through transdifferentiation from BM cells.

In a comprehensive review of the available data, Thorgeirsson et al. suggested that one or more types of hematopoietic cells may rarely acquire the hepatocyte phenotype in the liver (frequency $\sim 10^{-4}$). However, the nature of the hematopoietic cells involved and the mechanisms responsible for acquisition of a hepatocyte phenotype are still controversial. HSC do not appear to be direct precursors of hepatocytes; instead hepatocytes that carry a BM tag can be generated by fusion of hepatocytes with cells of the macrophage–monocyte lineage [32], which have been reported to be highly fusogenic [33]. Thorgeirsson et al. concluded that hematopoietic cells contribute little to hepatocyte formation under either physiological or pathological conditions, but may provide cytokines and growth factors that promote hepatocyte functions by paracrine mechanisms.

Thus, an important question was raised by Thorgeirsson et al., that is which specific type of hematopoietic stem cell may be able to support liver regeneration. Meanwhile, different subsets of HSC were analyzed with respect to their hepatic differentiation capacity [34–36], again yielding contradictory findings. The discussion around stem cell identity and definition of a pure population involves another issue, namely the request for the use of single cells as the ultimate test for multipotentiality of a given stem cell. In 2001 Krause et al. demonstrated multiorgan, multilineage engraftment by a single BM-derived stem cell using an elegant model of serial stem cell transplantation in mice [37]. Injection of single, selected BM stem cells generated a variable proportion of epithelial cells in various organs such as the lung, gastrointestinal tract, skin, and liver. Notably in the liver, only BM-derived cholangiocytes were detected, and no bone marrow derived hepatocytes [37].

According to Krause et al., the different engraftment frequencies in different organs observed in their study may be due to (1) the degree of tissue damage induced by the transplant, (2) the residual tissue-specific stem cell capacity within the organ, and/or (3) the normal rate of cell turnover in each organ [37]. These factors, however, might also explain some of the differences in the outcome of other studies, i.e., the formation of BM-derived liver cells in the presence of tissue injury [25, 26] while no or low numbers of such cells were detected in the absence of injury [27].

Recently, another interesting explanation of apparent transdifferentiation events in the liver was proposed. In a comparative study, Brulport et al. transplanted four different types of human extrahepatic precursor cells (cord blood derived, monocytes, BM, and pancreatic) into the livers of NOD/SCID mice. Initial results argued in favor of hepatic differentiation of the transplanted cells as they stained positive for human albumin and glycogen, given that the cells were negative for both markers before transplantation. However, cells with human nuclei (detected by in situ hybridization with human DNA-specific alu probes) did not show a hepatocyte-like morphology. In addition, they did not express cytochrome P450 3A4, a key marker of functional hepatocytes, suggesting that the engrafted human cells represented a mixed cell type potentially resulting from partial transdifferentiation. Surprisingly, a human albumin-positive cell type with hepatocyte-like morphology was found to contain a mouse, but not a human nucleus, therefore challenging the existence of human cell transdifferentiation. Although unproven, Brulport et al. suggest horizontal gene transfer as a likely mechanism, especially because small fragments of human nuclei were observed in mouse cells that originated from deteriorating transplanted cells. In conclusion, Brulport et al. provided evidence not favoring transdifferentiation, but rather suggesting a complex situation including partial differentiation of cord blood-derived donor cells and possibly horizontal gene transfer.

3.1.2 Mesenchymal Stem Cells

In addition to hematopoietic stem cells, the BM contains mesenchymal stem cells (MSC), another type of stem cell extensively studied for organ regeneration. MSC are typically enriched via isolation of the plastic adherent, fibroblast-like cell fraction. Despite their functional heterogeneity, MSC populations obtained from various tissues commonly express a number of surface receptors including CD29, CD44, CD49a–f, CD51, CD73, CD105, CD106, CD166, and Stro1 and lack expression of definitive hematopoietic lineage markers including CD11b, CD14, and CD45 [38]. Mesenchymal stem cells were also detected in the peripheral blood, most likely mobilized from the BM [39].

While their differentiation into adipocytes, chondrocytes and osteocytes as described by Prockop et al. [40] has become the gold standard for proving MSC differentiation capacity, reports on MSC (trans-)differentiation into other lineages such as hepatocytes are highly controversial. Contribution of MSC to the liver has been described in baboons by Devine et al. who infused MSC retrovirally tagged with enhanced GFP (eGFP) in adult animals following lethal total body irradiation [41]. The resulting data, 9–21 months later, suggested that MSC could contribute to the liver and possess the capacity to proliferate in a hepatic environment. In vitro differentiation into hepatocyte-like phenotypes has also been described for MSC derived from several species including mice [42], rats [43] and humans [44].

One problem concerning reports on the potential contribution of BM-derived cells to liver regeneration, is the lack of a comparable definition of the cell type used. Most of the early studies investigated whole BM preparations, while others defined certain subpopulations, such as recycling stem (RS-) cells [45, 46] or "human bone-marrow derived multipotent stem cells" (hBMSC) [47]. For example, Verfaillie's laboratory was able to demonstrate that postnatal BM-derived multipotent adult progenitor cells (MAPC) can differentiate into hepatocyte-like cells in vitro [48]. While MAPC copurify from the BM with MSC, they are considered a distinct population with a different phenotype. Human and rodent MAPC represent a CD44-negative, CD45-negative, HLA class I- and II-negative, as well as a cKit-negative subset of cells. When cultured on Matrigel with FGF-4 and HGF, they differentiated into epithelioid cells that expressed hepatocyte nuclear factor 3β

(HNF-3β), GATA4, cytokeratin 19 (CK19), transthyretin, and α-fetoprotein by day 7, and expressed CK18, HNF-4, and HNF-1α on days 14–28 [48]. Another in vitro study by Khurana et al. characterized the potential subpopulation of BM cells (BMC) involved in the repair of injured liver tissue to be a distinct subset of lineage (Lin)-negative BMC coexpressing CXCR4 and oncostatin M receptor β (OSMRβ), with/without stem cell antigen-1 (sca-1) [49].

Another problem became evident by the identification of cell fusion as the underlying mechanism for some of the earlier observations on MSC transdifferentiation, very similar to what has been outlined above for hematopoietic stem cells. Alvarez-Diego et al. described the cell fusion between MSC and resident liver cells detected by means of sophisticated genetic labeling [50]. For this study, mice expressing Cre recombinase ubiquitously under the control of a hybrid cytomegalovirus (CMV) enhancer β-actin promoter were used, and the conditional Cre reporter mouse line R26R. In this line, the LacZ reporter is exclusively expressed after the excision of a loxP-flanked (floxed) stop cassette by Cre mediated recombination, resulting in expression of the LacZ in fused cells.

Nevertheless, cell fusion not only accounts for misleading data on stem cell transdifferentiation, but can also have a therapeutic effect. Vassilopoulos et al. reported that transplanted BM regenerates liver by cell fusion in a model of tyrosinaemia type I [51]. Transplanted mice regained normal liver function and formed regenerating liver nodules with normal histology. Their hepatocytes expressed both donor and host genes, consistent with polyploid genome formation by fusion of host and donor cells.

Partial transdifferentiation was also observed, resulting in a chimeric phenotype with the expression of several lineage markers, but missing other markers fundamental to a bona fide functional cell type of a particular tissue. Lysy et al. demonstrated the persistence of a chimerical phenotype after hepatocyte differentiation of human BM-derived MSC, with the MSC partially preserving their mesenchymal phenotype [52]. Only after transplantation of MSC-derived hepatocyte-like cells into the liver of SCID mice did these cells lose their chimeric phenotype, but they conserved their hepatocyte-lineage markers, indicating that a hepatic environment in vivo is necessary for full maturation into functional hepatocytes.

To date, there is still not a common understanding of the processes occurring after transplantation of mesenchymal stem cells into the hepatic environment; thus further research will be needed to clarify the mechanism, in addition to the biological significance of MSC contribution to the liver.

3.2 Transdifferentiation into Myocytes

In contrast to studies on the hepatic differentiation of HSC, mostly investigating liver repopulation by circulating cells in vivo, studies on the conversion of HSC into different muscle cell types largely focused on stem cell transplantation via transmural injection directly into skeletal or heart muscle tissue in vivo.

In 1998, Ferrari et al. reported that BM cells can contribute to myogenesis in response to physiological stimuli [53]. However, according to Ferrari et al., the origin of the BM-derived myogenic cells, as well as their physiological role in the homeostasis of muscle tissue, could not be defined. Further studies concentrated on the identification of the myogenic cell type within the BM. In 2003, using a lineage tracing strategy, Corbel et al. showed that the progeny of a single HSC can both reconstitute the hematopoietic system and contribute to muscle regeneration [54]. Other reports identified immature myeloid cells as the predominant source of myogenic differentiation in vivo. Doyonnas et al. used fluorescence-activated cell sorter (FACS)-based protocols to test distinct hematopoietic fractions and showed that only fractions containing c-kit-positive immature myelomonocytic precursors were capable of contributing to muscle fibers after intramuscular injection [55]. In a similar approach, Abedi et al. transplanted animals with different populations of BMC from GFP transgenic mice, and the presence of GFP-positive muscle fibers were evaluated in cardiotoxin-injured *tibialis anterior* muscles [56]. GFP-positive muscle fibers were found mostly in animals that received either CD45-negative, Lin-negative, c-Kit-positive, Sca-1-positive or Flk-2-positive populations of BMC, suggesting that HSC rather than mesenchymal cells or more differentiated hematopoietic cells are responsible for the formation of GFP-positive muscle fibers. According to Adebi et al. and in contrast to Doyonnas et al., a CD11b-positive population of BMC was also associated with the emergence of GFP-positive skeletal muscle fibers.

While the contribution of HSC to skeletal muscle regeneration was confirmed by several groups, the exact phenotype and developmental stage of contributing cells, as well as the exact mechanism remains to be elucidated. Particularly, the question as to what extent cell fusion might play a role in this setting has not been answered. In contrast, the probability of adult stem cell contribution to cardiac muscle is still the subject of an ongoing debate.

3.2.1 Whole Bone Marrow and Hematopoietic Stem Cells

Initial reports on the possibility of BM-derived stem cells to regenerate cardiac myocytes after myocardial infarction in vivo were published by the group of Piero Anversa [57, 58] and others [59–61] while Eisenberg et al. proposed cardiac differentiation in vitro [62].

In 2001, Orlic et al. investigated whether ischemia damaged myocardium could be restored by transplanting BMC into infarcted mice [58]. Shortly after coronary ligation, Lin-negative/c-kit-positive cells were injected in the heart muscle wall bordering the infarct. This study claimed that donor cell-derived, newly formed myocardium occupied 68% of the infarcted portion of the ventricle 9 days after transplantation. In a similar study from the same group, a sex-mismatched mouse model with male eGFP-positive donor animals demonstrated that the engrafted cells were positive for eGFP, Y chromosome, and several myocyte-specific proteins including cardiac myosin and the transcription factors GATA-4, MEF2, and

Csx/Nkx2.5 [57]. The authors concluded that locally delivered BMC can generate de novo myocardium, ameliorating the outcome of coronary artery disease by improving several hemodynamic parameters [57, 58].

Coculture experiments with adult mouse BM cells and embryonic heart tissue seemed to confirm that hematopoietic progenitor cells are able both to integrate into cardiac tissue and to differentiate into cardiomyocytes [62]. Remarkably, Eisenberg et al. reported that macrophages cocultured with cardiac explants were also able to integrate into contractile heart tissue and undergo cardiac differentiation. Another cell population from the BM, the so-called Side Population (SP) cells, or highly purified CD34-negative/c-kit-positive/sca-1-positive cells, have also been reported to differentiate into cardiac lineages and improve cardiac function after transplantation into infarcted myocardium [60, 61]. According to Agbulut et al. BM-derived cells that can contribute to cardiac differentiation are present in total unpurified BM, but not in the sca-1-positive hematopoietic progenitor cell population [59]. However, the very small number of transdifferentiated cells (5.6 ± 2.3 cells per 3×10^{-2} mm^3 of mouse heart tissue at 7 days after transplantation of 6×10^6 cells) raised concern regarding their functional efficiency.

These early reports on transdifferentiation were challenged by contradictory data. Nygren et al. reported that BM-derived hematopoietic cells generate cardiomyocytes at a low frequency through cell fusion, but not transdifferentiation [63]. While they were able to confirm earlier reports on efficient engrafting of unfractionated BMC and a purified population of hematopoietic stem and progenitor cells to the injured myocardium, they also found this engraftment to be transient. In addition, all engrafted cells expressed the pan-hematopoietic marker CD45, coexpressed myeloid blood lineage markers (Gr-1/Mac-1) failed to express cardiac-specific markers. In contrast, BM-derived cardiomyocytes were observed outside the infarcted myocardium at a low frequency and were derived exclusively through cell fusion.

These results are in line with the observations of Murry et al., who used both cardiomyocyte-restricted and ubiquitously expressed reporter transgenes to track the fate of HSC transplants into normal and injured adult mouse hearts [64]. Their results indicated that HSC do not readily acquire a cardiac phenotype raising a cautionary note for clinical studies of infarct repair. The notion that hematopoietic cells may engraft to the myocardium without transdifferentiation into cardiomyocytes was further corroborated by Balsam et al. by showing that HSC adopt mature hematopoietic fates in ischemic myocardium [65]. Cells were isolated from transgenic mice constitutively expressing GFP driven by the chicken β-actin promoter and injected directly into ischemic myocardium of wild-type mice. Abundant GFP-positive cells were detected in the myocardium after 10 days, but by 30 days few cells were detectable. These GFP-positive cells did not express cardiac tissue-specific markers; rather, most of the donor cells expressed the hematopoietic marker CD45 and myeloid marker Gr-1, suggesting that even in the microenvironment of the injured heart, HSC adopt only hematopoietic fates. In contrast to widely publicized reports of HSC plasticity, Weissman et al. failed to reproduce transdifferentiation of HSC to lineages comprising skeletal muscle, heart, brain or gut.

They concluded that rare cell fusion events and incomplete purifications of HSC contaminated with tissue-committed stem cells were likely explanations for the other published results [66].

In contrast to the negative findings concerning the transdifferentiation capacity of HSC into cardiomyocytes, the Anversa group published further data in favor of this phenomenon. In 2005, Kajstura et al. reported that BMC differentiated into cardiac cell lineages after infarction, independent of cell fusion [67]. In this publication, using the same mouse model as described in the reports of Orlic et al., transdifferentiation into cardiac myocytes was demonstrated by immunohistology followed by morphological measurements of infarcted and regenerated areas in addition to Y-chromosome FISH analysis. Finding no evidence of angiogenesis or myocyte proliferation in remote parts of the heart, the authors excluded a paracrine effect of injected BMC in myocardial recovery. Kajstura et al. attribute the obvious discrepancy between their findings and others to (1) technical differences in experimental protocols, (2) identity of the applied donor cell(s), and (3) details in tissue preparation and immunocytochemical analysis of the myocardium. However, Kajstura et al. did not provide data on long-term engraftment beyond 10 days. In addition, the fact that Kajstura et al. did not observe angiogenesis or proliferation after cell transplantation, does not unequivocally exclude paracrine effects, e.g., on cardiomyocyte survival.

How animated the controversy on the subject has become by now can be estimated from the following statement of Kajstura et al.: "The assumption made by Balsam et al. [65] and Murry et al. [64] that the technical approach that they have used in the identification and measurement of myocardial structures is superior to that used in our laboratory does not reflect any scientific reality but the emotional disbelief that bone marrow cells can adopt myocardial cell lineages and repair the injured heart."[67]

Nevertheless, it should be noted that some of the criticism concerning methods and conclusions described by the Anversa group might be justified. For example, Kajstura et al. report the difficulty of cell transplantation into the infarcted myocardium with a 50% probability of correct injection. To control for this, rhodamine particles were added to the cell suspension used for transplantation. It was stated that "the unsuccessfully injected mice (no rhodamine particles) were considered the most appropriate control animals for the successfully treated mice" [67]. This practice obviously neglects general (nonspecific) effects of cell transplantation into the myocardium, in particular local inflammatory processes that can be expected after the usually injection-related death of transplanted cells. In addition, the improvement in heart function after stem cell transplantation reported by Orlic et al. [57] leaves room for discussion as acquisition of functional data in small animals is extremely difficult and should be interpreted with caution.

The accuracy of Y-chromosome FISH analysis may be another issue. Kajstura et al. reported that this method underestimated the frequency of positive cells by nearly 50%. Other studies reported visualization of 62% of nuclei in a male mouse due to partial nuclear sampling as the plane of each section does not always cut through the Y-chromosome [37]. Thus, FISH data can show significant variations.

On the other hand, it has been argued that at least in the human system, data is available which demonstrate the existence of male cells in a female's heart, totally unrelated to any cell transplantation, which in turn might lead to false-positive results. This phenomenon is attributable to the persistence of fetal cell microchimerism following the birth of male children, a fact that should be considered when using sex-mismatched transplantation models [68, 69].

Thus, the phenomenon of cardiac transdifferentiation of HSC is still controversially discussed and should be addressed diligently and with an open mind in the future.

3.2.2 Mesenchymal Stem Cells

Most of the early reports on cardiac differentiation of MSC focused on the effect of 5′-azacytidine on MSC marker expression in vitro [70–72] and on the outcome of subsequent MSC transplantation into the infarcted myocardium [73–75] with contradictory results. Although some studies claimed improvement of heart function after stem cell transplantation [75, 76], different explanations have been proposed including transdifferentiation [75], scar formation [77], improved revascularization [78] and/or cell fusion [50]. In contrast, and even though their influence on cardiac function has not been evaluated yet, calcification and/or ossification after MSC transplantation into the infarcted myocardium as demonstrated by Breitbach et al. show that these cells can also adapt fates with potentially deleterious effects in the engrafted tissue [79].

Wakitani et al. were among the first to describe a myogenic differentiation of BM-derived mesenchymal stem cells after treatment with the DNA demethylating compound 5′-azacytidine [72]. Rat BM-derived MSC were exposed to 5′-azacytidine for 24 h resulting in long, multinucleated myotubes with spontaneous contractions. Later studies using immortalized murine MSC, demonstrated not only the formation of myogenic structures, but the resulting cells displayed spontaneous beating, as well as the expression of several cardiac marker proteins, specific characteristics of cardiac myocytes [71]. Likewise, cardiac differentiation of murine MSC was described after cocultivation with rat cardiomyocytes [80]. However, it should be noted that the expression of certain cardiac marker genes alone, does not provide evidence for cardiac transdifferentiation. Other evidence, including the absence of markers from other lineages should be demonstrated, in addition to functionality of the resulting cell type.

Importantly, the DNA demethylating agent 5′-azacytidine does not induce specific genes, but effects global gene expression, suggesting that partial "reprogramming" rather than transdifferentiation of the MSC may occur. Recently, 5′-azacytidine has been used to enhance the reprogramming efficiency of mouse and human somatic stem cells by ectopic expression of transcription factors, thus generating induced pluripotent stem (iPS) cells, by approximately tenfold [81].

Further studies aiming at the differentiation of MSC isolated from rat bone marrow yielded contradictory results. In contrast to Wakitani et al., other studies were not able to generate spontaneously contracting cells after 5′-azacytidin or 5′-

aza-2-deoxycytidin treatment of MSC. Furthermore, the resulting cells did not express cardiac marker proteins such as cardiac myosin heavy chain, connexin 43 or troponin [70]. Experiments in rats [74] and pigs [75] using marrow stromal cells showed an improved heart function after transplantation of 5′-azacytidine-treated cells in an infarct model, as well as induced angiogenesis in the scar. However, improvement of cardiac function was also observed after transplantation of untreated BM stromal cells [73], as well as the formation of fibrotic scar tissue [77]. In the following years similar findings have been described after transplantation of human BM-derived cells [47, 76, 82].

Moreover, the mechanism of tissue engraftment and improvement of cardiac function is controversial. On one hand, cellular effects could play a decisive role if the applied cells led to an improvement by differentiation into functional cardiomyocytes. On the other hand, there are also reports on the fusion of transplanted stem cells with cardiomyocytes [50, 83], which may account for false-positive data on transdifferentiation. Nevertheless, fusion may also have a therapeutic effect as described for liver damage above [51]. Importantly, injected stem cells may exert paracrine effects potentially influencing the survival and/or proliferation of endogenous myocardial cells thereby reducing scar formation. Additionally, paracrine effects could result in stabilization of the infarcted area leading to an improvement of cardiac function. As the expression and secretion of cytokines, i.e., FGF, VEGF and angiopoetin, are upregulated in MSC under hypoxic conditions [84], enhanced vascularization by these cytokines is also plausible. In fact, the differentiation of MSC into endothelial phenotypes [78], as well as induction of cardiac nerve sprouting after MSC injection in a pig model of myocardial infarction [85] have been described.

Therefore, neither improvement of cardiac function nor homing of the transplanted cells to the myocardium as such, can provide clear evidence for the transdifferentiation of MSC into cardiomyocytes. For that reason, there is a clear need to investigate the cellular events following transplantation in order to analyze further cell fate, i.e., engraftment and transdifferentiation. Müller-Ehmsen et al. showed effective engraftment, but poor mid-term persistence of mononuclear (MNC) and mesenchymal BMC in acute and chronic rat myocardial infarction in a sex-mismatch setting [86]. The percentage of intramyocardially transplanted MNC or BMC in the heart decreased rapidly, independent from the donor cell type, donor cell number, and the application time (0–7 days post myocardial infarction). Besides the heart, transplanted cells were found predominantly in the lung and more rarely in liver and kidney. In other organs, donor cells were either absent or detected few in number.

Although Rota et al. worked with a similar animal model using transgenic mice for transplantation of BMC to the myocardium in a sex-mismatch setting, they obtained completely different results. According to their comprehensive study using sophisticated methods for donor cell detection and phenotype analysis, it was found that BMC adopt a cardiomyogenic fate in vivo [22]. Rota et al. reported that BMC engraft, both survive and grow within the spared myocardium following infarction by forming junctional complexes with resident myocytes. BMC and endogenous cardiomyocytes expressed connexin 43 and N-cadherin at their interface, as determined by immunofluorescence staining using primary antibodies directly labeled by quantum dots to enable discrimination from autofluorescence. BMC subsequently transdifferentiated into cardiomyogenic and vascular phenotypes. This process seemed to occur independently of cell fusion (only diploid DNA and a maximum of two sex chromosomes were detected within the cells) and ameliorated structurally and functionally the outcome of the heart after infarction [22].

Most of the data presented in this study relate to rather early time-points after transplantation (up to 48 h) and some of the data on long-term engraftment have been challenged by other studies. Rota et al. using two-photon laser scanning fluorescence microscopy (TPLSM) demonstrated that some donor-derived cells develop electrical stimulation-evoked rhod-2 transients in synchrony with host cardiomyocytes 30 days following transplantation [22]. However, Scherschel et al. claim that control experiments demonstrating sufficient in situ *z*-axis spatial resolution to discriminate between signals originating in donor and host cells under the experimental conditions employed were lacking [87]. Based on previous reports [88, 89], they conclude that it is highly possible that the rhod-2 transients observed in donor-derived cells in the study arose as a consequence of fluorescence contamination from juxtaposed host cardiomyocytes, and do not represent intrinsic cardiomyogenic activity in the donor cell.

Ghodsizad et al. detected neither transdifferentiation nor fusion of cord blood derived mesenchymal cells after transplantation into the acutely ischemic lateral wall of the left ventricle [90]. They applied an alternative somatic cell type, human cord-blood derived unrestricted somatic stem cells (USSC), in a porcine model of acute myocardial infarction. Although a remarkable improvement of cardiac function was demonstrated using transesophageal echocardiography, sex- and species-specific FISH/immunostaining failed to detect engrafted donor cells 8 weeks postinfarction. Since differentiation, apoptosis, and macrophage mobilization at the infarct site were excluded as underlying mechanisms, paracrine effects are most likely to account for the observed functional effects of the USSC treatment. One possible reason for the failure of long-term engraftment might originate from the fact that a xenogeneic model was used for this study. As immunodeficient pigs are unavailable to date, to mimic the setting of small animal experiments in SCID mice, an immunosuppressive regimen has to be used in this setting. However, it is important to note that an effective immunosuppression in the human-to-pig xenotransplantation setting is difficult to achieve and a rapid rejection of the xenograft might have occurred despite the medication.

In summary, the outcome following stem cell transplantation into the infarcted heart seems to depend strongly on the donor cell type(s) and particularly on the animal model used in the respective study.

3.2.3 Endothelial Progenitor Cells

The blood is also a source for another progenitor cell type that has been tested for heart repair. Circulating endothelial progenitor cells (EPC) and endothelial cells have been proposed for transdifferentiation into cardiomyocytes [91, 92]. However, these reports have been challenged by others that attributed these findings either to cell fusion [93], inappropriate viral labeling of transplanted donor cells [94], or concerns regarding donor cell detection and stringency of data analysis [95]. In addition, serious doubts on the cell type identity of EPC have been raised [96–98].

The identification of human EPC disproved the assumption that a postnatal vascularization depended exclusively on the proliferation and migration of terminally differentiated endothelial cells. EPC were identified based on their expression of CD34 and flk-1, as well as their adherence to tissue culture plastic surfaces. In contrast to leukocytes, they are CD45-negative and express further endothelial marker proteins, e.g., Tie-2 and CD117 [99].

After transplantation of labeled EPC into ischemic tissue of mice and rabbits, the cells were incorporated into neovascularized areas of capillaries and smaller arteries [99]. Thus, in the adult organism EPC may ameliorate reduced perfusion as in myocardial infarction and lead to improved cardiac function [100].

First reports on a cardiac transdifferentiation of endothelial cells were published by Condorelli et al. in 2001. Endothelial cells of various origins were labeled first using adenoviral or lentiviral vectors and subsequently cocultivated with neonatal rat cardiomyocytes or transplanted into ischemic areas of an infarcted mouse heart. In up to 10% of the labeled cells, the expression of cardiac marker proteins was detected by immunofluorescence staining. Such double staining as an indicator of transdifferentiation of endothelial cells was observed only after direct cell–cell contact of endothelial cells and cardiomyocytes [92]. However, these results might potentially be due to the transfer of viral vectors from one cell type to another as was proposed by Blomer et al. [94] and others [101, 102].

In contrast to Condorelli et al., Welikson et al. reported in 2006 that human umbilical vein endothelial cells (HUVEC) fuse with cardiomyocytes, but do not activate cardiac gene expression [93]. Analysis with a Cre/lox recombination assay indicated that virtually all HUVEC containing cardiac markers had indeed fused with cardiomyocytes.

A similar controversy exists on the cardiac differentiation potential of endothelial progenitor cells. In 2003, cardiac differentiation of circulating human endothelial progenitor cells after cocultivation with neonatal rat cardiomyocytes was described by Badorff et al. [91]. Within these cultures, an increase in cell size was demonstrated for the 1,1′-dioctadecyl-3,3,3′,3′-tetramethylindocarbocyanine (DiI)–labeled EPC and immunofluorescence staining determined that approximately 10% of these labeled cells expressed cardiac marker proteins. Notably, double staining was observed only after direct cell–cell contact. Dye transfer between EPC and cardiomyocytes demonstrated the formation of gap junctions between the two cell types. Control experiments were carried out using fixed cardiomyocytes to exclude the possibility of cell fusion as an underlying reason for the double labeling.

To date, cardiac differentiation of EPC as described by Badorff et al. has not been confirmed by other groups and the phenotype of the cells used in the study is controversial. Different studies defined EPC as VEGFR2-positive/CD133-positive/CD34-positive subpopulations of MNC [103], or as CD34-positive/VEGFR2-positive [104] or CD133-positive/VEGFR2-positive cells originating from the BM and mobilizing as the need arises [105]. However, CD34-negative cell populations have also been identified which have differentiated into EPC and endothelial cells [106]. While the cells described by Badorff et al. were assumed to have an endothelial phenotype due to the uptake of acetylated low density lipoprotein (LDL) and binding of the lectin Ulex europaeus agglutinin-1 (UEA-1), more recent data suggest that these cells represent almost exclusively monocytes/macrophages [95, 98]. Only a small percentage of the cells express endothelial markers; therefore, they have been termed "circulating angiogenic cells" (CAC) [96]. The CD14-positive/CD34-negative cell population within the expanded EPC might exert a proangiogenic effect by releasing paracrine factors [107]. In addition, CD14-positive cells release cytokines that may be important signals for wound healing [108, 109]. Recently, it has been confirmed that blood-derived monocytes [98], as well as other immune cells [97] can mimic EPC due to LDL uptake and lectin binding abilities in addition to colony forming capacities.

In contrast to Badorff et al., a study by our group performing coculture experiments with DiI-labeled huEPC and neonatal rat cardiomyocytes (NRCM) did not support transdifferentiation of huEPC into functionally active cardiomyocytes. Gruh et al. analyzed the cocultivated cells by means of flow cytometry, 3D confocal laser microscopy, species-specific RT-PCR for the expression of human cardiac marker genes, and electron microscopy [95]. Although FACS analysis and conventional wide-field fluorescence microscopy suggested the existence of DiI-positive human cardiomyocytes in cocultures, we obtained no convincing evidence of cardiac differentiation of huEPC. Rather, DiI-positive cardiomyocytes were identified as necrotic NRCM or NRCM-derived vesicles with high levels of autofluorescence, or alternatively, as NRCM lying on top of or below labeled huEPC or huEPC fragments. Accordingly, no expression of human Nkx2.5, GATA-4, or cardiac troponin I was detected. Although it cannot be excluded that slightly different culture conditions may have prevented transdifferentiation in our own experiments, our data highlight technical limitations of FACS analysis and conventional 2D immunofluorescence, as well as confocal microscopy for the analysis of stem cell differentiation in coculture settings.

3.3 Transdifferentiation into Neuronal Cells

3.3.1 Hematopoietic Stem Cells

First reports on the contribution of HSC to the brain described the differentiation into microglia and macroglia in adult mice [110], and were later confirmed in several studies [111, 112]. In contrast, the contribution of HSC to other cell types in the brain is controversial and initial reports on neuronal differentiation of HSC [113–115] could not be confirmed by others [116, 117]. These discrepancies have led to a discussion on the validity of different approaches used for cell tracing in transplantation experiments [118].

To test the ability of adult HSC to contribute to the central nervous system, Eglitis et al. transplanted adult female mice with donor BMC genetically marked with either a retroviral tag or by using male donor cells [110]. Using in situ hybridization histochemistry, a continuing influx of BM-derived hematopoietic cells into the brain was detected. These cells were widely distributed throughout regions in the brain, including the cortex, hippocampus, thalamus, brain stem, and cerebellum. When in situ hybridization histochemistry was combined with immunohistochemical staining using lineage-specific markers, some BM-derived cells were positive for the microglial marker F4/80. Other BM-derived cells expressed the astroglial marker glial fibrillary acidic protein (GFAP). From these results, Eglitis et al. concluded that some microglia and astroglia arise from a precursor that is a normal constituent of adult BM. This idea became widely accepted [111, 112]; however it was followed by controversial discussions regarding the contribution of HSC to other cell types in the brain.

In 2000, Brazelton et al. reported the expression of neuronal phenotypes from BM-derived cells, following intravascular delivery of genetically marked adult mouse BM into lethally irradiated adult mice. These cells persisted in the brain for at least 6 months after transplantation, as assessed by flow cytometry and showed typical neuronal gene expression profiles (NeuN, 200-kilodalton neurofilament, and class III beta-tubulin) demonstrated by confocal microscopy [113].

In the same year, Mezey et al. showed that transplanted adult BMC migrated into the brain and differentiated into cells that expressed neuron-specific antigens [114]. Later, the same group also investigated whether HSC contribute to neuronal cells in humans. To this effect, they examined postmortem brain samples from females who had received BM transplants from male donors [115]. Using a combination of neuron-specific antibodies for immunocytochemistry and FISH histochemistry, cells containing Y-chromosomes were detected in several brain regions. Most of these cells were identified as nonneuronal (e.g., endothelial cells); however, neurons in the hippocampus and cerebral cortex were detected. The distribution of the labeled cells was not homogeneous with clusters of Y-chromosome-positive cells, suggesting that single progenitor cells underwent clonal expansion and differentiation. Mezey et al. concluded that adult human BMC can enter the brain and generate neurons in a manner similar to rodent cells.

In contrast to these data, Castro et al. report the failure of BMC to transdifferentiate into neural cells in vivo, both after transplantation of BM-derived side population cells, as well as unfractionated BM [116]. None of the recipients had donor-derived neural-like cells in the brain and cervical spinal cords, regardless of injury. Comments on this report by Mezey et al. point out that this discrepancy might be due to the different methodologies used for cell tracing [118]. While Mezey et al. used immunocytochemistry in combination with FISH histochemistry for Y-chromosome-positive cells in a sex-mismatch model, Castro et al. used genetically labeled donor cells from a Rosa-LacZ mouse strain expressing the LacZ reporter gene under transcriptional control of the Rosa26 promoter. The latter approach, however, depends on uniform ubiquitous transgene expression in the tissues analyzed, as well as on error-prone detection methods [119]. Therefore, it is not unlikely that the study by Castro et al. underestimated the actual number of donor-derived cells in their model.

Interestingly, another study using a reporter gene approach for labeling of HSC was also not able to detect transdifferentiation of BM-derived cells into neuronal lineages. In 2006, Roybon et al. investigated whether highly purified mouse adult HSC, characterized by lineage marker depletion and expression of the cell surface markers Sca1 and c-Kit (Lin-negativ/Sca1-positive/c-Kit-positive), can be stimulated to adopt a neuronal fate [117]. In this study, transgenic mice expressing GFP under control of the chicken β-actin promoter were used. First, Roybon et al. tried to induce neural differentiation in vitro with protocols that have been successfully used to differentiate either neuronal or embryonic stem cells or multipotent adult progenitor cells from BM into neuronal cells. As a result, up to 50% of the cells expressed the neural progenitor marker nestin. However, electrophysiological recordings on neuron-like cells showed that these cells were incapable of generating action potentials. Therefore, at least in vitro, HSC did not seem to be able to differentiate into functional neuronal cell types. According to Roybon et al., neither cocultivation with neural precursors nor transplantation into the striatum or cerebellum of wild-type mice, resulted in HSC-derived cells with a true neuronal phenotype. Rather, the applied HSC differentiated into macrophage/microglia or died.

One major point of criticism concerning the findings of Castro et al. raised by Mezey et al. was that blue LacZ-positive microglia, which like other monocyte/macrophage cells originate from HSC, were absent from the brains of the transplanted animals. In contrast, Roybon et al. did find GFP-positive microglia after HSC transplantation. Thus, their method seems valid for the detection of transdifferentiated neuronal cells in principle, in return raising doubts on the data presented by Brazelton et al. and Mezey et al. In conclusion, further studies using sophisticated methods are mandatory to unambiguously prove or disprove the contribution of BM-derived HSC to functional neuronal cell types in vivo.

3.3.2 Mesenchymal Stem Cells

It was reported by several groups that stem cells isolated from the BM were capable of differentiation towards neural like cells (reviewed in [120]). Most studies based their conclusions on an evaluation of changes in cell morphology, i.e., the formation of neurite-like structures, and on the detection of neuronal-cell specific marker gene expression, mostly detected by immunohistology. However, other studies demonstrated that neuronal marker expression was already present in undifferentiated MSC [121] and is induced in response to stress [122, 123]. In addition, these studies questioned the validity of morphological analyses of neuronal transdifferentiation in vitro. While some studies attributed the beneficial effects of MSC transplantation to the brain as a result of transdifferentiation [124], immunological effects have also been considered [125].

Early reports on neuronal transdifferentiation of MSC were contradicted, for example by in vitro experiments based on protocols by Woodbury et al. [126], that used exposure to certain chemicals as a neural differentiation stimulus for MSC. Investigation with time-lapse video recording showed that the formation of neurites is not the result of an outgrowth of dendrite- and axon-like structures, but merely a result of cell shrinkage and retraction of the cell edge in response to stress [122, 123, 127]. In addition, some neural marker proteins have been found to be expressed in undifferentiated MSC [121]. Furthermore, exposure of MSC to stress causes an increase in expression levels of the neural markers neuronal nuclei (NeuN), neuron-specific enolase (NSE) [123], neurofilament 200 (NF200) and tau [122].

In addition, for MAPC from the BM, Raedt et al. reported a baseline expression of neural markers beta III tubulin and NF200. Furthermore, the application of several protocols for neural differentiation did not result in an increase in expression levels as determined using real-time PCR and immunohistochemistry [128].

Nevertheless, in vivo experiments using MSC for transplantation into the brain yielded positive results. In 2006, Arnhold et al. investigated the therapeutic potential of MSC by stereotactic engraftment into the lateral ventricle of adult rats [124]. They reported that human BM stromal cells display certain neural characteristics and integrate into the subventricular compartment after injection into the liquor system and took up a close host graft interaction without any degenerative influence on the host cells. Arnhold et al. reported morphological, as well as immunohistochemical evidence for a transdifferentiation of MSC within the host tissue.

In contrast, Gerdoni et al. obtained different results investigating the therapeutic effect of MSC transplantation to the brain in experimental autoimmune encephalomyelitis [125]. MSC-treated mice showed a significantly milder disease and fewer relapses compared to control mice. This was also accompanied with a decreased number of inflammatory infiltrates, reduced demyelination, and axonal loss. However, no evidence of GFP-labeled neural cells was detected inside the brain parenchyma, thus not supporting the hypothesis of MSC transdifferentiation. In contrast, the analysis of in vivo T- and B-cell responses and antibody titers suggested that the beneficial effect of MSC in experimental autoimmune encephalomyelitis is mainly the result of an interference with the pathogenic autoimmune response.

In fact, it is conceivable that any stem cell transplantation may lead to a reaction that could be characterized as a "proregenerative inflammation." In this setting, the induced lesion, as well as the transplanted cells can trigger the attraction of immune cells to the site of transplantation and result in a proregenerative cytokine release.

4 Critical Aspects of Differentiation Experiments

4.1 Cell Type Identity

Identifying the stem cell type used in experiments investigating transdifferentiation is critical. For a comparative analysis of stem cell plasticity, especially when being performed by different groups, an unambiguous definition of the cell's phenotype is crucial. However, besides inconsistencies in the protocols for isolation and cultivation of the described cells, the rapidly expanding knowledge on stem cell populations and subpopulations complicates an objective comparison of the existing data. While early reports investigated the fate of "adult BMC" [114] or "BM stromal cells" [73], others used different subpopulations. These were classified either by the expression of single marker proteins like "CD34-positive BMC"[35], "purified BM Sca-1-positive cells" [59], or differentiation potential as for "multipotent adult progenitor cells (MAPC)" [106] or "human BM-derived multipotent stem cells" (hBMSC) [47]. To address inconsistencies, Horwitz et al. suggested a clarification of the nomenclature for MSC in an International Society for Cellular Therapy position statement [129]. Herein, the authors propose that the plastic-adherent cells currently described as mesenchymal stem cells be termed multipotent mesenchymal stromal cells, while the term mesenchymal stem cells should be reserved for a subset of these (or other) cells that demonstrate stem cell activity by clearly stated criteria. These include demonstrations of long-term survival with self-renewal capacity and tissue-repopulation with multilineage differentiation. For both cell populations, the acronym MSC may be used, however, investigators should unequivocally define the acronym in their work.

The analysis of transdifferentiation processes is especially complicated in the case of mixed populations or when investigating in vivo migration and homing to sites of injury. Besides engraftment of a single cell type potentially leading to the regeneration of damaged tissue, synergistic effects might play a key role. This might be conceivable following transplantation of BMC with different cell subtypes exerting proangiogenic, antiapoptotic and/or antiinflammatory effects. For example, the expression and secretion of cytokines like FGF, VEGF and angiopoetin in MSC [84], could potentially modulate the transdifferentiation capacity of other cell types. The complications resulting from the use of mixed cell populations can be circumvented by the clonal transplantation of single cells as performed, for example, by Krause et al. [37]. However, this approach has certain limitations and may be difficult to perform for many cell types, as in vivo cell survival and proliferation capacity following a single cell transplantation are usually low.

One of the most prominent examples of a controversial cell type identification is the ongoing debate regarding "endothelial progenitor cells" (EPC) or "circulating angiogenic progenitor cells" (CAC). In recent years, difficulties in discriminating between EPC and cells of monocytic/macrophage origin became more and more evident [91, 96, 99, 103]. It was demonstrated that blood derived monocytes [98], as well as immune cells [97] can mimic EPC; thus questioning the validity of earlier reports.

Obviously, not only the potential cell source for transdifferentiation can be controversial, but surely also the cell type resulting from this event. The question being: which criteria need to be met by the resulting cell to be considered a hepatocyte, cardiomyocyte or neuronal cell? When Lysy et al. investigated the hepatic differentiation of MSC, the resulting cells displayed expression of several hepatocyte markers such as albumin, alpha-fetoprotein, cytokeratin 18, representing at least "hepatocyte-like" cells [52]. However, it was also demonstrated that these cells partially retained mesenchymal markers, suggesting that the cells were not "fully" differentiated. Consequently, it seems to be crucial to define the conditions that have to be fulfilled by cells to be considered a fully differentiated and most of all functional cell type.

4.2 Cell Labeling

As studies on transdifferentiation frequently involve more than one cell type, e.g., in cocultivation approaches or transplantation settings, an optimal cell labeling method has to be applied for an interpretable read-out of the experiment.

The first possibility for cell labeling is with fluorescent dyes that bind to cellular components covalently or noncovalently. For example, in a study investigating the cardiac differentiation potential of endothelial progenitor cells, cells were labeled through the uptake of DiI-LDL prior to cocultivation with neonatal rat cardiomyocytes [91]. This approach has several drawbacks: (1) dyes are diluted upon further cell division, (2) once labeled, dead cells will retain the label and (3) fluorescent cell debris can be taken up by other cells, e.g., macrophages, or stick to other cells leading to false positive results.

Some of these problems can be overcome using genetic labeling, most commonly with reporter genes such as LacZ and GFP. These reporter genes have been used in combination with ubiquitous promoters, for example to investigate the capacity of BMC to transdifferentiate into neural cells after transplantation to the brain. As described above, Castro et al. used genetically labeled donor cells from a Rosa-LacZ mouse strain expressing the LacZ reporter gene under transcriptional control of the Rosa26 promoter [116], while Roybon et al. used cells expressing GFP under control of the chicken β-actin promoter [117]. Both studies did not provide evidence for transdifferentiation events, in contrast to Mezey et al., who used immunocytochemistry in combination with FISH histochemistry for Y-chromosome-positive cells in a sex-mismatch model [115]. It is known, that a reporter gene assay depends on uniform ubiquitous transgene expression in the analyzed cells; therefore, it is crucial that the transcriptional activity of a given promoter is on a similar level in both undifferentiated and differentiated (stem) cells. As was demonstrated for murine embryonic stem cells, promoter activity may vary significantly throughout the process of differentiation [130]. This issue should be considered a possible explanation for discrepancies in the outcome of adult stem cell differentiation experiments using genetic labeling.

In addition to the mere labeling of cells by ubiquitous reporter gene expression, conditional genetic labeling techniques have added greatly to the knowledge on stem cell transdifferentiation. Tissue specific promoters can be used to switch on reporter gene expression only in case of differentiation towards a certain cell type.

Rota et al. used the reporters eGFP and a c-myc-tagged nuclear-targeted-Akt transgene, both driven by the cardiac-specific α-myosin-heavy-chain (α-MHC) promoter, to investigate the cardiomyogenic fate of BMC [22]. However, for this approach, the cell specificity of the promoter has to be carefully analyzed, as leaky or unspecific expression in other cell types may occur, especially in case of higher copy numbers of the transgenes within the cells [131]. Recently, sophisticated genetic labeling has been used for the detection of cell fusion. As described above, Alvarez-Dolado et al. used a conditional Cre/lox recombination, enabling detection of fused cells by X-gal staining for LacZ expression [50].

Another important issue is how the transgene is transferred to the cells. When transgenic cell lines or animals are not available, the gene transfer has to be performed directly before the experiment, by using either standard transfection methods or viral vectors. Both with plasmid transfection and nonintegrating viral vectors, e.g., adenoviruses, the problem of signal dilution can occur in dividing cells. In contrast, integrating viruses like lentiviral vectors, have turned out to be an efficient method for stable gene transfer for both in vitro and in vivo studies. However, we have identified an important weakness of this method in cases where cells need immediate transplantation after preparation, e.g., to prevent cell death, differentiation or dedifferentiation [94]. Although these cells are usually washed several times following viral transduction, there may be the risk of viral vector shuttle via transplanted cells resulting in undesired in vivo transduction of recipient cells. We explored a potential viral shuttle via ex vivo lentivirally transduced cardiomyocytes in vitro, following transplantation into the brain and peripheral muscle. By this, we demonstrated that even after extensive washing, infectious viral vector particles can be detected in cell suspensions. As a result, the lentiviral vector particles stably transduced resident cells of the recipient central nervous system and muscle in vivo.

This phenomenon can also be seen using other cell types, as was confirmed by further studies demonstrating that retroviral particles adhere nonspecifically, or "hitchhike," to the surface of T-cells [132]. After transplantation, secondary transduction has been observed due to the adherence of vector particles to hematopoietic target cells [102] or endothelial cells [101]. In some cases, for example in a study by Condorelli et al., these findings might be one of the possible reasons for the discrepancies among studies investigating stem cell differentiation in transplantation models and cocultivation systems [92].

4.3 Imaging Techniques

Transplantation models and cocultivation systems suffer from another difficulty, as the identification of transdifferentiated cells can be complex. Methods based on immunohistology have to be carefully evaluated with respect to the specificity of the obtained signals, inclusion of all necessary controls and exclusion of staining artifacts. As described above, the detection of the LacZ transgene relies on error-prone detection methods, including the risk of unspecific staining after prolonged incubation [119]. Detection of GFP or using immunofluorescence approaches can also be impaired by high levels of tissue or cell-specific background fluorescence [133]. Importantly, high levels of autofluorescence can frequently be observed in necrotic or apoptotic cells leading to false interpretations, in particular in transplantation or coculture-based transdifferentiation experiments. Laflamme et al. reported that apart from the normal autofluorescence in striated heart muscle, this fluorescence increases after myocardial injury due to accumulated lipofuscin, blood-derived pigments and other intrinsic fluorochromes such as flavins and reduced nicotinamide adenine dinucleotide (NADH) [134]. While early reports using GFP-labeled cells for transplantation might have overlooked this fact, recent publications used GFP-specific antibodies and/or validation of the emission spectrum to unequivocally identify GFP-expressing cells [135]. Increasing levels of autofluorescence in the course of a coculture experiment, as has been demonstrated by our group using flow cytometry analyses as shown in Fig. 2 [95], can potentially lead to misinterpretation of the obtained data as might be the case in

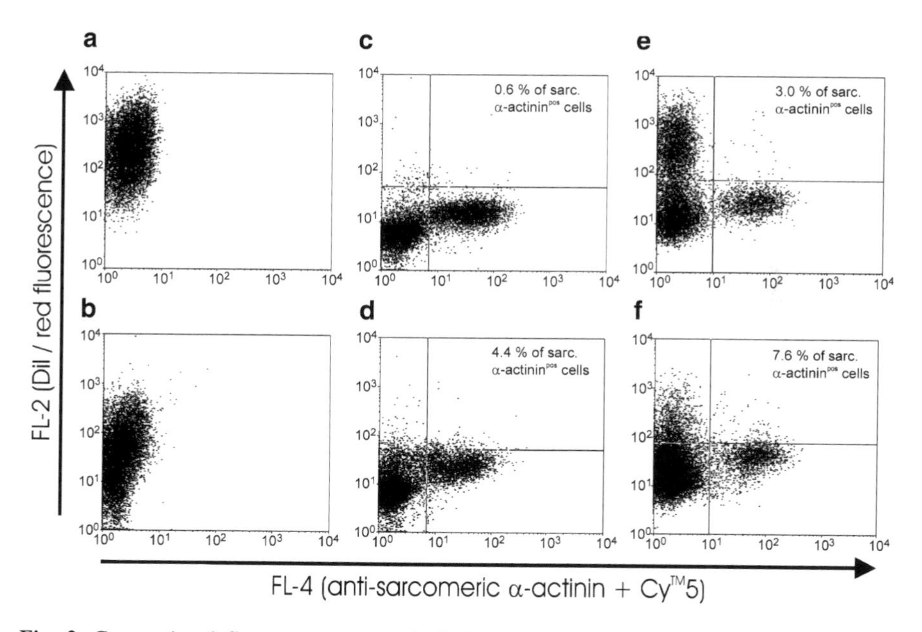

Fig. 2 Conventional flow cytometry analysis is not suitable to identify DiI-positive human cardiomyocytes within cocultures of human endothelial cell progenitors (huEPC) and neonatal rat cardiomyocytes (NRCM). Analyses of cocultures (**e**,**f**), as well as of monocultures of huEPC (**a**,**b**) and NRCM (**c**,**d**) at day 2 (**a**,**c**,**e**) and day 6 (**b**,**d**,**f**) demonstrate a significant increase in sarcomeric α-actinin-positive cardiomyocytes displaying red DiI-like fluorescence. Modified after Gruh et al. [95]

reports on the alleged cardiac differentiation potential of endothelial progenitor cells [91].

When using conventional two-dimensional image analysis for the evaluation of double or multiple immunostaining, sometimes a genuine colocalization within the same cell is hard to discern from an overlay of signals from two neighboring cells and three-dimensional confocal imaging should be preferred instead [95]. However, even in the case of three-dimensional analysis, data interpretation can be difficult.

In some cases, another dimension has to be included: the monitoring of the cell's fate over time. As described above, time-lapse video recording was able to reveal that morphological changes associated with a presumed neuronal differentiation of mesenchymal stem cells were actually not a result of outgrowing neurites, but of cell shrinkage in response to stress signals [122, 123, 127].

Immunohistology and immunofluorescence can also be error-prone and common problems include a weak signal from otherwise specific antibodies and/or nonuniform staining. For example, in our xenogeneic pig model studies, we have observed this problem when using an antibody detecting a human mitochondrial antigen. We found that this antibody led to nonuniform staining when used to detect different human cell types. On one hand, it conferred strong labeling of cardiomyocytes while on the other, human fibroblasts showed insufficient staining. While this finding is in line with expected differences in metabolic activity and numbers of mitochondria per cell in the two cell types, it prevented the use of this antibody in our study. Alternatively, another antibody directed against human nuclear antigen (HuNu) was used [90].

The high background of unspecific staining can also be a problem and unfortunately, published data often lack critical controls such as images of the appropriate isotype staining control. As long as images show the detection of structural proteins resulting in distinct staining patterns, e.g., cardiomyocyte specific staining of contractile proteins that shows clearly visible cross-striations, this might not be problematic. Otherwise, it is difficult to discern diffuse staining of cytoplasmic proteins from background levels. It is therefore advised to include these controls either in the original publication or as online supporting material. Moving forward, both editors and reviewers should be made aware that these controls would add to the reliability, and thus quality of the published data.

Unfortunately, appropriate isotype controls are not always available. This is true when using rabbit serum for staining. The correct control would be preimmunized serum obtained from the same animal. Thus, experiments using unpurified serum should at least include negative control staining with other, nontarget cells; and the specificity of immunostaining strategies with unpurified antibodies should be interpreted with caution.

4.4 Data Interpretation

The interpretation of transdifferentiation experiments can be difficult when too many conclusions are drawn from too little data. Early reports on the presumed neural differentiation of mesenchymal stem cells provide an important example. In this case, initial reports concentrated on the apparent morphology of the cells, as well as the detection of a limited number of markers [120]. Only later did data disprove the initial reports by demonstrating that some neural marker proteins are already expressed in undifferentiated MSC [121] and that stress causes an increase in the expression level of several neural markers [122, 123].

Similarly, later studies investigating differentiation of transplanted stem cells set out to analyze not only colocalization of donor-derived and tissue-specific markers, but also considered cell fusion as an alternative mechanism. In 2003, Wang et al. demonstrated that cell fusion was the principal source of BM-derived hepatocytes by investigating the ploidy of the presumably transdifferentiated donor cells [31]. Subsequent studies using the same assays, for example a study published by Sato et al., tried to elucidate the cellular components of human BM that potently differentiated into hepatocytes. Sato et al. stated that cell fusion was not likely involved, as both human and rat chromosomes were independently identified by FISH [136]. However, fusion as an underlying mechanism for the detection of double labeled, presumably transdifferentiated cells, cannot be excluded from earlier reports, as this possibility was not explicitly investigated.

Early in vivo data on the transdifferentiation of adult stem cells concentrated on the therapeutic effects following stem cell transplantation. Improved heart function and increased angiogenesis in the scar were observed after transplantation of 5-azacytidine-treated marrow stromal cells in an infarct model [74, 75]. Although some labeled bone marrow-derived cells within the infarct region stained positively for a cardiac marker protein, it remains unclear to what extent transdifferentiation into cardiomyocytes is the reason for the improvement or whether this may be due to other cardio-protective effects as described above [77, 78, 84, 85]. Therefore, functional improvement alone, does not provide evidence of transdifferentiation and leaves room for different interpretations with respect to the impact of individual effects triggered by stem cell transplantation.

4.5 Biological Significance

Most of the experiments investigating stem cell transdifferentiation represent a highly artificial setting with limited biological significance in vivo. This holds true especially for transplantation experiments, whereby stem cells are transferred from one part of the body to another, and sometimes to a rather remote compartment. These settings may not resemble naturally occurring stem cell mobilization and/or recruitment processes; and therefore, might not be ideal to mimic and investigate in vivo regeneration. However, the importance of stem cell transplantation, as a future method with great clinical significance, should not be overshadowed by the complications of replicating the exact events occurring in nature.

Indeed, the detection of a therapeutic effect does not prove stem cell transdifferentiation per se, even unequivocally confirmed stem cell transdifferentiation into another somatic cell type does not guarantee a therapeutic effect. When Wu et al. investigated whether human BMC could contribute to liver regeneration in vivo, they detected cells from extrahepatic sources that had homed to the tissue, ultimately transdifferentiating into hepatocytes. However, these cells did not increase in number, thus a robust repopulation of the tissue was not observed [137].

It should be noted that, apart from the role of transdifferentiation of stem cells in tissue regeneration following injury, and/or in homeostasis, this process might also have an impact on pathogenesis. For example, it has been proposed recently that BM-derived circulating precursor cells participate in the development of human lung fibrosis and lesion formation, especially in *bronchiolitis obliterans* [138].

In reality, the biological significance of transdifferentiation, with respect to its meaning, is still poorly understood. The question remains to be determined whether transdifferentiation reflects a natural process, i.e., an inherent ability of a given cell to switch its fate under certain conditions, or an artificial change in its expression profile, as might be the case for differentiation processes induced by treatment with 5-azacytidine. As this agent confers the demethylation of DNA leading to a random induction of gene expression, subsequent changes could be interpreted as an artificially-induced reprogramming, a rather hard reset of the cellular differentiation program.

Lastly, the incidence of transdifferentiation and/or cell fusion might also play a role in determining biological significance. To date, only rare events have been described, and although interesting, the findings might be irrelevant for therapeutic purposes in vivo, due to the low frequency of occurrence.

5 Conclusions

New data on the plasticity of stem cells of various lineages have emerged. These data, in addition to the developing new field of adult stem cell differentiation, are not without controversy. Today, most of the reported discrepancies cannot be explained satisfactorily due to several reasons. For example, many studies lack a common starting point, i.e., it remains unclear whether the exact same cell population was analyzed. In addition, the methodology for precise analyses of differentiation events is still rapidly evolving. As a reaction to criticism concerning early and sometimes too enthusiastic reports on the transdifferentiation of stem cells and its envisaged therapeutic potential, sophisticated methods have been developed or adapted, e.g., in the area of cell labeling, imaging and tracing. However, to prove unequivocally stem cell transdifferentiation, there is a clear need to prove the functionality of the resulting cell type. It is not sufficient to show that a given cell *looks* like a transdifferentiated cell solely due to the expression of specific marker proteins, but to answer the question as to whether it *acts* accordingly.

As the field of stem cell biology progresses, it will be crucial to analyze further not only *if* a certain stem cell type differentiates into a certain phenotype, whether or not expected, but also to investigate in detail *how* this process works. This will include the identification of key factors inducing cell fate switches, and the molecular mechanisms and chronological sequence of the conversion itself. This includes the question of how *we*, as investigators, *force* a given cell to transdifferentiate into a desired cell type, for example by the over-expression of cell-type specific transcription factors, regardless of the in vivo and/or in vitro significance of the particular conversion.

By focusing on these mechanisms, insight into the original question will be addressed: Do stem cells undergo direct differentiation towards a more specialized somatic cell, or, must they be reprogrammed or "de"-differentiated, thereby first changing into a more common ancestor to then be "trans"-differentiated into specialized cells ultimately involving the same pathways as in organ development.

References

1. Takahashi K, Tanabe K, Ohnuki M, Narita M, Ichisaka T, Tomoda K, and Yamanaka S (2007) Induction of pluripotent stem cells from adult human fibroblasts by defined factors. *Cell* 131:861–872
2. Takahashi K, Yamanaka S (2006) Induction of pluripotent stem cells from mouse embryonic and adult fibroblast cultures by defined factors. *Cell* 126:663–676
3. Guan K, Nayernia K, Maier LS, Wagner S, Dressel R, Lee JH, Nolte J, Wolf F, Li M, Engel W, Hasenfuss G (2006) Pluripotency of spermatogonial stem cells from adult mouse testis. *Nature* 440:1199–1203
4. Maccarty WC, Caylor HD (1922) Metaplasia in ovarian dermoids and cystadenomas: report of three cases. *Ann Surg* 76:238–245
5. Thiery JP (2003) Epithelial-mesenchymal transitions in development and pathologies. *Curr Opin Cell Biol* 15:740–746
6. Takeichi M (1991) Cadherin cell adhesion receptors as a morphogenetic regulator. *Science* 251:1451–1455
7. Takeichi M (1993) Cadherins in cancer: implications for invasion and metastasis. *Curr Opin Cell Biol* 5:806–811
8. Lamb HD (1934) The pathogenesis of some intra-ocular osseus tissue. True metaplasia in the eye. *Trans Am Ophthalmol Soc* 32:294–311
9. Tseng SC (1989) Concept and application of limbal stem cells. *Eye* 3(2):141–157
10. Pittack C, Jones M, Reh TA (1991) Basic fibroblast growth factor induces retinal pigment epithelium to generate neural retina in vitro. *Development* 113:577–588
11. Wei ZG, Wu RL, Lavker RM, Sun TT (1993) In vitro growth and differentiation of rabbit bulbar, fornix, and palpebral conjunctival epithelia. Implications on conjunctival epithelial transdifferentiation and stem cells. *Invest Ophthalmol Vis Sci* 34:1814–1828
12. Opas M, Dziak E (1998) Direct transdifferentiation in the vertebrate retina. *Int J Dev Biol* 42:199–206
13. Kim JT, Lee EH, Chung KH, Kang IC, Lee DH, Joo CK (2004) Transdifferentiation of cultured bovine lens epithelial cells into myofibroblast-like cells by serum modulation. *Yonsei Med J* 45:380–391

14. Bachem MG, Sell KM, Melchior R, Kropf J, Eller T, Gressner AM (1993) Tumor necrosis factor alpha (TNF alpha) and transforming growth factor beta 1 (TGF beta 1) stimulate fibronectin synthesis and the transdifferentiation of fat-storing cells in the rat liver into myofibroblasts. *Virchows Arch B Cell Pathol Incl Mol Pathol* 63:123–130
15. Fan JM, Ng YY, Hill PA, Nikolic-Paterson DJ, Mu W, Atkins RC, Lan HY (1999) Transforming growth factor-beta regulates tubular epithelial-myofibroblast transdifferentiation in vitro. *Kidney Int* 56:1455–1467
16. Funderburgh JL, Funderburgh ML, Mann MM, Corpuz L, Roth MR (2001) Proteoglycan expression during transforming growth factor beta -induced keratocyte-myofibroblast transdifferentiation. *J Biol Chem* 276:44173–44178
17. Boros LG, Torday JS, Paul Lee WN, Rehan VK (2002) Oxygen-induced metabolic changes and transdifferentiation in immature fetal rat lung lipofibroblasts. *Mol Genet Metab* 77:230–236
18. Buyon JP, Clancy RM (2005) Autoantibody-associated congenital heart block: TGFbeta and the road to scar. *Autoimmun Rev* 4:1–7
19. Untergasser G, Gander R, Lilg C, Lepperdinger G, Plas E, Berger P (2005) Profiling molecular targets of TGF-beta1 in prostate fibroblast-to-myofibroblast transdifferentiation. *Mech Ageing Dev* 126:59–69
20. Real C, Glavieux-Pardanaud C, Vaigot P, Le-Douarin N, Dupin E (2005) The instability of the neural crest phenotypes: Schwann cells can differentiate into myofibroblasts. *Int J Dev Biol* 49:151–159
21. Eberhard D, Tosh D (2008) Transdifferentiation and metaplasia as a paradigm for understanding development and disease. *Cell Mol Life Sci* 65:33–40
22. Rota M, Kajstura J, Hosoda T, Bearzi C, Vitale S, Esposito G, Iaffaldano G, Padin-Iruegas ME, Gonzalez A, Rizzi R, Small N, Muraski J, Alvarez R, Chen X, Urbanek K, Bolli R, Houser SR, Leri A, Sussman MA, Anversa P (2007) Bone marrow cells adopt the cardiomyogenic fate in vivo. *Proc Natl Acad Sci U S A* 104:17783–17788
23. Koestenbauer S (2006) *J Reprod Med Endocrinol.* 3(5):324–330
24. Eisenberg LM, Eisenberg CA (2003) Stem cell plasticity, cell fusion, and transdifferentiation. *Birth Defects Res C Embryo Today* 69:209–218
25. Petersen BE, Bowen WC, Patrene KD, Mars WM, Sullivan AK, Murase N, Boggs SS, Greenberger JS, Goff JP (1999) Bone marrow as a potential source of hepatic oval cells. *Science* 284:1168–1170
26. Lagasse E, Connors H, Al-Dhalimy M, Reitsma M, Dohse M, Osborne L, Wang X, Finegold M, Weissman IL, Grompe M (2000) Purified hematopoietic stem cells can differentiate into hepatocytes in vivo. *Nat Med* 6:1229–1234
27. Theise ND, Nimmakayalu M, Gardner R, Illei PB, Morgan G, Teperman L, Henegariu O, Krause DS (2000) Liver from bone marrow in humans. *Hepatology* 32:11–16
28. Wagers AJ, Sherwood RI, Christensen JL, Weissman IL (2002) Little evidence for developmental plasticity of adult hematopoietic stem cells. *Science* 297:2256–2259
29. Dahlke MH, Popp FC, Larsen S, Schlitt HJ, Rasko JE (2004) Stem cell therapy of the liver–fusion or fiction? *Liver Transpl* 10:471–479
30. Menthena A, Deb N, Oertel M, Grozdanov PN, Sandhu J, Shah S, Guha C, Shafritz DA, Dabeva MD (2004) Bone marrow progenitors are not the source of expanding oval cells in injured liver. *Stem Cells* 22:1049–1061
31. Wang X, Willenbring H, Akkari Y, Torimaru Y, Foster M, Al-Dhalimy M, Lagasse E, Finegold M, Olson S, Grompe M (2003) Cell fusion is the principal source of bone-marrow-derived hepatocytes. *Nature* 422:897–901
32. Thorgeirsson SS, Grisham JW (2006) Hematopoietic cells as hepatocyte stem cells: a critical review of the evidence. *Hepatology* 43:2–8
33. Camargo FD, Chambers SM, Goodell MA (2004) Stem cell plasticity: from transdifferentiation to macrophage fusion. *Cell Prolif* 37:55–65
34. Brulport M, Schormann W, Bauer A, Hermes M, Elsner C, Hammersen FJ, Beerheide W, Spitkovsky D, Hartig W, Nussler A, Horn LC, Edelmann J, Pelz-Ackermann O, Petersen J,

Kamprad M, von Mach M, Lupp A, Zulewski H, Hengstler JG (2007) Fate of extrahepatic human stem and precursor cells after transplantation into mouse livers. *Hepatology* 46:861–870

35. Fiegel HC, Lioznov MV, Cortes-Dericks L, Lange C, Kluth D, Fehse B, Zander AR (2003) Liver-specific gene expression in cultured human hematopoietic stem cells. *Stem Cells* 21:98–104
36. Tanabe Y, Tajima F, Nakamura Y, Shibasaki E, Wakejima M, Shimomura T, Murai R, Murawaki Y, Hashiguchi K, Kanbe T, Saeki T, Ichiba M, Yoshida Y, Mitsunari M, Yoshida S, Miake J, Yamamoto Y, Nagata N, Harada T, Kurimasa A, Hisatome I, Terakawa N, Murawaki Y, Shiota G (2004) Analyses to clarify rich fractions in hepatic progenitor cells from human umbilical cord blood and cell fusion. *Biochem Biophys Res Commun* 324:711–718
37. Krause DS, Theise ND, Collector MI, Henegariu O, Hwang S, Gardner R, Neutzel S, Sharkis SJ (2001) Multi-organ, multi-lineage engraftment by a single bone marrow-derived stem cell. *Cell* 105:369–377
38. Phinney DG, Prockop DJ (2007) Concise review: mesenchymal stem/multipotent stromal cells: the state of transdifferentiation and modes of tissue repair–current views. *Stem Cells* 25:2896–2902
39. Kuznetsov SA, Mankani MH, Gronthos S, Satomura K, Bianco P, Robey PG (2001) Circulating skeletal stem cells. *J Cell Biol* 153:1133–1140
40. Prockop DJ (1997) Marrow stromal cells as stem cells for nonhematopoietic tissues. *Science* 276:71–74
41. Devine SM, Cobbs C, Jennings M, Bartholomew A, Hoffman R (2003) Mesenchymal stem cells distribute to a wide range of tissues following systemic infusion into nonhuman primates. *Blood* 101:2999–3001
42. Shi XL, Qiu YD, Wu XY, Xie T, Zhu ZH, Chen LL, Li L, Ding YT (2005) In vitro differentiation of mouse bone marrow mononuclear cells into hepatocyte-like cells. *Hepatol Res* 31:223–231
43. Kang XQ, Zang WJ, Song TS, Xu XL, Yu XJ, Li DL, Meng KW, Wu SL, Zhao ZY (2005) Rat bone marrow mesenchymal stem cells differentiate into hepatocytes in vitro. *World J Gastroenterol* 11:3479–3484
44. Lee KD, Kuo TK, Whang-Peng J, Chung YF, Lin CT, Chou SH, Chen JR, Chen YP, Lee OK (2004) In vitro hepatic differentiation of human mesenchymal stem cells. *Hepatology* 40:1275–1284
45. Colter DC, Sekiya I, Prockop DJ (2001) Identification of a subpopulation of rapidly self-renewing and multipotential adult stem cells in colonies of human marrow stromal cells. *Proc Natl Acad Sci U S A* 98:7841–7845
46. Prockop DJ, Sekiya I, Colter DC (2001) Isolation and characterization of rapidly self-renewing stem cells from cultures of human marrow stromal cells. *Cytotherapy* 3:393–396
47. Yoon YS, Wecker A, Heyd L, Park JS, Tkebuchava T, Kusano K, Hanley A, Scadova H, Qin G, Cha DH, Johnson KL, Aikawa R, Asahara T, Losordo DW (2005) Clonally expanded novel multipotent stem cells from human bone marrow regenerate myocardium after myocardial infarction. *J Clin Invest* 115:326–338
48. Schwartz RE, Reyes M, Koodie L, Jiang Y, Blackstad M, Lund T, Lenvik T, Johnson S, Hu WS, Verfaillie CM (2002) Multipotent adult progenitor cells from bone marrow differentiate into functional hepatocyte-like cells. *J Clin Invest* 109:1291–1302
49. Khurana S, Mukhopadhyay A (2007) Characterization of the potential subpopulation of bone marrow cells involved in the repair of injured liver tissue. *Stem Cells* 25:1439–1447
50. Alvarez-Dolado M, Pardal R, Garcia-Verdugo JM, Fike JR, Lee HO, Pfeffer K, Lois C, Morrison SJ, Alvarez-Buylla A (2003) Fusion of bone-marrow-derived cells with Purkinje neurons, cardiomyocytes and hepatocytes. *Nature* 425:968–973
51. Vassilopoulos G, Wang PR, Russell DW (2003) Transplanted bone marrow regenerates liver by cell fusion. *Nature* 422:901–904
52. Lysy PA, Campard D, Smets F, Malaise J, Mourad M, Najimi M, Sokal EM (2008) Persistence of a chimerical phenotype after hepatocyte differentiation of human bone marrow mesenchymal stem cells. *Cell Prolif* 41:36–58

53. Ferrari G, Cusella-De Angelis G, Coletta M, Paolucci E, Stornaiuolo A, Cossu G, Mavilio F (1998) Muscle regeneration by bone marrow-derived myogenic progenitors. *Science* 279:1528–1530
54. Corbel SY, Lee A, Yi L, Duenas J, Brazelton TR, Blau HM, Rossi FM (2003) Contribution of hematopoietic stem cells to skeletal muscle. *Nat Med* 9:1528–1532
55. Doyonnas R, LaBarge MA, Sacco A, Charlton C, Blau HM (2004) Hematopoietic contribution to skeletal muscle regeneration by myelomonocytic precursors. *Proc Natl Acad Sci U S A* 101:13507–13512
56. Abedi M, Foster BM, Wood KD, Colvin GA, McLean SD, Johnson KW, Greer DA (2007) Haematopoietic stem cells participate in muscle regeneration. *Br J Haematol* 138:792–801
57. Orlic D, Kajstura J, Chimenti S, Bodine DM, Leri A, Anversa P (2001) Transplanted adult bone marrow cells repair myocardial infarcts in mice. *Ann N Y Acad Sci* 938:221–229; discussion 229–230
58. Orlic D, Kajstura J, Chimenti S, Jakoniuk I, Anderson SM, Li B, Pickel J, McKay R, Nadal-Ginard B, Bodine DM, Leri A, Anversa P (2001) Bone marrow cells regenerate infarcted myocardium. *Nature* 410:701–705
59. Agbulut O, Menot ML, Li Z, Marotte F, Paulin D, Hagege AA, Chomienne C, Samuel JL, Menasche P (2003) Temporal patterns of bone marrow cell differentiation following transplantation in doxorubicin-induced cardiomyopathy. *Cardiovasc Res* 58:451–459
60. Goodell MA, Jackson KA, Majka SM, Mi T, Wang H, Pocius J, Hartley CJ, Majesky MW, Entman ML, Michael LH, Hirschi KK (2001) Stem cell plasticity in muscle and bone marrow. *Ann N Y Acad Sci* 938:208–218; discussion 218–220
61. Jackson KA, Majka SM, Wang H, Pocius J, Hartley CJ, Majesky MW, Entman ML, Michael LH, Hirschi KK, Goodell MA (2001) Regeneration of ischemic cardiac muscle and vascular endothelium by adult stem cells. *J Clin Invest* 107:1395–1402
62. Eisenberg LM, Burns L, Eisenberg CA (2003) Hematopoietic cells from bone marrow have the potential to differentiate into cardiomyocytes in vitro. *Anat Rec A Discov Mol Cell Evol Biol* 274:870–882
63. Nygren JM, Jovinge S, Breitbach M, Sawen P, Roll W, Hescheler J, Taneera J, Fleischmann BK, Jacobsen SE (2004) Bone marrow-derived hematopoietic cells generate cardiomyocytes at a low frequency through cell fusion, but not transdifferentiation. *Nat Med* 10:494–501
64. Murry CE, Soonpaa MH, Reinecke H, Nakajima H, Nakajima HO, Rubart M, Pasumarthi KB, Virag JI, Bartelmez SH, Poppa V, Bradford G, Dowell JD, Williams DA, Field LJ (2004) Haematopoietic stem cells do not transdifferentiate into cardiac myocytes in myocardial infarcts. *Nature* 428:664–668
65. Balsam LB, Wagers AJ, Christensen JL, Kofidis T, Weissman IL, Robbins RC (2004) Haematopoietic stem cells adopt mature haematopoietic fates in ischaemic myocardium. *Nature* 428:668–673
66. Weissman IL (2005) Normal and neoplastic stem cells. *Novartis Found Symp* 265:35–50; discussion 50–34, 92–37
67. Kajstura J, Rota M, Whang B, Cascapera S, Hosoda T, Bearzi C, Nurzynska D, Kasahara H, Zias E, Bonafe M, Nadal-Ginard B, Torella D, Nascimbene A, Quaini F, Urbanek K, Leri A, Anversa P (2005) Bone marrow cells differentiate in cardiac cell lineages after infarction independently of cell fusion. *Circ Res* 96:127–137
68. Bianchi DW, Johnson KL, Salem D (2002) Chimerism of the transplanted heart. *N Engl J Med* 346:1410–1412; author reply 1410–1412
69. Johnson KL, McAlindon TE, Mulcahy E, Bianchi DW (2001) Microchimerism in a female patient with systemic lupus erythematosus. *Arthritis Rheum* 44:2107–2111
70. Liu Y, Song J, Liu W, Wan Y, Chen X, Hu C (2003) Growth and differentiation of rat bone marrow stromal cells: does 5-azacytidine trigger their cardiomyogenic differentiation? *Cardiovasc Res* 58:460–468
71. Makino S, Fukuda K, Miyoshi S, Konishi F, Kodama H, Pan J, Sano M, Takahashi T, Hori S, Abe H, Hata J, Umezawa A, Ogawa S (1999) Cardiomyocytes can be generated from marrow stromal cells in vitro. *J Clin Invest* 103:697–705

72. Wakitani S, Saito T, Caplan AI (1995) Myogenic cells derived from rat bone marrow mesenchymal stem cells exposed to 5-azacytidine. *Muscle Nerve* 18:1417–1426
73. Olivares EL, Ribeiro VP, Werneck de Castro JP, Ribeiro KC, Mattos EC, Goldenberg RC, Mill JG, Dohmann HF, dos Santos RR, de Carvalho AC, Masuda MO (2004) Bone marrow stromal cells improve cardiac performance in healed infarcted rat hearts. *Am J Physiol Heart Circ Physiol* 287:H464–H470
74. Tomita S, Li RK, Weisel RD, Mickle DA, Kim EJ, Sakai T, Jia ZQ (1999) Autologous transplantation of bone marrow cells improves damaged heart function. *Circulation* 100:II247–II256
75. Tomita S, Mickle DA, Weisel RD, Jia ZQ, Tumiati LC, Allidina Y, Liu P, Li RK (2002) Improved heart function with myogenesis and angiogenesis after autologous porcine bone marrow stromal cell transplantation. *J Thorac Cardiovasc Surg* 123:1132–1140
76. Wollert KC, Drexler H (2005) Clinical applications of stem cells for the heart. *Circ Res* 96:151–163
77. Wang JS, Shum-Tim D, Chedrawy E, Chiu RC (2001) The coronary delivery of marrow stromal cells for myocardial regeneration: pathophysiologic and therapeutic implications. *J Thorac Cardiovasc Surg* 122:699–705
78. Silva GV, Litovsky S, Assad JA, Sousa AL, Martin BJ, Vela D, Coulter SC, Lin J, Ober J, Vaughn WK, Branco RV, Oliveira EM, He R, Geng YJ, Willerson JT, Perin EC (2005) Mesenchymal stem cells differentiate into an endothelial phenotype, enhance vascular density, and improve heart function in a canine chronic ischemia model. *Circulation* 111:150–156
79. Breitbach M, Bostani T, Roell W, Xia Y, Dewald O, Nygren JM, Fries JW, Tiemann K, Bohlen H, Hescheler J, Welz A, Bloch W, Jacobsen SE, Fleischmann BK (2007) Potential risks of bone marrow cell transplantation into infarcted hearts. *Blood* 110:1362–1369
80. Xu W, Zhang X, Qian H, Zhu W, Sun X, Hu J, Zhou H, Chen Y (2004) Mesenchymal stem cells from adult human bone marrow differentiate into a cardiomyocyte phenotype in vitro. *Exp Biol Med (Maywood)* 229:623–631
81. Huangfu D, Maehr R, Guo W, Eijkelenboom A, Snitow M, Chen AE, Melton DA (2008) Induction of pluripotent stem cells by defined factors is greatly improved by small-molecule compounds. *Nat Biotechnol* 26:795–797
82. Perin EC, Dohmann HF, Borojevic R, Silva SA, Sousa AL, Silva GV, Mesquita CT, Belem L, Vaughn WK, Rangel FO, Assad JA, Carvalho AC, Branco RV, Rossi MI, Dohmann HJ, Willerson JT (2004) Improved exercise capacity and ischemia 6 and 12 months after transendocardial injection of autologous bone marrow mononuclear cells for ischemic cardiomyopathy. *Circulation* 110:II213–218
83. Zhang S, Wang D, Estrov Z, Raj S, Willerson JT, Yeh ET (2004) Both cell fusion and transdifferentiation account for the transformation of human peripheral blood CD34-positive cells into cardiomyocytes in vivo. *Circulation* 110:3803–3807
84. Kinnaird T, Stabile E, Burnett MS, Lee CW, Barr S, Fuchs S, Epstein SE (2004) Marrow-derived stromal cells express genes encoding a broad spectrum of arteriogenic cytokines and promote in vitro and in vivo arteriogenesis through paracrine mechanisms. *Circ Res* 94:678–685
85. Pak HN, Qayyum M, Kim DT, Hamabe A, Miyauchi Y, Lill MC, Frantzen M, Takizawa K, Chen LS, Fishbein MC, Sharifi BG, Chen PS, Makkar R (2003) Mesenchymal stem cell injection induces cardiac nerve sprouting and increased tenascin expression in a Swine model of myocardial infarction. *J Cardiovasc Electrophysiol* 14:841–848
86. Muller-Ehmsen J, Krausgrill B, Burst V, Schenk K, Neisen UC, Fries JW, Fleischmann BK, Hescheler J, Schwinger RH (2006) Effective engraftment but poor mid-term persistence of mononuclear and mesenchymal bone marrow cells in acute and chronic rat myocardial infarction. *J Mol Cell Cardiol* 41:876–884
87. Scherschel JA, Soonpaa MH, Srour EF, Field LJ, Rubart M (2008) Adult bone marrow-derived cells do not acquire functional attributes of cardiomyocytes when transplanted into peri-infarct myocardium. *Mol Ther* 16:1129–1137

88. Booth MJ, Wilson T (2001) Refractive-index-mismatch induced aberrations in single-photon and two-photon microscopy and the use of aberration correction. *J Biomed Opt* 6:266–272
89. Niesner R, Andresen V, Neumann J, Spiecker H, Gunzer M (2007) The power of single and multibeam two-photon microscopy for high-resolution and high-speed deep tissue and intravital imaging. *Biophys J* 93:2519–2529
90. Ghodsizad A, Niehaus M, Kogler G, Martin U, Wernet P, Bara C, Khaladj N, Loos A, Makoui M, Thiele J, Mengel M, Karck M, Klein HM, Haverich A, Ruhparwar A (2009), Transplanted human cord blood-derived unrestricted somatic stem cells improve left-ventricular function and prevent left-ventricular dilation and scar formation after acute myocardial infarction. *Heart* 95:27–35
91. Badorff C, Brandes RP, Popp R, Rupp S, Urbich C, Aicher A, Fleming I, Busse R, Zeiher AM, Dimmeler S (2003) Transdifferentiation of blood-derived human adult endothelial progenitor cells into functionally active cardiomyocytes. *Circulation* 107:1024–1032
92. Condorelli G, Borello U, De Angelis L, Latronico M, Sirabella D, Coletta M, Galli R, Balconi G, Follenzi A, Frati G, Cusella De Angelis MG, Gioglio L, Amuchastegui S, Adorini L, Naldini L, Vescovi A, Dejana E, Cossu G (2001) Cardiomyocytes induce endothelial cells to trans-differentiate into cardiac muscle: implications for myocardium regeneration. *Proc Natl Acad Sci U S A* 98:10733–10738
93. Welikson RE, Kaestner S, Reinecke H, Hauschka SD (2006) Human umbilical vein endothelial cells fuse with cardiomyocytes but do not activate cardiac gene expression. *J Mol Cell Cardiol* 40:520–528
94. Blomer U, Gruh I, Witschel H, Haverich A, Martin U (2005) Shuttle of lentiviral vectors via transplanted cells in vivo. *Gene Ther* 12:67–74
95. Gruh I, Beilner J, Blomer U, Schmiedl A, Schmidt-Richter I, Kruse ML, Haverich A, Martin U (2006) No evidence of transdifferentiation of human endothelial progenitor cells into cardiomyocytes after coculture with neonatal rat cardiomyocytes. *Circulation* 113:1326–1334
96. Rehman J, Li J, Orschell CM, March KL (2003) Peripheral blood "endothelial progenitor cells" are derived from monocyte/macrophages and secrete angiogenic growth factors. *Circulation* 107:1164–1169
97. Rohde E, Bartmann C, Schallmoser K, Reinisch A, Lanzer G, Linkesch W, Guelly C, Strunk D (2007) Immune cells mimic the morphology of endothelial progenitor colonies in vitro. *Stem Cells* 25:1746–1752
98. Rohde E, Malischnik C, Thaler D, Maierhofer T, Linkesch W, Lanzer G, Guelly C, Strunk D (2006) Blood monocytes mimic endothelial progenitor cells. *Stem Cells* 24:357–367
99. Asahara T, Murohara T, Sullivan A, Silver M, van der Zee R, Li T, Witzenbichler B, Schatteman G, Isner JM (1997) Isolation of putative progenitor endothelial cells for angiogenesis. *Science* 275:964–967
100. Kawamoto A, Gwon HC, Iwaguro H, Yamaguchi JI, Uchida S, Masuda H, Silver M, Ma H, Kearney M, Isner JM, Asahara T (2001) Therapeutic potential of ex vivo expanded endothelial progenitor cells for myocardial ischemia. *Circulation* 103:634–637
101. Burghoff S, Ding Z, Godecke S, Assmann A, Wirrwar A, Buchholz D, Sergeeva O, Leurs C, Hanenberg H, Muller HW, Bloch W, Schrader J (2008) Horizontal gene transfer from human endothelial cells to rat cardiomyocytes after intracoronary transplantation. *Cardiovasc Res* 77:534–543
102. Pan YW, Scarlett JM, Luoh TT, Kurre P (2007) Prolonged adherence of human immunodeficiency virus-derived vector particles to hematopoietic target cells leads to secondary transduction in vitro and in vivo. *J Virol* 81:639–649
103. Peichev M, Naiyer AJ, Pereira D, Zhu Z, Lane WJ, Williams M, Oz MC, Hicklin DJ, Witte L, Moore MA, Rafii S (2000) Expression of VEGFR-2 and AC133 by circulating human CD34(+) cells identifies a population of functional endothelial precursors. *Blood* 95:952–958

104. Vasa M, Fichtlscherer S, Aicher A, Adler K, Urbich C, Martin H, Zeiher AM, Dimmeler S (2001) Number and migratory activity of circulating endothelial progenitor cells inversely correlate with risk factors for coronary artery disease. *Circ Res* 89:E1–7
105. Gill M, Dias S, Hattori K, Rivera ML, Hicklin D, Witte L, Girardi L, Yurt R, Himel H, Rafii S (2001) Vascular trauma induces rapid but transient mobilization of VEGFR2(+)AC133(+) endothelial precursor cells. *Circ Res* 88:167–174
106. Reyes M, Dudek A, Jahagirdar B, Koodie L, Marker PH, Verfaillie CM (2002) Origin of endothelial progenitors in human postnatal bone marrow. *J Clin Invest* 109:337–346
107. Rehman J, Li J, Parvathaneni L, Karlsson G, Panchal VR, Temm CJ, Mahenthiran J, March KL (2004) Exercise acutely increases circulating endothelial progenitor cells and monocyte-/macrophage-derived angiogenic cells. *J Am Coll Cardiol* 43:2314–2318
108. Hubner G, Brauchle M, Smola H, Madlener M, Fassler R, Werner S (1996) Differential regulation of pro-inflammatory cytokines during wound healing in normal and glucocorticoid-treated mice. *Cytokine* 8:548–556
109. Rappolee DA, Mark D, Banda MJ, Werb Z (1988) Wound macrophages express TGF-alpha and other growth factors in vivo: analysis by mRNA phenotyping. *Science* 241:708–712
110. Eglitis MA, Mezey E (1997) Hematopoietic cells differentiate into both microglia and macroglia in the brains of adult mice. *Proc Natl Acad Sci U S A* 94:4080–4085
111. Djukic M, Mildner A, Schmidt H, Czesnik D, Bruck W, Priller J, Nau R, Prinz M (2006) Circulating monocytes engraft in the brain, differentiate into microglia and contribute to the pathology following meningitis in mice. *Brain* 129:2394–2403
112. Tanaka R, Komine-Kobayashi M, Mochizuki H, Yamada M, Furuya T, Migita M, Shimada T, Mizuno Y, Urabe T (2003) Migration of enhanced green fluorescent protein expressing bone marrow-derived microglia/macrophage into the mouse brain following permanent focal ischemia. *Neuroscience* 117:531–539
113. Brazelton TR, Rossi FM, Keshet GI, Blau HM (2000) From marrow to brain: expression of neuronal phenotypes in adult mice. *Science* 290:1775–1779
114. Mezey E, Chandross KJ, Harta G, Maki RA, McKercher SR (2000) Turning blood into brain: cells bearing neuronal antigens generated in vivo from bone marrow. *Science* 290:1779–1782
115. Mezey E, Key S, Vogelsang G, Szalayova I, Lange GD, Crain B (2003) Transplanted bone marrow generates new neurons in human brains. *Proc Natl Acad Sci U S A* 100:1364–1369
116. Castro RF, Jackson KA, Goodell MA, Robertson CS, Liu H, Shine HD (2002) Failure of bone marrow cells to transdifferentiate into neural cells in vivo. *Science* 297:1299
117. Roybon L, Ma Z, Asztely F, Fosum A, Jacobsen SE, Brundin P, Li JY (2006) Failure of transdifferentiation of adult hematopoietic stem cells into neurons. *Stem Cells* 24:1594–1604
118. Mezey E, Nagy A, Szalayova I, Key S, Bratincsak A, Baffi J, Shahar T (2003) Comment on "Failure of bone marrow cells to transdifferentiate into neural cells in vivo". *Science* 299:1184; author reply 1184
119. Trainor PA, Zhou SX, Parameswaran M, Quinlan GA, Gordon M, Sturm K, Tam PP (1999) Application of lacZ transgenic mice to cell lineage studies. *Methods Mol Biol* 97:183–200
120. Phinney DG, Isakova I (2005) Plasticity and therapeutic potential of mesenchymal stem cells in the nervous system. *Curr Pharm Des* 11:1255–1265
121. Woodbury D, Reynolds K, Black IB (2002) Adult bone marrow stromal stem cells express germline, ectodermal, endodermal, and mesodermal genes prior to neurogenesis. *J Neurosci Res* 69:908–917
122. Bertani N, Malatesta P, Volpi G, Sonego P, Perris R (2005) Neurogenic potential of human mesenchymal stem cells revisited: analysis by immunostaining, time-lapse video and microarray. *J Cell Sci* 118:3925–3936
123. Lu P, Blesch A, Tuszynski MH (2004) Induction of bone marrow stromal cells to neurons: differentiation, transdifferentiation, or artifact? *J Neurosci Res* 77:174–191
124. Arnhold S, Klein H, Klinz FJ, Absenger Y, Schmidt A, Schinkothe T, Brixius K, Kozlowski J, Desai B, Bloch W, Addicks K (2006) Human bone marrow stroma cells display certain

neural characteristics and integrate in the subventricular compartment after injection into the liquor system. *Eur J Cell Biol* 85:551–565

125. Gerdoni E, Gallo B, Casazza S, Musio S, Bonanni I, Pedemonte E, Mantegazza R, Frassoni F, Mancardi G, Pedotti R, Uccelli A (2007) Mesenchymal stem cells effectively modulate pathogenic immune response in experimental autoimmune encephalomyelitis. *Ann Neurol* 61:219–227
126. Woodbury D, Schwarz EJ, Prockop DJ, Black IB (2000) Adult rat and human bone marrow stromal cells differentiate into neurons. *J Neurosci Res* 61:364–370
127. Neuhuber B, Gallo G, Howard L, Kostura L, Mackay A, Fischer I (2004) Reevaluation of in vitro differentiation protocols for bone marrow stromal cells: disruption of actin cytoskeleton induces rapid morphological changes and mimics neuronal phenotype. *J Neurosci Res* 77:192–204
128. Raedt R, Pinxteren J, Van Dycke A, Waeytens A, Craeye D, Timmermans F, Vonck K, Vandekerckhove B, Plum J, Boon P (2007) Differentiation assays of bone marrow-derived Multipotent Adult Progenitor Cell (MAPC)-like cells towards neural cells cannot depend on morphology and a limited set of neural markers. *Exp Neurol* 203:542–554
129. Horwitz EM, Le Blanc K, Dominici M, Mueller I, Slaper-Cortenbach I, Marini FC, Deans RJ, Krause DS, Keating A (2005) Clarification of the nomenclature for MSC: The International Society for Cellular Therapy position statement. *Cytotherapy* 7:393–395
130. Wang R, Liang J, Jiang H, Qin LJ, Yang HT (2008) Promoter-dependent EGFP expression during embryonic stem cell propagation and differentiation. *Stem Cells Dev* 17:279–289
131. Gruh I, Wunderlich S, Winkler M, Schwanke K, Heinke J, Blomer U, Ruhparwar A, Rohde B, Li RK, Haverich A, Martin U (2008) Human CMV immediate-early enhancer: a useful tool to enhance cell-type-specific expression from lentiviral vectors. *J Gene Med* 10:21–32
132. Cole C, Qiao J, Kottke T, Diaz RM, Ahmed A, Sanchez-Perez L, Brunn G, Thompson J, Chester J, Vile RG (2005) Tumor-targeted, systemic delivery of therapeutic viral vectors using hitchhiking on antigen-specific T cells. *Nat Med* 11:1073–1081
133. Gao G, Johansson U, Rundquist I, Ollinger K (1994) Lipofuscin-induced autofluorescence of living neonatal rat cardiomyocytes in culture. *Mech Ageing Dev* 73:79–86
134. Laflamme MA, Murry CE (2005) Regenerating the heart. *Nat Biotechnol* 23:845–856
135. van Laake LW, Passier R, Monshouwer-Kloots J, Verkleij AJ, Lips DJ, Freund C, den Ouden K, Ward-van Oostwaard D, Korving J, Tertoolen LG, van Echteld CJ, Doevendans PA, Mummery CL (2007) Human embryonic stem cell-derived cardiomyocytes survive and mature in the mouse heart and transiently improve function after myocardial infarction. *Stem Cell Res.* 1:9–24
136. Sato Y, Araki H, Kato J, Nakamura K, Kawano Y, Kobune M, Sato T, Miyanishi K, Takayama T, Takahashi M, Takimoto R, Iyama S, Matsunaga T, Ohtani S, Matsuura A, Hamada H, Niitsu Y (2005) Human mesenchymal stem cells xenografted directly to rat liver are differentiated into human hepatocytes without fusion. *Blood* 106:756–763
137. Wu T, Cieply K, Nalesnik MA, Randhawa PS, Sonzogni A, Bellamy C, Abu-Elmagd K, Michalopolous GK, Jaffe R, Kormos RL, Gridelli B, Fung JJ, Demetris AJ (2003) Minimal evidence of transdifferentiation from recipient bone marrow to parenchymal cells in regenerating and long-surviving human allografts. *Am J Transplant* 3:1173–1181
138. Brocker V, Langer F, Fellous TG, Mengel M, Brittan M, Bredt M, Milde S, Welte T, Eder M, Haverich A, Alison MR, Kreipe H, Lehmann U (2006) Fibroblasts of recipient origin contribute to bronchiolitis obliterans in human lung transplants. *Am J Respir Crit Care Med* 173:1276–1282

Adv Biochem Engin/Biotechnol (2009) 114: 107-128
DOI: 10.1007/10_2008_37

Published online: 19 June 2009

Stem Cells for Cardiac Regeneration by Cell Therapy and Myocardial Tissue Engineering

Jun Wu, Faquan Zeng, Richard D. Weisel, and Ren-Ke Li

Abstract Congestive heart failure, which often occurs progressively following a myocardial infarction, is characterized by impaired myocardial perfusion, ventricular dilatation, and cardiac dysfunction. Novel treatments are required to reverse these effects – especially in older patients whose endogenous regenerative responses to currently available therapies are limited by age. This review explores the current state of research for two related approaches to cardiac regeneration: cell therapy and tissue engineering. First, to evaluate cell therapy, we review the effectiveness of various cell types for their ability to limit ventricular dilatation and promote functional recovery following implantation into a damaged heart. Next, to assess tissue engineering, we discuss the characteristics of several biomaterials for their potential to physically support the infarcted myocardium and promote implanted cell survival following cardiac injury. Finally, looking ahead, we present recent findings suggesting that hybrid constructs combining a biomaterial with stem and supporting cells may be the most effective approaches to cardiac regeneration.

Keywords Cell therapy, Congestive heart failure, Myocardial infarction, Stem cells, Tissue engineering

Contents

J. Wu, F. Zeng, R.D. Weisel, and R.-K. Li (✉)
Division of Cardiovascular Surgery, Toronto General Research Institute, University of Toronto, Toronto, Ontario, Canada
e-mail: renkeli@uhnres.utoronto.ca

Abbreviations

BMC	Bone marrow cell
CHF	Congestive heart failure
CSC	Cardiac stem cell
ES cell	Embryonic stem cell
HSC	Hematopoietic stem cell
MI	Myocardial infarction
MSC	Mesenchymal stem cell
PEG-PVL	Poly(δ-valerolactone)-*block*-poly(ethylene glycol)-*block*-poly (δ-valerolactone)

1 Congestive Heart Failure

Congestive heart failure (CHF), most often following a myocardial infarction (MI) [1], is the result of a structural or functional disorder that impairs the ability of the heart to pump sufficient blood to meet the metabolic demands of the body. Despite optimal medical care, many survivors of an acute MI with an ejection fraction less than 40% will experience progressive ventricular dysfunction [2, 3]. Heart transplantation has been a therapeutic option for many decades. However, because a shortage of donor organs limits the feasibility of this procedure, transplantation is unlikely to provide a viable option for the large number of patients who face a 35% 2 year mortality rate. In those who do receive organs, the return to "normal activity" is often limited by infection or rejection despite immunosuppressive treatments.

Recent studies found that the human heart has some capacity for self-repair after an injury [4–6]. This repair process, termed "cardiac regeneration", appears to involve the participation of stem cells (both bone marrow- and heart-derived) [7–9], and is partially responsible for restoring heart function after an MI. However, the potential for endogenous regeneration is diminished with age so that cardiac regeneration and functional restoration are limited following an MI in aged patients. In this population, irreversible cardiomyocyte loss and fibrosis within the myocardial scar lead to progressive ventricular remodeling, and eventually to ventricular dilatation and CHF. Novel therapies are urgently required to treat and prevent CHF, particularly in older, debilitated individuals.

2 Stem Cell Therapy to Promote Cardiac Regeneration

Cell therapy is a novel treatment to prevent ventricular dilatation and cardiac dysfunction in patients who have suffered an MI. The concept of repopulating and regenerating the injured myocardium by implanting muscle or stem cells into the damaged tissue originated in the early 1990s [10–13], during which time a variety of somatic and stem cells were tested in infarcted animal hearts. The experimental data clearly demonstrated that implanted cells survived after implantation, restored cardiac function, increased regional perfusion and prevented progressive ventricular dilation. Animals that received implanted cells had better cardiac function than media-injected controls.

Although the mechanisms responsible for the restoration of cardiac function by the implanted cells remain unclear, the original concept was that the implanted cells improved ventricular function by increasing the number of functioning muscle cells through repopulation or transdifferentiation. For example, implantation of fetal cardiomyocytes could repopulate the myocardial scar with beating muscle cells, limiting scar expansion and preventing cardiac dysfunction [11]. Skeletal muscle cells injected into the damaged myocardial tissue also survived and restored cardiac function [14]. Bone marrow cells (BMCs), which either homed to or were injected directly into the injured myocardium, appeared to form new muscle cells – albeit in very small numbers – and induced the formation of new capillaries and arterioles [8, 12]. However, subsequent studies [15, 16] suggested that improved cardiac function was probably the result of paracrine effects induced by the implanted cells, including accelerated angiogenesis, decreased deleterious matrix remodeling, and increased recruitment of circulating stem cells to the damaged tissue. An alternate explanation was that the few "new" cardiomyocytes observed after BMC implantation might have arisen from the fusion of implanted cells with persisting cardiomyocytes [17, 18].

Because few, if any, BMCs underwent myogenic "transdifferentiation" following cell implantation [18–20], it was difficult to understand how so few engrafted cells could restore function to the damaged heart. Multiple biological factors released or induced by the implanted cells have since been identified within the infarct area. Angiogenic factors increase the blood vessel density in the ischemic area [10]. The release of cytokines after cell implantation could also prevent host cardiomyocyte apoptosis and direct stem cell homing to the damaged area [21]. Protease inhibitors released by the implanted cells within the infarct and the remote (normal) myocardium inhibit matrix degradation and ventricular dilatation [22]. Although the exact mechanisms remain ambiguous, most pre-clinical cell transplantation studies reported improvements in ventricular function [23], suggesting that cell therapy offers a unique opportunity to modify the remodeling process and enhance the endogenous mechanisms that promote cardiac regeneration.

Over the last 6 years, the encouraging pre-clinical findings led to the initiation of numerous clinical cell therapy trials [24] designed to test the efficacy of skeletal myoblasts or bone marrow precursor cells injected into infarcted myocardium or arteries [25, 26]. However, the dramatic benefits observed in the pre-clinical animal experiments have not been replicated clinically. While the direct injection of muscle cells into the infarcted myocardium was shown to be clinically safe and some of the implanted cells survived, data from clinical trials did not demonstrate a fundamental improvement in cardiac function. For example, in a phase II study, myoblast implantation did not prevent ventricular dysfunction [27]. The implantation of BMCs into an infarcted artery resulted in only transient improvements in ventricular function in the BOOST trial [28]. The Repair AMI trial – a randomized clinical trial involving the implantation of autologous stem cells from the bone marrow into the infarct-related coronary artery – achieved a statistically significant, but modest, improvement in cardiac function (2.9% increase in ejection fraction compared to controls). The medical benefits of such a small increase compared to the potential interventional risks of cell injection are questionable. Combining results from all clinical trials demonstrated only minimal benefits when BMCs were implanted in a randomized, double-blind manner. In addition, the rationale for these clinical trials has been questioned because the implanted bone marrow-derived cells may not have the ability to become cardiomyocytes. Numerous subsequent studies have questioned the "plasticity" (potential for transdifferentiation) of hematopoietic stem cells (HSCs – a fraction of BMCs most commonly used for clinical trials) [18–20].

The discrepancy between the results of the pre-clinical studies and the clinical trials could be due to the effects of co-morbidities such as diabetes, smoking, hypertension and diffuse vascular disease. Further, the age of the patients at the time of MI could be particularly relevant, because the majority of patients treated for MI are older adults, while pre-clinical studies were carried out in young animals. In particular, the regenerative capacity of endogenous stem cells becomes limited in aged individuals [29, 30]. In other words, the number and potency of endogenous stem cells available to contribute to homing, inflammatory responses, and the cellular micro-environment for implanted cell survival is reduced in older compared to younger individuals. Achieving cardiac functional improvement in aged individuals will require that we augment not only the capacity for engraftment/survival and regeneration of the donor cells, but also the recipients' own intrinsic capacity for regeneration. Ongoing research should focus on understanding how the benefits of cell transplantation are impaired by aging so that the next generation of bio-interventions may be designed to correct these defects. For instance, if diminished function of older donor cells is the primary defect, then the donor cells might be treated, or "rejuvenated", prior to implantation. Alternatively, strategies could be directed to induce tolerance in allogenic cells from young donors. If, however, the defect is related to an abnormal response of the recipient to cell implantation, then treatments could be aimed at restoring the host's regenerative capacity. For example, an attenuated bone marrow response to ischemia (including a suppressed inflammatory response) could be corrected using strategies to restore the bone marrow response to approximate that of a young individual.

3 Optimal Cells for Cardiac Repair by Cell Therapy

Since the concept of cell therapy to restore cardiac function was introduced a decade ago, much research has focused on identifying the optimal cells for implantation. Donor cells must be effective, easy to obtain, abundant, and unaffected by immunorejection.

Autologous cells, such as cardiomyocytes [11], BMCs [10], skeletal muscle cells [14] and smooth muscle cells [23], have been widely used in pre-clinical studies. Although these cells can prevent ventricular dysfunction when implanted after an MI, clinical cell therapy trials have been unable to replicate the new tissue formation and neovascularization reported in the pre-clinical studies. Therefore, we must re-examine the characteristics of the various candidate cell types:

3.1 Autologous Somatic Cells

The original hypothesis for cell therapy was that donor cells would restore function to the damaged heart by repopulating the muscle cells within the infarct area and increasing the number of functioning muscle units. Pre-clinical studies demonstrated that cardiomyocytes derived from fetal and neonatal hearts survived within the damaged myocardium and did form new cardiac muscle, which effectively prevented ventricular dilation and improved heart function [11]. The research data showed similar results following the implantation of adult skeletal myoblasts and smooth muscle cells derived from biopsies [23]. Although all three muscle cell types are excellent candidates for cell therapy, adult human donor cardiomyocytes are not available clinically, and ethical issues obstruct the use of fetal and neonatal cardiomyocytes. Since skeletal myoblasts (satellite cells) can be easily isolated from skeletal muscle, Menasche and colleagues implanted autologous skeletal myoblasts into patients with congestive heart failure. Their phase I clinical trial demonstrated the safety and potential efficacy of the cells [31]. However, the myoblasts produced only a limited clinical benefit in a multicenter phase II trial [27]. The repopulation of muscle cells within the damaged myocardial tissue results from a combination of factors. Among them, patient age might be particularly important, as we have demonstrated that growth rates differ significantly in cells isolated from young and aged rats [32]. The regenerative capacity of adult muscle cells may become diminished with age (or other co-morbidities) in patients with CHF. Because aging limits the effectiveness of somatic cell therapy, future studies will need to evaluate new methods to restore the regenerative capacity of cells isolated from aged patients.

3.2 Autologous Stem Cells

The new biomedical field of regenerative medicine has emerged from recent advances in stem cell biology. Stem cells are generally considered to be capable of both self-renewal and differentiation [33]. By definition, a self-renewing cell should be capable of generating sufficient cells for cell therapy. Perhaps the most attractive characteristic of the stem cells is their multipotency. The microenvironment is reported to induce stem cell differentiation. The rationale for using autologous stem cells is that a patient's own stem cells can be implanted into their myocardial tissue, where the cells will rebuild the dysfunctional heart by differentiating into muscle cells, blood vessel cells and matrix cells. Autologous stem cells have been actively evaluated due to the identified limitations of autologous somatic muscle cells for cell therapy.

3.2.1 Hematopoietic Stem Cells (HSCs)

HSCs produce blood cells and are responsible for the regeneration of blood forming tissue [34]. They are characterized by the expression of c-kit, thy-1, sca-1 (mouse) and CD34 (human), are lineage negative (Lin^-) [35, 36], and can be isolated from the bone marrow and peripheral blood. A number of in vivo studies demonstrated that HSCs (Lin^-/c-kit^+) can transdifferentiate into myogenic cells, producing new cardiomyocytes, smooth muscle cells and vascular endothelial cells after implantation into the infarcted myocardium [8]; tissue regeneration was associated with improvements in regional perfusion and cardiac function. As reviewed in Sect. 2, HSCs have been employed in several clinical trials over the last 6–7 years because they are easily obtained from blood and bone marrow, and are effective for cardiac functional restoration in vivo. Unfortunately, the randomized trials reported only limited beneficial effects, or failure. Effective cell therapy has only been demonstrated in animal models in which young recipients received cells from young, healthy donors, while clinical results were obtained when aged cells were implanted into aged patients. Dimmeler and colleagues have attributed this discrepancy to diminished "stemness" in the stem cells of aged patients. Similarly, several recent studies have questioned the "plasticity" of HSCs [37]. As a consequence, novel and creative techniques will be required to develop cell therapies that employ HSCs.

3.2.2 Mesenchymal Stem Cells (MSCs)

MSCs are multipotent stem cells in bone marrow and adipose tissue [38–40]. These cells are attractive candidates for cell therapy because they are easy to access and lack the cell surface expression of MHC [41], the receptor responsible for initiating immune rejection. In addition, MSCs can be expanded in culture and may differentiate into myogenic cells [42–45]. For example, Makino and colleagues [46] showed that MSCs became beating cardiomyocytes when cultured in the presence of 5-azacytidine, a chromosomal demethylating agent that removes epigenetic restrictions on cell differentiation pathways. The resulting beating cells resembled adult cardiomyocytes in terms of transcription factors, protein expression, electron microscopic structure, and electromechanical activity. A number of in vivo studies have demonstrated that MSCs implanted into the infarcted myocardium can survive within the implanted area, increase regional blood vessel density and prevent scar expansion [13]. Improvements in cardiac function have subsequently been confirmed using a clinically relevant porcine model [47], and are also supported by several pre-clinical studies that used other models [48, 49]. Adipose-derived MSCs have also been successfully used to repair infarcted regions in the rat heart [50].

The positive findings have sparked several clinical applications in which MSCs are implanted into damaged myocardial tissue to restore cardiac function. However, as with autologous somatic cells and HSCs, the clinical application of MSCs produced only limited improvement in left ventricular ejection fraction [51].

3.2.3 Cardiac Stem Cells (CSCs)

The traditional view of the adult heart as a terminally differentiated organ without the potential for self-renewal has changed dramatically. The heart has recently been found to contain CSCs positive for various stem cell markers, including lineage negative c-kit [52–55], Sca-1 [56, 57], isl1 [58] and cardiosphere-forming cells [59, 60]. The c-kit positive CSCs, identified using anti-c-kit antibodies, are small and round, self-renewing, clonogenic, and multipotent, giving rise to cardiomyocytes, smooth muscle cells and endothelial cells. CSC injection into the infarct border zone of an ischemic rat heart resulted in new myocardium containing cardiomyocytes, capillaries and arterioles [61]. Intracoronary delivery after 4 hours of reperfusion following 90 min of coronary occlusion in rats limited infarct size, attenuated LV remodeling and ameliorated myocardial dysfunction at 5 weeks after MI [62]. A second CSC isolation technique utilizes an antibody against stem cell antigen 1 (Sca-1). Sca-1 positive cells lack markers for blood lineage cells or c-kit, and do not differentiate spontaneously in vitro; however, when exposed to 5'-azacytidine, a small fraction of these cells demonstrated biochemical evidence of cardiomyocyte differentiation [63]. Further, when injected intravenously after ischemia/reperfusion in mice, Sca-1 positive cells homed to the injured myocardium and differentiated into cardiomyocytes, with or without fusion with host cardiomyocytes [56]. In Sca-1 positive CSCs recently isolated from human heart tissue, the efficiency of cardiomyocyte formation and maturation in vitro was greatly enhanced by the addition of the growth factor TGFb1 during differentiation [64]. Chien and colleagues [58, 65] identified a third group of cardiac progenitor cells, called islet-1 (isl1) positive cells, in the heart tissue of newborn rats, mice, and humans. These cells are defined by the presence of an isl1 protein without concomitant expression of c-kit or Sca-1. Unlike c-kit and Sca-1, isl1 is a transcription factor, and isl1-expressing cells isolated from neonatal mice have the capacity to mature into beating cardiomyocytes [58]. Recently traced in the mouse heart, isl1 positive cells produced not only cardiomyocytes, but also smooth muscle and endothelial cells [65]. However, the scarcity of isl1 positive cells, and the fact that their presence is restricted to very young animals and humans, may limit their application. Finally, CSCs can be isolated from the myocardial tissue using cardiosphere-forming technology. Sphere-forming cells have recently been derived from subcultures of human atrial or ventricular biopsy specimens, and from murine hearts. These cells are clonogenic, express stem and endothelial progenitor cell antigens/markers, and appear to have the properties of adult CSCs. They are capable of long-term self-renewal and can differentiate in vitro and after transplantation in vivo to yield the major specialized cell types of the heart: myocytes and vascular cells [59]. CSCs are a new and challenging stem cell source, and much research will be required to determine their potential for human cell therapy.

3.3 Allogenic Stem Cells

The accumulated evidence has demonstrated that both autologous stem and somatic cells have limited capacities to repair damaged myocardial tissue and restore cardiac function after an MI in patients. For aged individuals with congestive heart failure, the major obstacle is likely the diminished regenerative capacity – of their cells due to age or other co-morbidities. Since optimal cardiac regeneration likely requires the introduction of young, healthy muscle cells or bone marrow-derived stem cells, allogenic cells have been proposed as candidates for cell therapy. Another point in favour of allogenic cells is that they can be pre-prepared for transplantation, and then "shelved" as products until they are required. The feasibility of allogenic MSCs for cardiac repair has been closely examined because MSCs may be immunoprivileged [41], and some studies suggest that they are not rejected after implantation [42–45]. For example, infarcted pig hearts functioned better after the implantation of allogenic MSCs than without cell therapy [66]. In a short- and long-term study, Dai and colleagues found that allogenic MSCs survived for 6 months after implantation into damaged hearts. However, although the surviving MSCs differentiated into muscle cells that improved cardiac function at 4 weeks [67], the benefits of the surviving cells did not persist to 6 months. Therefore, the extent to which MSCs are useful for long-term cell therapy will require further evaluation.

3.4 Embryonic Stem (ES) Cells

ES cells, derived from the inner cell mass of blastocyst-stage embryos, were first isolated from in vitro cultures of mouse blastocysts [68], and were isolated in humans in the late 1990s [69]. They are pluripotent, and can differentiate into all derivatives of the three primary germ layers (ectoderm, endoderm, mesoderm) [70]. Protocols for ES cell isolation and propagation have been well-established, while techniques to induce cardiogenic differentiation in ES cells are still being developed. With their extensive capacity for self-renewal and potential for myogenic differentiation, ES cells may be a promising candidate cell source for cell therapy.

For cardiac regeneration, an undisputed advantage of ES cells is their ability to differentiate into all cardiac cell types [69, 71, 72]. Scientifically, ES cells are perhaps the most promising cell source to generate genuine cardiomyocytes. Klug and colleagues [73] reported on the potential use of ES-derived myocytes for cardiac repair in dystrophic mice. In that study, they described the presence of ES-derived cardiomyocyte grafts for as long as 7 weeks after implantation. Implanted ES-derived cardiomyocytes survived in the injured rat myocardium [74] and improved global cardiac function and myocardial contractility. Human ES cells, which have cardiac structures that include intercalated discs, sarcomeric organization, and electromechanical integration, are currently receiving much attention for their potential to generate human cardiomyocytes [75–77]. The consistent formation of myocardial grafts in the infarcted rat heart has already been demonstrated following the implantation of human ES cells. The engrafted human myocardium attenuated ventricular dilatation and preserved regional and global contractile function relative to non-cardiac human ES cell-derivatives or vehicle (controls) at 4 weeks after coronary artery ligation [78]. However, in a comparable study, the improvement in cardiac function observed at 4 weeks after transplantation of human ES cell-derived cardiomyocytes into mice that had undergone an MI was not sustained at 12 weeks compared with mice receiving human ES cell-derived non-cardiomyocytes [79]. Although the disagreement between these observations may be related to species incompatibilities, shorter-term studies must be interpreted with caution, and longer-term follow-up is necessary before the efficacy of myocardial cell transplantation can be clearly determined.

Despite their obvious advantages for use in regenerative medicine, ES cells present a number of clinical challenges. First is the continuing ethical debate about the harvesting of human embryos to create the cells. A second major obstruction is the need for new techniques to isolate, expand and purify cardiac committed progenitor cells from the ES population in sufficient numbers for clinical use. A recent report [80] demonstrated that undifferentiated human ES cells did not produce cardiomyocytes after transplantation into normal or infarcted hearts. More recently however, Keller and colleagues [81] defined the sequence required to induce reproducibly beating cardiomyocytes from ES cells by carefully isolating the stages of development of the cardiovascular lineages in human ES cell differentiation cultures. By induction with combinations of activin A, bone morphogenetic protein 4 (BMP4), basic fibroblast growth factor (bFGF, also known as FGF2), vascular endothelial growth factor (VEGF, also known as VEGFA) and dickkopf homolog 1 (DKK1) in serum-free media, they produced human ES cell-derived embryoid bodies that generated a KDR^{low}/c-kit^{neg} $(CD117)^{neg}$ population with cardiac, endothelial and vascular smooth muscle potential in vitro and, after transplantation, in vivo. Third, the effectiveness of ES cell transplantation may be limited by rejection. Although one study [82] reported no significant immune response, subsequent studies [83] reported a significant T-lymphocyte and dendritic cell population around transplanted allogenic ES cells, suggesting an immune response associated with ES cell differentiation in vivo. Finally, because ES cells are pluripotent, their implantation into the heart could result in teratoma formation. While pre-commitment of undifferentiated ES cells has been shown to limit teratoma formation [84], the inclusion within an ES cell graft of a single undifferentiated cell might produce a teratoma. Taken together, these studies raise exciting prospects for the use of ES cells in regenerative and tissue engineering therapies aimed at repairing and "rebuilding" the hearts of patients with congenital or acquired myocardial disease.

4 Optimal Biomaterials for Cardiac Tissue Engineering

Ventricular dilatation is a major contributor to congestive heart failure. Cell therapy provides a novel opportunity to repopulate muscle cells within the infarct region, strengthen the damaged myocardial tissue and prevent ventricular dilatation and cardiac dysfunction. However, multiple studies have demonstrated that the number of implanted cells is very low within the infarct or ischemic region of the dysfunctional heart at the experimental endpoints. This is possibly due to detrimental effects of the ischemic environment on the implanted cells. Since the number of implanted cells within the infarcted heart is positively correlated with the degree of cardiac functional restoration achieved [85], numerous studies have attempted to determine the optimal means to improve the microenvironment and promote the survival of implanted cells. For example, angiogenic factors, such as VEGF [86] and bFGF [87], were injected into the ischemic myocardial tissue to improve regional perfusion prior to cell implantation, and gene modification or pre-conditioning to increase heat-shock proteins were used to increase the resistance of the implanted cells to an ischemic environment [88, 89]. These techniques have significantly increased the number of implanted cells at the implanted area and the associated improvements in cardiac function. However, the proportion of implanted cells retained in the implanted area of the heart is still extremely low.

Recently, molecular imaging techniques have been employed for the non-invasive tracking of metabolically active cells in living animals [66, 90–92] (Fig. 1) and humans [93, 94]; these approaches offer effective tools to elucidate the spatial and temporal distributions of implanted cells in living subjects. Several groups have used these techniques to demonstrate that cell leakage from the implanted area results in a significant loss of donor cells immediately after implantation. In fact, few implanted cells actually remain within the damaged tissue. Post-implantation cell retention is a major determinant of the effectiveness of cell therapy. Fortunately, advances in the development of biomaterials for myocardial tissue engineering suggest that combining cell transplantation with biomaterial technologies could significantly improve the potential of these interventions to achieve cardiac regeneration.

Two types of biomaterials have been used to repair the heart after injury: injectable biomaterials and surgically implanted biomaterial grafts.

4.1 Injectable Biomaterials

In situ polymer biodegradable biomaterials include injectable fibrin glue, matrigel, collagen, alginate gels and self-assembling peptides (summarized in Table 1). These materials are injected in liquid form, and assume a gel structure in the heart [95, 96]. The greatest advantage of an injectable biomaterial is the opportunity to combine it with a mixture of cells and biologically active molecules prior to implantation [97]. The resulting extracellular matrixes can not only reduce physical infarct expansion,

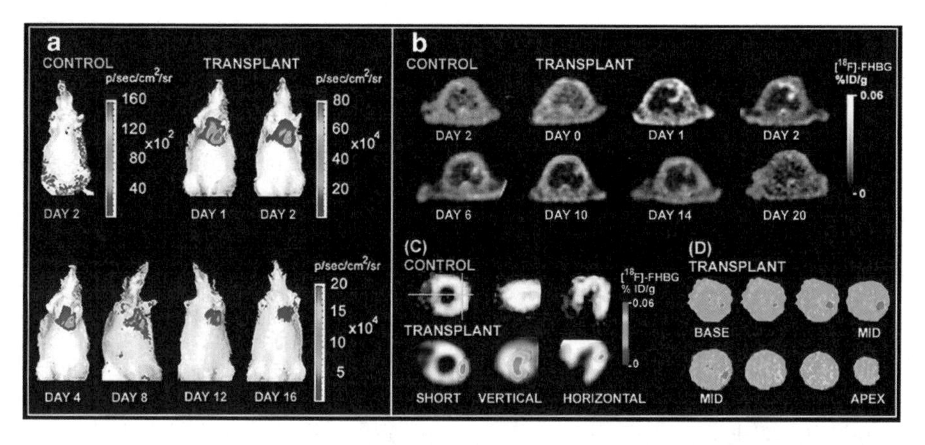

Fig. 1 a–d Molecular imaging of cardiac cell transplantation in living animals. **a** Optical imaging in a representative rat (Transplant) transplanted with embryonic cardiomyoblasts expressing the Fluc reporter gene (emits significant cardiac bioluminescence activity) at days 1, 2, 4, 8, 12, and 16 ($P < 0.05$ vs control). Background signal only (day 2) is shown in Control rat. **b** MicroPET imaging in a representative rat (Transplant) transplanted with cardiomyoblasts expressing the HSV1-sr39tk reporter gene, whereby the location(s), magnitude, and survival duration were monitored by longitudinal imaging of [18F]-FHBG reporter activity (*gray scale*). **c** Detailed tomographic views of cardiac microPET images shown in short, vertical, and horizontal axes for Control and Transplant animals. At day 2, a representative transplant rat exhibited significant activity at the lateral wall, as indicated by the magnitude of [18F]-FHBG reporter activity (*color scale*) overlaid on the [13N]-NH3 perfusion image (*gray scale*). In contrast, the control rat exhibited homogeneous [13N]-NH3 perfusion, but background [18F]-FHBG reporter activity. **d** Autoradiography from the same transplant rat confirmed trapping of [18F] radioactivity by transplanted cells at the lateral wall at a finer spatial resolution (50 μm)

Table 1 Selected injectable biomaterials (gels) used to create scaffolds for myocardial tissue engineering

Gel type	Gel composition	Gelling process	MTE applications	References
Protein-based	Fibrin glue	Fibrin + thrombin	Scar limitation, cell retention	102–105
	Matrigel	Temperature-sensitive (liquid at 4°C; gels at 37°C)	Cell retention	106–108
Natural polymer-based	Protein nono-fibers	Polymer self-aggregation	Cell retention and cytokine carrier	109–111
	Alginate	Alginate + calcium chloride	Cell retention and cytokine carrier	112
Synthetic polymer-based	PEG-PVL	Temperature-sensitive (liquid at (at room temp; gels at 37°C, pH 7–8)	Cell retention	unpublished
	PIPAAm	Temperature-sensitive (gels at 37°C)	Cell sheets	58,114

but also closely mimic the myocardial microenvironment by facilitating the sustained, local release of growth factors and cytokines that enhance implanted cell survival and induce neovessel formation within the ischemic myocardium [98]. Most important, gelation of the biomaterial after injection into the myocardial tissue can significantly reduce the leakage of implanted cells and biological materials.

4.1.1 Protein-Based Hydrogel

A gel is defined as a three-dimensional network swollen by a solvent, a major component of the gel system. Gels can be classified into 2 categories based on the formation of their molecular network: physical gels (formed by secondary forces), and chemical gels (formed by covalent bonds) [99]. Some physical gels are heat reversible, also called "thermoreversible" [100], and others will gel in response to pH changes, ionic cross-linking, solvent exchange or crystallization. The most commonly used protein-based gels used for cardiovascular applications include fibrin glue, matrigel, collagen, and amphiphilic peptides.

Fibrin glue is composed of 2 separate solutions: fibrinogen and thrombin. When mixed together, these agents mimic the last stages of the clotting cascade to form a fibrin clot, or a protein gel into which cells can be incorporated. When injected into the ischemic myocardium of adult rats, the compound dramatically increased regional blood vessel density [96]. Fibrin glue, in combination with several cell types – including skeletal muscle cells [97, 101, 102], bone marrow derived mononuclear cells [103], and endothelial cells [104] – has been injected into the damaged myocardium after an MI. Compared to the delivery of cells alone, this combined treatment increased the number of implanted cells at the implanted area [101] – likely because the mixture of fibrin glue and cells increased cell retention and induced greater neovascularization within the ischemic myocardium. In turn, these effects reduced infarct expansion beyond the limitation observed following the injection of biomaterial alone.

Matrigel is a protein mixture secreted by mouse tumor cells that resembles the complex extracellular environment found in many tissues. It exists as a liquid when chilled to 4 °C, but its proteins self-assemble into a gel at 37 °C (body temperature). Kofidis and colleagues demonstrated that injecting a mixture of matrigel and mouse ES cells into infarcted myocardium improved both implanted cell survival and heart function relative to the injection of matrigel or cells alone [105, 106]. A similar study demonstrated that ventricular geometry and cardiac function were improved relative to controls when cardiomyocytes were mixed with matrigel before they were delivered into infarcted myocardial tissue [107].

Amphiphilic peptides form *nanofibers* (7–100 nm in diameter) through molecular self-assembly (protein folding) via hydrogen bonding, van der Waals interaction, ionic bonds or hydrophobic interactions in solution at low pH or low osmolarity [108]. Exposed to physiological ionic strength and pH conditions, the peptides interact and self-assemble into stable nanofibers that interweave rapidly to form a hydrated network [108]. A hydrogel scaffold created using the properties of self-assembling peptide nanofibers was used to mimic the intramyocardial microenvironment [95, 109]. After the peptides were injected into the infarcted myocardium, progenitor cells expressing endothelial markers and vascular smooth-muscle cells were recruited into the nanofiber microenvironments. Local delivery of stromal cell derived factor-1 (SDF-1) with the nanofibers further promoted stem cell recruitment and improved cardiac function [110]. Enhancing the chemotaxis of stem cells by local chemokine delivery with an injectable biomaterial is a promising new strategy for tissue regeneration.

4.1.2 Natural Polymer-Based Hydrogel

Biodegradable alginates are gelatinous substances obtained from seaweed or other natural sources whose gel-forming properties are determined by their component proportions of galuronic and mannuronic acids. *Sodium alginate* is a naturally occurring polysaccharide that is easily polymerized into a matrix. Alginate with calcium chloride solution co-injected into the infarcted myocardium formed gel, stimulated angiogenesis and efficiently attenuated infarct expansion, heart dilatation, and cardiac dysfunction [110, 111]. Alginates offer an acellular option to facilitate neovascularization and self-repair within the infarcted myocardium. The combination of these natural polymers with appropriated cells may in future provide a new strategy for cardiac regeneration.

4.1.3 Synthetic Polymer-Based Hydrogel

While the temperature-controlled gel-forming properties of matrigel are desirable for use in combination with cell therapy for cardiac regeneration, proteins in the matrigel can stimulate severe inflammatory reactions that induce significant side effects within the host after implantation. To overcome these biological disadvantages of matrigel, we recently synthesized a temperature-sensitive aliphatic polyester hydrogel, poly(δ-valerolactone)-*block*-poly(ethylene glycol)-*block*-poly(δ-valerolactone) *(PEG-PVL)*. This biomaterial is a liquid at room temperature, but forms a gel at 37 °C in aqueous solution. Our preliminary data indicate that the polymer solution forms a hydrogel after injection into the subcutaneous tissue of adult rats (at 37 °C). Progressive angiogenesis observed in hydro-gel (PEG-PVL) implants harvested at 1, 2, 3 and 4 weeks after implantation suggests that this novel, synthetic, biodegradable and biocompatible polymer could be useful for the treatment of cardiovascular diseases.

Poly(*N*-isopropylacrylamide) *(PIPAAm)* is another temperature-responsive and non-biodegradable polymer. By increasing the temperature to 37 °C, *PIPAAm* will aggregate to form PIPAAm gel. Although it has not been used for cell injection, this biomaterial was used in the creation of cardiac tissue for cardiac repair [112, 113].

4.2 Cardiac Grafts

Biodegradable biomaterials can be used to create sheets for cell growth, matrices upon which cells can form a tissue structure, or a spongy material used to generate a three-dimensional structure that supports cell growth on its surface.

Okano and colleagues [50, 113] created monolayered, MSC cell sheets by growing the cells on top of a culture dish coated with poly(*N*-isopropylacrylamide), which forms a gel at 37 °C. The sheets were then detached from the dish surface by lowering the temperature to below 32 °C, at which point the dish surface became (reversibly) hydrophilic and ceased to support adhesion. Following their implantation onto the epicardial surface of the infarcted myocardium [50], the grafted cell sheets showed evidence of in situ growth characterized by increased thickness, vascularization, and some cardiomyocyte differentiation. More importantly, the grafts improved left ventricular geometry and function within 4 weeks of implantation [102]. In that study, the self-propagating, multipotential and angiogenic characteristics of MSCs led to a thickening of the scar area and a neovascularization of the surrounding tissue, which reduced ventricular wall stress and facilitated cardiac functional improvement.

Since cell sheets may not be strong enough to prevent expansion of a large myocardial infarct, mixed structures comprised of cells and biomaterials have been engineered and studied. Dr. Leor and colleagues reported that fetal rat cardiomyocytes seeded into porous scaffolds composed of alginate biomaterials retained their viability within the scaffolds and formed multicellular beating cell clusters after 24 hours [114]. Following implantation of these constructs into the infarcted myocardium, some of the cells appeared to differentiate into mature myocardial fibers. The implanted grafts were supplied by intensive neovascularization, which evidently contributed to the survival of the seeded cells. The biografts attenuated left ventricular dilatation and prevented functional deterioration. Similarly, heart tissue engineered by culturing a mixture of neonatal rat cardiomyocytes and liquid collagen plus matrigel [115] became heavily vascularized by 14 days after implantation, and retained a well-organized myocardial structure. The contractile function of the graft tissue was also preserved in vivo.

In the late 1990s, our research group created a three-dimensional muscle tissue in vitro by growing fetal cardiomyocytes and/or smooth muscle cells on the surface of a gelatin sponge [116, 117]. The sponge contracted regularly and spontaneously both in culture and in vivo, and histological studies demonstrated that the seeded cardiomyocytes grew and formed junctions characteristic of cardiac tissue [116] (Fig. 2). When we used the cell-seeded grafts for surgical repair in adult rats [117], we found that they were suitable for reconstructing both a right ventricular outflow tract defect and a left ventricular transmural defect [118]. The grafts reduced chamber volumes and improved post-MI left ventricular function. Bioengineered muscle grafts may therefore be superior to synthetic materials for the surgical repair of ventricular scar tissue.

Cardiac tissue has been engineered from numerous somatic cell types, such as cardiomyocytes, smooth muscle cells and skeletal muscle cells. For example, a research team led by Drs. Zimmermann and Eschenhagen made significant

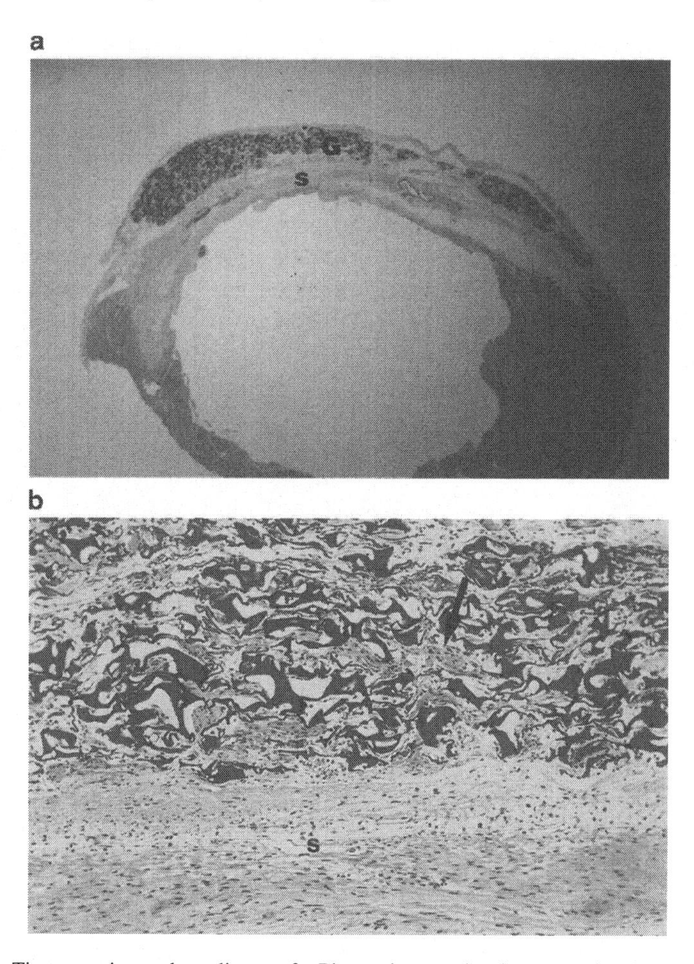

Fig. 2 a-b Tissue-engineered cardiac graft. Photomicrographs demonstrating hematoxylin and eosin staining (a: magnification = 40×; b: magnification = 100×) of a tissue-engineered graft (G; composed of cell-seeded gelatin sponge) at 5 weeks after suturing to the surface of a left ventricular scar (s). Seeded cells (*arrow* in b) filled the interstices of the gelatin mesh and formed tissue resembling myocardium. The graft (G) was adherent to the surface of the scar

contributions to the field of myocardial tissue engineering by developing a novel cardiac muscle construct called engineered heart tissue (EHT) [119–122] using a combination of neonatal cardiomyocytes and an artificial extracellular matrix made from collagen type I and matrigel. These all-natural EHTs demonstrated contractility in vitro and contributed to improved cardiac geometry and function after implantation in a rat MI model; however, EHTs may not be applicable for clinical use due to their limited availability, and ethical concerns related to the use of fetal cardiomyocytes and the biomaterial's high immunogenicity [122]. Further, because tissue formation requires large numbers and multiple types of cells, a major weakness of these techniques is the limited number of somatic cells that can be isolated for cardiac constructs.

5 Stem Cell-Based Cardiac Tissue Engineering

Because they are more potent and proliferative than somatic cells, stem cells are a promising alternative cell source for cardiac tissue engineering. For example, bone marrow stem cells have been used in the construction of in vivo tissue-engineered vascular autografts [123], and to seed a biodegradable scaffold used to repair cardiac defects. The main advantage of these stem cells is their multipotency. Of the seeded cells, some will differentiate into endothelial cells that adhere to the scaffold to prevent thrombosis, while others will proliferate and differentiate into smooth muscle cells, and still others will produce VEGF and angiopoietin-1 within the graft to stimulate angiogenesis.

Tissue-engineered vascular autographs engineered using bone marrow stem cells are particularly useful for cardiovascular surgical applications in humans, especially in children, who require biocompatible materials with growth potential. In 2005, Shin'oka and colleagues reported the first clinical series of tissue-engineered vascular autografts for repair of congenital heart defects [124]. Forty-two patients received an autologous BMC-derived biodegradable patch or conduit composed of an l-lactide and ε-caprolactone co-polymer reinforced with a woven PGA fabric. Mid-term follow-up (mean of 490 days, maximum of 31 months) confirmed the procedure was both feasible and safe, producing only a single mortality unrelated to graft function, and no morbidity or prosthesis-related complications. Follow-up angiography and computed tomography revealed two patch stenoses but no dilation or rupture of the implanted autografts. All conduits were patent without signs of calcification and demonstrated increases in graft diameter, suggesting growth as the children matured.

Autologous bone marrow stem cells were used in combination with a hybrid biodegradable polymer scaffold to tissue-engineer a venous vascular patch in the dog [125]. Eight weeks after implantation, the patches remained patent with no signs of thrombosis, stenoses, or dilatation. Histological, immunohistochemical, and scanning electron microscopic analyses of the excised vascular patches revealed regeneration of the endothelium and smooth muscle cells, and the presence of collagen. Immunofluorescent double staining confirmed that the implanted cells not only survived after implantation, but also contributed to tissue regeneration observed within the implanted patches. A similar autologous conduit was constructed from bone marrow-derived vascular cells in an ovine model [126].

ES cells can also be used to create cardiac grafts. In mice, ES cell-seeded bioabsorbable patches grafted onto the post-MI myocardium repaired the infarcted tissue and improved cardiac function [127]. In another innovative study, a vascularized, three-dimensional tissue-engineered human cardiac muscle was created in vitro using cardiomyocytes and endothelial cells plus fibroblasts derived from human ES cells [128]. The cells were seeded into porous biodegradable sponges composed of poly-(L-lactic acid) and poly(lactic acid-co-glycolic acid) (50%:50%). The authors confirmed cardiac-specific molecular, ultrastructural, and functional properties within the generated myocardial constructs, with synchronous activity mediated by action potential propagation through gap junctions. Such cell-based cardiac grafts offer great possibilities for the next generation of engineered heart tissues [129]. However, functional integration, connection to the vascular system and electrophysiological coupling of donor grafts to the recipient heart tissue will present challenges and will require investigation.

6 Future Research

Cell therapy and tissue engineering (or the combination) offer unique advantages as the starting points of new therapies that may someday be used to treat CHF patients. We have seen many advances in the field, but these successes have raised further questions. To date, no single repair technique has been shown to generate tissue with all the desirable characteristics of engineered myocardial tissue, including consistent and meaningful contractility, stable electrophysiological properties, vascularization, and an autologous cell source.

Before we can engineer the "ideal" cardiac tissue for clinical use, we need to identify the optimal cells, biomaterials and techniques that will compose the tissue. Cardiac grafts seeded with skeletal myoblasts, cardiomyocytes, or smooth muscle cells can effectively repair damaged myocardial tissue and improve cardiac function; however, the relatively small number of cells that can be obtained from aged patients prevents the generation of sufficiently large grafts for clinical use. Highly- proliferative and multipotent bone marrow stem cells can generate a cardiac graft with multiple cell types, but the tissue generated from BMC-engineered grafts is non-beating. To this end, ES cells have been used to engineer contractile cardiomyocytes and contractile cardiac grafts that have been used in cardiac repair with encouraging results. However, a main concern with the use of ES cells for cardiac repair is the continuing ethical debate regarding the harvesting of human embryos. Another major limitation is the potential for immunorejection, since the cardiac grafts must be generated using allogenic ES cells. Alternatively, ES cell lines (currently collected at stem cell banks world wide) that match the MHC-pattern of a large number of recipients in the population might be used in combination with a mild immunosuppressive treatment. Another approach is to use nuclear transfer to produce individualized ES cells [130, 131].

A recent advance in stem cell biology offers the opportunity to produce fully differentiated cardiomyocytes from adult cells [132–136]. Induced pluripotent stem cells (iPS cells) are similar to ES cells but are derived from adult cells. IPS cells can be generated from fully-differentiated somatic cells (such as skin fibroblasts) by exposure to defined factors such as Oct3/4, Sox2, c-Myc and Klf4 [134]. Mouse and human fibroblasts can also be reprogrammed into ES-like cells by introducing transcription factors such as OCT4, SOX2, NANOG and LIN28 [136]. The human iPS cells have normal karyotypes and telomerase activity, express cell surface markers and genes that characterize human ES cells, and maintain the developmental potential to differentiate into advanced derivatives of all 3 primary germ layers. The cardiomyocytes generated from these induced cells have the potential to regenerate the injured myocardium [132].

This new cell source may eliminate concerns over immunorejection by facilitating the induction of multipotential ES cells from individuals with heart disease. New advances in stem cell biology are bringing cell therapy closer to clinical application.

Acknowledgements This work was supported by grants to R-KL from the Canadian Institutes of Health Research (RMF82498, MOP14795). R-KL is a Career Investigator of the Heart and Stroke Foundation of Canada, and holds a Canada Research Chair in cardiac regeneration. The authors thank Heather McDonald Kinkaid for her editorial assistance.

References

1. Thom T, Haase N, Rosamond W, Howard VJ, Rumsfeld J, Manolio T, Zheng ZJ, Flegal K, O'Donnell C, Kittner S, Lloyd-Jones D, Goff DC Jr, Hong Y, Adams R, Friday G, Furie K, Gorelick P, Kissela B, Marler J, Meigs J, Roger V, Sidney S, Sorlie P, Steinberger J, Wasserthiel-Smoller S, Wilson M, Wolf P (2006) Circulation 113:e85
2. McMurray J, Pfeffer MA (2002) Circulation 105:2099
3. St John SM, Pfeffer MA, Moye L, Plappert T, Rouleau JL, Lamas G, Rouleau J, Parker JO, Arnold MO, Sussex B, Braunwald E (1997) Circulation 96:3294
4. Deb A, Wang S, Skelding KA, Miller D, Simper D, Caplice NM (2003) Circulation 107:1247
5. Jackson KA, Majka SM, Wang H, Pocius J, Hartley CJ, Majesky MW, Entman ML, Michael LH, Hirschi KK, Goodell MA (2001) J Clin Invest 107:1395
6. Laflamme MA, Myerson D, Saffitz JE, Murry CE (2002) Circ Res 90:634
7. Leri A, Kajstura J, Anversa P (2005) Physiol Rev 85:1373
8. Orlic D, Kajstura J, Chimenti S, Jakoniuk I, Anderson SM, Li B, Pickel J, McKay R, Nadal-Ginard B, Bodine DM, Leri A, Anversa P (2001) Nature 410:701
9. Schuster MD, Kocher AA, Seki T, Martens TP, Xiang G, Homma S, Itescu S (2004) Am J Physiol Heart Circ Physiol 287:H525
10. Kocher AA, Schuster MD, Szabolcs MJ, Takuma S, Burkhoff D, Wang J, Homma S, Edwards NM, Itescu S (2001) Nat Med 7:430
11. Li RK, Jia ZQ, Weisel RD, Mickle DA, Zhang J, Mohabeer MK, Rao V, Ivanov J (1996) Ann Thorac Surg 62:654
12. Orlic D, Kajstura J, Chimenti S, Limana F, Jakoniuk I, Quaini F, Nadal-Ginard B, Bodine DM, Leri A, Anversa P (2001) Proc Natl Acad Sci U S A 98:10344
13. Tomita S, Li RK, Weisel RD, Mickle DA, Kim EJ, Sakai T, Jia ZQ (1999) Circulation 100:II-247
14. Taylor DA, Atkins BZ, Hungspreugs P, Jones TR, Reedy MC, Hutcheson KA, Glower DD, Kraus WE (1998) Nat Med 4:929
15. Formigli L, Perna AM, Meacci E, Cinci L, Margheri M, Nistri S, Tani A, Silvertown J, Orlandini G, Porciani C, Zecchi-Orlandini S, Medin J, Bani D (2007) J Cell Mol Med 11:1087
16. Uemura R, Xu M, Ahmad N, Ashraf M (2006) Circ Res 98:1414
17. Alvarez-Dolado M, Pardal R, Garcia-Verdugo JM, Fike JR, Lee HO, Pfeffer K, Lois C, Morrison SJ, Alvarez-Buylla A (2003) Nature 425:968
18. Nygren JM, Jovinge S, Breitbach M, Sawen P, Roll W, Hescheler J, Taneera J, Fleischmann BK, Jacobsen SE (2004) Nat Med 10:494
19. Balsam LB, Wagers AJ, Christensen JL, Kofidis T, Weissman IL, Robbins RC (2004) Nature 428:668
20. Murry CE, Soonpaa MH, Reinecke H, Nakajima H, Nakajima HO, Rubart M, Pasumarthi KB, Virag JI, Bartelmez SH, Poppa V, Bradford G, Dowell JD, Williams DA, Field LJ (2004) Nature 428:664

21. Fazel S, Cimini M, Chen L, Li S, Angoulvant D, Fedak P, Verma S, Weisel RD, Keating A, Li RK (2006) J Clin Invest 116:1865
22. Fedak PW, Szmitko PE, Weisel RD, Altamentova SM, Nili N, Ohno N, Verma S, Fazel S, Strauss BH, Li RK (2005) J Thorac Cardiovasc Surg 130:1430
23. Li RK, Jia ZQ, Weisel RD, Merante F, Mickle DA (1999) J Mol Cell Cardiol 31:513
24. Fazel S, Tang GH, Angoulvant D, Cimini M, Weisel RD, Li RK, Yau TM (2005) Ann Thorac Surg 79:S2238
25. Assmus B, Schachinger V, Teupe C, Britten M, Lehmann R, Dobert N, Grunwald F, Aicher A, Urbich C, Martin H, Hoelzer D, Dimmeler S, Zeiher AM (2002) Circulation 106:3009
26. Britten MB, Abolmaali ND, Assmus B, Lehmann R, Honold J, Schmitt J, Vogl TJ, Martin H, Schachinger V, Dimmeler S, Zeiher AM (2003) Circulation 108:2212
27. Menasche P, Alfieri O, Janssens S, McKenna W, Reichenspurner H, Trinquart L, Vilquin JT, Marolleau JP, Seymour B, Larghero J, Lake S, Chatellier G, Solomon S, Desnos M, Hagege AA (2008) Circulation 117:1189
28. Meyer GP, Wollert KC, Lotz J, Steffens J, Lippolt P, Fichtner S, Hecker H, Schaefer A, Arseniev L, Hertenstein B, Ganser A, Drexler H (2006) Circulation 113:1287
29. Heeschen C, Lehmann R, Honold J, Assmus B, Aicher A, Walter DH, Martin H, Zeiher AM, Dimmeler S (2004) Circulation 109:1615
30. Hill JM, Zalos G, Halcox JP, Schenke WH, Waclawiw MA, Quyyumi AA, Finkel T (2003) N Engl J Med 348:593
31. Menasche P, Hagege AA, Scorsin M, Pouzet B, Desnos M, Duboc D, Schwartz K, Vilquin JT, Marolleau JP (2001) Lancet 357:279
32. Zhang H, Fazel S, Tian H, Mickle DA, Weisel RD, Fujii T, Li RK (2005) Am J Physiol Heart Circ Physiol 289:H2089
33. Smith A (2006) Nature 441:1060
34. Weissman IL (2000) Science 287:1442
35. Bradfute SB, Graubert TA, Goodell MA (2005) Exp Hematol 33:836
36. Murray L, DiGiusto D, Chen B, Chen S, Combs J, Conti A, Galy A, Negrin R, Tricot G, Tsukamoto A (1994) Blood Cells 20:364
37. Dimmeler S, Vasa-Nicotera M (2003) J Am Coll Cardiol 42:2081
38. Gaustad KG, Boquest AC, Anderson BE, Gerdes AM, Collas P (2004) Biochem Biophys Res Commun 314:420
39. Rangappa S, Fen C, Lee EH, Bongso A, Sim EK (2003) Ann Thorac Surg 75:775
40. Zuk PA, Zhu M, Ashjian P, De Ugarte DA, Huang JI, Mizuno H, Alfonso ZC, Fraser JK, Benhaim P, Hedrick MH (2002) Mol Biol Cell 13:4279
41. Le BK, Tammik C, Rosendahl K, Zetterberg E, Ringden O (2003) Exp Hematol 31:890
42. Pittenger MF, Mackay AM, Beck SC, Jaiswal RK, Douglas R, Mosca JD, Moorman MA, Simonetti DW, Craig S, Marshak DR (1999) Science 284:143
43. Pittenger MF, Martin BJ (2004) Circ Res 95:9
44. Shake JG, Gruber PJ, Baumgartner WA, Senechal G, Meyers J, Redmond JM, Pittenger MF, Martin BJ (2002) Ann Thorac Surg 73:1919
45. Toma C, Pittenger MF, Cahill KS, Byrne BJ, Kessler PD (2002) Circulation 105:93
46. Makino S, Fukuda K, Miyoshi S, Konishi F, Kodama H, Pan J, Sano M, Takahashi T, Hori S, Abe H, Hata J, Umezawa A, Ogawa S (1999) J Clin Invest 103:697
47. Tomita S, Mickle DA, Weisel RD, Jia ZQ, Tumiati LC, Allidina Y, Liu P, Li RK (2002) J Thorac Cardiovasc Surg 123:1132
48. Memon IA, Sawa Y, Miyagawa S, Taketani S, Matsuda H (2005) J Thorac Cardiovasc Surg 130:646
49. Zhang DZ, Gai LY, Liu HW, Jin QH, Huang JH, Zhu XY (2007) Chin Med J (Engl) 120:300
50. Miyahara Y, Nagaya N, Kataoka M, Yanagawa B, Tanaka K, Hao H, Ishino K, Ishida H, Shimizu T, Kangawa K, Sano S, Okano T, Kitamura S, Mori H (2006) Nat Med 12:459
51. Janssens S, Dubois C, Bogaert J, Theunissen K, Deroose C, Desmet W, Kalantzi M, Herbots L, Sinnaeve P, Dens J, Maertens J, Rademakers F, Dymarkowski S, Gheysens O, Van CJ,

Bormans G, Nuyts J, Belmans A, Mortelmans L, Boogaerts M, Van de WF (2006) Lancet 367:113
52. Beltrami AP, Barlucchi L, Torella D, Baker M, Limana F, Chimenti S, Kasahara H, Rota M, Musso E, Urbanek K, Leri A, Kajstura J, Nadal-Ginard B, Anversa P (2003) Cell 114:763
53. Dawn B, Stein AB, Urbanek K, Rota M, Whang B, Rastaldo R, Torella D, Tang XL, Rezazadeh A, Kajstura J, Leri A, Hunt G, Varma J, Prabhu SD, Anversa P, Bolli R (2005) Proc Natl Acad Sci U S A 102:3766
54. Urbanek K, Torella D, Sheikh F, De Angelis A, Nurzynska D, Silvestri F, Beltrami CA, Bussani R, Beltrami AP, Quaini F, Bolli R, Leri A, Kajstura J, Anversa P (2005) Proc Natl Acad Sci U S A 102:8692
55. Urbanek K, Rota M, Cascapera S, Bearzi C, Nascimbene A, De Angelis A, Hosoda T, Chimenti S, Baker M, Limana F, Nurzynska D, Torella D, Rotatori F, Rastaldo R, Musso E, Quaini F, Leri A, Kajstura J, Anversa P (2005) Circ Res 97:663
56. Oh H, Bradfute SB, Gallardo TD, Nakamura T, Gaussin V, Mishina Y, Pocius J, Michael LH, Behringer RR, Garry DJ, Entman ML, Schneider MD (2003) Proc Natl Acad Sci U S A 100:12313
57. Oh H, Chi X, Bradfute SB, Mishina Y, Pocius J, Michael LH, Behringer RR, Schwartz RJ, Entman ML, Schneider MD (2004) Ann N Y Acad Sci 1015:182
58. Laugwitz KL, Moretti A, Lam J, Gruber P, Chen Y, Woodard S, Lin LZ, Cai CL, Lu MM, Reth M, Platoshyn O, Yuan JX, Evans S, Chien KR (2005) Nature 433:647
59. Messina E, De AL, Frati G, Morrone S, Chimenti S, Fiordaliso F, Salio M, Battaglia M, Latronico MV, Coletta M, Vivarelli E, Frati L, Cossu G, Giacomello A (2004) Circ Res 95:911
60. Smith RR, Barile L, Cho HC, Leppo MK, Hare JM, Messina E, Giacomello A, Abraham MR, Marban E (2007) Circulation 115:896
61. Beltrami AP, Barlucchi L, Torella D, Baker M, Limana F, Chimenti S, Kasahara H, Rota M, Musso E, Urbanek K, Leri A, Kajstura J, Nadal-Ginard B, Anversa P (2003) Cell 114:763
62. Dawn B, Stein AB, Urbanek K, Rota M, Whang B, Rastaldo R, Torella D, Tang XL, Rezazadeh A, Kajstura J, Leri A, Hunt G, Varma J, Prabhu SD, Anversa P, Bolli R (2005) Proc Natl Acad Sci U S A 102:3766
63. van VP, Roccio M, Smits AM, van Oorschot AA, Metz CH, van Veen TA, Sluijter JP, Doevendans PA, Goumans MJ (2008) Neth Heart J 16:163
64. Goumans MJ, de Boer TP, Smits AM, van Laake W, van Vliet P, Metz CH (2008). Stem Cell Res 1:138
65. Moretti A, Caron L, Nakano A, Lam JT, Bernshausen A, Chen Y, Qyang Y, Bu L, Sasaki M, Martin-Puig S, Sun Y, Evans SM, Laugwitz KL, Chien KR (2006) Cell 127:1151
66. Amado LC, Schuleri KH, Saliaris AP, Boyle AJ, Helm R, Oskouei B, Centola M, Eneboe V, Young R, Lima JA, Lardo AC, Heldman AW, Hare JM (2006) J Am Coll Cardiol 48:2116
67. Dai W, Hale SL, Martin BJ, Kuang JQ, Dow JS, Wold LE, Kloner RA (2005) Circulation 112:214
68. Evans MJ, Kaufman MH (1981) Nature 292:154
69. Thomson JA, Itskovitz-Eldor J, Shapiro SS, Waknitz MA, Swiergiel JJ, Marshall VS, Jones JM (1998) Science 282:1145
70. Bradley A, Evans M, Kaufman MH, Robertson E (1984) Nature 309:255
71. Ali NN, Xu X, Brito-Martins M, Poole-Wilson PA, Harding SE, Fuller SJ (2004) Basic Res Cardiol 99:382
72. Kehat I, Kenyagin-Karsenti D, Snir M, Segev H, Amit M, Gepstein A, Livne E, Binah O, Itskovitz-Eldor J, Gepstein L (2001) J Clin Invest 108:407
73. Klug MG, Soonpaa MH, Koh GY, Field LJ (1996) J Clin Invest 98:216
74. Min JY, Yang Y, Converso KL, Liu L, Huang Q, Morgan JP, Xiao YF (2002) J Appl Physiol 92:288
75. Caspi O, Gepstein L (2006) Eur Heart J Suppl 8:E43

76. Kehat I, Khimovich L, Caspi O, Gepstein A, Shofti R, Arbel G, Huber I, Satin J, Itskovitz-Eldor J, Gepstein L (2004) Nat Biotechnol 22:1282
77. Xu C, Police S, Rao N, Carpenter MK (2002) Circ Res 91:501
78. Laflamme MA, Chen KY, Naumova AV, Muskheli V, Fugate JA, Dupras SK, Reinecke H, Xu C, Hassanipour M, Police S, O'Sullivan C, Collins L, Chen Y, Minami E, Gill EA, Ueno S, Yuan C, Gold J, Murry CE (2007) Nat Biotechnol 25:1015
79. van Laake LW, Passier R, Doevendans PA, Mummery CL (2008) Circ Res 102:1008
80. Leor J, Gerecht S, Cohen S, Miller L, Holbova R, Ziskind A, Shachar M, Feinberg MS, Guetta E, Itskovitz-Eldor J (2007) Heart 93:1278
81. Yang L, Soonpaa MH, Adler ED, Roepke TK, Kattman SJ, Kennedy M, Henckaerts E, Bonham K, Abbott GW, Linden RM, Field LJ, Keller GM (2008) Nature 453:524
82. Menard C, Hagege AA, Agbulut O, Barro M, Morichetti MC, Brasselet C, Bel A, Messas E, Bissery A, Bruneval P, Desnos M, Puceat M, Menasche P (2005) Lancet 366:1005
83. Kofidis T, deBruin JL, Tanaka M, Zwierzchoniewska M, Weissman I, Fedoseyeva E, Haverich A, Robbins RC (2005) Eur J Cardiothorac Surg 28:461
84. Behfar A, Perez-Terzic C, Faustino RS, Arrell DK, Hodgson DM, Yamada S, Puceat M, Niederlander N, Alekseev AE, Zingman LV, Terzic A (2007) J Exp Med 204:405
85. Pouzet B, Vilquin JT, Hagege AA, Scorsin M, Messas E, Fiszman M, Schwartz K, Menasche P (2001) Ann Thorac Surg 71:844
86. Yau TM, Kim C, Ng D, Li G, Zhang Y, Weisel RD, Li RK (2005) Ann Thorac Surg 80:1779
87. Sakakibara Y, Nishimura K, Tambara K, Yamamoto M, Lu F, Tabata Y, Komeda M (2002) J Thorac Cardiovasc Surg 124:50
88. Liu TB, Fedak PW, Weisel RD, Yasuda T, Kiani G, Mickle DA, Jia ZQ, Li RK (2004) Am J Physiol Heart Circ Physiol 287:H2840
89. Nakamura Y, Yasuda T, Weisel RD, Li RK (2006) Am J Physiol Heart Circ Physiol 291:H939
90. Cao F, Lin S, Xie X, Ray P, Patel M, Zhang X, Drukker M, Dylla SJ, Connolly AJ, Chen X, Weissman IL, Gambhir SS, Wu JC (2006) Circulation 113:1005
91. Li Z, Wu JC, Sheikh AY, Kraft D, Cao F, Xie X, Patel M, Gambhir SS, Robbins RC, Cooke JP, Wu JC (2007) Circulation 116:I-46
92. Wu JC, Chen IY, Sundaresan G, Min JJ, De A, Qiao JH, Fishbein MC, Gambhir SS (2003) Circulation 108:1302
93. Hofmann M, Wollert KC, Meyer GP, Menke A, Arseniev L, Hertenstein B, Ganser A, Knapp WH, Drexler H (2005) Circulation 111:2198
94. Kang WJ, Kang HJ, Kim HS, Chung JK, Lee MC, Lee DS (2006) J Nucl Med 47:1295
95. Davis ME, Hsieh PC, Grodzinsky AJ, Lee RT (2005) Circ Res 97:8
96. Huang NF, Yu J, Sievers R, Li S, Lee RJ (2005) Tissue Eng 11:1860
97. Christman KL, Vardanian AJ, Fang Q, Sievers RE, Fok HH, Lee RJ (2004) J Am Coll Cardiol 44:654
98. Davis ME, Motion JP, Narmoneva DA, Takahashi T, Hakuno D, Kamm RD, Zhang S, Lee RT (2005) Circulation 111:442
99. Higgs PG, Ball RC (1990) Physical networks: polymers and gels. Elsevier, New York
100. Guenet JM (1992) Thermoreversible gelation of polymers and biopolymers. Academic Press, New York
101. Christman KL, Fok HH, Sievers RE, Fang Q, Lee RJ (2004) Tissue Eng 10:403
102. Christman KL, Lee RJ (2006) J Am Coll Cardiol 48:907
103. Ryu JH, Kim IK, Cho SW, Cho MC, Hwang KK, Piao H, Piao S, Lim SH, Hong YS, Choi CY, Yoo KJ, Kim BS (2005) Biomaterials 26:319
104. Chekanov V, Akhtar M, Tchekanov G, Dangas G, Shehzad MZ, Tio F, Adamian M, Colombo A, Roubin G, Leon MB, Moses JW, Kipshidze NN (2003) Pacing Clin Electrophysiol 26:496
105. Kofidis T, de Bruin JL, Hoyt G, Lebl DR, Tanaka M, Yamane T, Chang CP, Robbins RC (2004) J Thorac Cardiovasc Surg 128:571
106. Kofidis T, Lebl DR, Martinez EC, Hoyt G, Tanaka M, Robbins RC (2005) Circulation 112:I-173
107. Zhang P, Zhang H, Wang H, Wei Y, Hu S (2006) Artif Organs 30:86

108. Zhang S (2003) Nat Biotechnol 21:1171
109. Davis ME, Hsieh PC, Takahashi T, Song Q, Zhang S, Kamm RD, Grodzinsky AJ, Anversa P, Lee RT (2006) Proc Natl Acad Sci U S A 103:8155
110. Segers VF, Tokunou T, Higgins LJ, MacGillivray C, Gannon J, Lee RT (2007) Circulation 116:1683
111. Leor J, Miller L, Feinberg MS, Shachar S, Landa N, Holbova R (2004) A novel injectable alginate scaffold promotes angiogenesis and preserves left ventricular geometry and function after extensive myocardial infarction in rat. Circulation 110(Suppl):III-279
112. Cohen S, Leor J (2004) Sci Am 291:44
113. Shimizu T, Yamato M, Kikuchi A, Okano T (2003) Biomaterials 24:2309
114. Leor J, Patterson M, Quinones MJ, Kedes LH, Kloner RA (1996) Circulation 94:II-332
115. Zimmermann WH, Didie M, Wasmeier GH, Nixdorff U, Hess A, Melnychenko I, Boy O, Neuhuber WL, Weyand M, Eschenhagen T (2002) Circulation 106:I-151
116. Li RK, Jia ZQ, Weisel RD, Mickle DA, Choi A, Yau TM (1999) Circulation 100:II-63
117. Li RK, Yau TM, Weisel RD, Mickle DA, Sakai T, Choi A, Jia ZQ (2000) J Thorac Cardiovasc Surg 119:368
118. Matsubayashi K, Fedak PW, Mickle DA, Weisel RD, Ozawa T, Li RK (2003) Circulation 108(Suppl 1):II219
119. Zimmermann WH, Fink C, Kralisch D, Remmers U, Weil J, Eschenhagen T (2000) Biotechnol Bioeng 68:106
120. Zimmermann WH, Eschenhagen T (2005) Cardiovasc Res 68:344
121. Zimmermann WH, Didie M, Doker S, Melnychenko I, Naito H, Rogge C, Tiburcy M, Eschenhagen T (2006) Cardiovasc Res 71:419
122. Zimmermann WH, Melnychenko I, Wasmeier G, Didie M, Naito H, Nixdorff U, Hess A, Budinsky L, Brune K, Michaelis B, Dhein S, Schwoerer A, Ehmke H, Eschenhagen T (2006) Nat Med 12:452
123. Matsumura G, Hibino N, Ikada Y, Kurosawa H, Shin'oka T (2003) Biomaterials 24:2303
124. Shin'oka T, Matsumura G, Hibino N, Naito Y, Watanabe M, Konuma T, Sakamoto T, Nagatsu M, Kurosawa H (2005) J Thorac Cardiovasc Surg 129:1330
125. Cho SW, Jeon O, Lim JE, Gwak SJ, Kim SS, Choi CY, Kim DI, Kim BS (2006) J Vasc Surg 44:1329
126. Roh JD, Brennan MP, Lopez-Soler RI, Fong PM, Goyal A, Dardik A, Breuer CK (2007) J Pediatr Surg 42:198
127. Ke Q, Yang Y, Rana JS, Chen Y, Morgan JP, Xiao YF (2005) Sheng Li Xue Bao 57:673
128. Caspi O, Lesman A, Basevitch Y, Gepstein A, Arbel G, Habib IH, Gepstein L, Levenberg S (2007) Circ Res 100:263
129. Zimmermann WH, Eschenhagen T (2007) Trends Cardiovasc Med 17:134
130. Hochedlinger K, Jaenisch R (2003) N Engl J Med 349:275
131. Jaenisch R (2004) N Engl J Med 351:2787
132. Narazaki G, Uosaki H, Teranishi M, Okita K, Kim B, Matsuoka S, Yamanaka S, Yamashita JK (2008) Circulation 118:498
133. Okita K, Ichisaka T, Yamanaka S (2007) Nature 448:313
134. Takahashi K, Yamanaka S (2006) Cell 126:663
135. Takahashi K, Tanabe K, Ohnuki M, Narita M, Ichisaka T, Tomoda K, Yamanaka S (2007) Cell 131:861
136. Yu J, Vodyanik MA, Smuga-Otto K, Antosiewicz-Bourget J, Frane JL, Tian S, Nie J, Jonsdottir GA, Ruotti V, Stewart R, Slukvin II, Thomson JA (2007) Science 318:1917

Adv Biochem Engin/Biotechnol (2009) 114: 129-172
DOI: 10.1007/10_2008_8

Published online: 01 October 2008

Stem Cells and Scaffolds for Vascularizing Engineered Tissue Constructs

E. Luong and S. Gerecht

Abstract The clinical impact of tissue engineering depends upon our ability to direct cells to form tissues with characteristic structural and mechanical properties from the molecular level up to organized tissue. Induction and creation of functional vascular networks has been one of the main goals of tissue engineering either in vitro, for the transplantation of prevascularized constructs, or in vivo, for cellular organization within the implantation site. In most cases, tissue engineering attempts to recapitulate certain aspects of normal development in order to stimulate cell differentiation and functional tissue assembly. The induction of tissue growth generally involves the use of biodegradable and bioactive materials designed, ideally, to provide a mechanical, physical, and biochemical template for tissue regeneration. Human embryonic stem cells (hESCs), derived from the inner cell mass of a developing blastocyst, are capable of differentiating into all cell types of the body. Specifically, hESCs have the capability to differentiate and form blood vessels de novo in a process called vasculogenesis. Human ESC-derived endothelial progenitor cells (EPCs) and endothelial cells have substantial potential for microvessel formation, in vitro and in vivo. Human adult EPCs are being isolated to understand the fundamental biology of how these cells are regulated as a population and to explore whether these cells can be differentiated and reimplanted as a cellular therapy in order to arrest or even reverse damaged vasculature. This chapter focuses on advances made toward the generation and engineering of functional vascular tissue, focusing on both the scaffolds – the synthetic and biopolymer materials – and the cell sources – hESCs and hEPCs.

Keywords Blood vessels, Endothelial progenitor cells, Human embryonic stem cells, Scaffolds, Vascular development

E. Luong and S. Gerecht (✉)
Department of Chemical and Biomolecular Engineering, Johns Hopkins University,
3400 N. Charles St., Baltimore, MD, 21218, USA
e-mail: gerecht@jhu.edu

Contents

Abbreviations

acLDL acetylated low-density lipoprotein;
BM bone marrow
CFU-EC EC colony-forming units;
EB embryonic body;
EC endothelial cells
ECFC endothelial colony-forming cell;
ECM extracellular matrix;
EPC endothelial progenitor cell;
FGF2 fibroblast growth factor 2;
GF growth factor;
HA hyaluronic acid;
hESC human embryonic stem cell;
HUVEC human umbilical cord vein endothelial cell;
KDR kinase insert domain-containing receptor;
MB mononuclear blood;
PDGF platelet-derived growth factor;
PEG polyethylene glycol;
PEGDA poly(ethylene glycol) diacrylate;

PGA polyglycolic acid;
PLA polylactic acid;
PLLA poly-l-lactic acid;
SMC smooth muscle cell;
SSEA stage-specific embryonic antigen;
VEGF vascular endothelial growth factor;
VEGFR2 vascular endothelial growth factor receptor 2;
vWF von Willebrand factor

1 Introduction

In the embryo, vasculogenesis is the process in which blood vessel formation occurs by the differentiation of vascular endothelial cells (ECs) from angioblastic precursors, which in turn give rise to a primitive vascular plexus [1–3]. The derivation of hESCs and their vascular differentiation potential have positioned these cells as an appealing source for advancing research into human vascular development [4] and for cellular therapies of the vasculature, including vascular tissue engineering. In recent decades, postnatal vasculogenesis has been purported to be an important mechanism for angiogenesis, via marrow-derived circulating human endothelial progenitor cells (hEPCs). On the basis of this paradigm, hEPCs have been extensively studied as biomarkers of cardiovascular disease and for their use in cell-based therapies for the repair of damaged blood vessels [5].

Angiogenesis can be defined as the sprouting of capillaries from existing vessels leading to the formation of new microvessels in previously avascular tissues. The molecular basis for angiogenesis is easily characterized by viewing the process as a stepwise progression [2, 6]. The initial vasodilation of existing vessels is accompanied by increases in permeability and degradation of the surrounding matrix, which allows activated and proliferating ECs to migrate and form lumens [2, 6, 7].

This chapter will explore advances in the derivation of EPCs and ECs from hESCs and will consider the various sources and regenerative potential of adult hEPCs. Specifically, it will do so in the context of hESC derivatives and hEPCs and their ability to generate networks within biomaterials in vitro that result in enhanced network formation in vivo [8]. Recent efforts and developments in synthetic and biopolymer materials for vascular tissue engineering will also be discussed at length.

2 Vascular Development and Disease

2.1 The Hemangioblast

Following differentiation of the endothelium from the mesoderm during embryogenesis, the vascular system is the first functional organ to appear in the embryo

[1, 9, 10]. Development of the vascular system begins with an aggregate of cells resulting in the formation of a blood island in the yolk sac [1, 9, 11]. These blood islands consist of an outer layer of angioblasts and a center containing hematopoietic precursors. The angioblasts will give rise to endothelial precursor cells, whereas the hematopoietic precursors will give rise to blood cells. Because the blood island gives rise to both endothelial and hematopoietic precursors, a common precursor was theorized, the hemangioblast [12–15].

Attempts to identify the hemangioblast in the developing embryo faced obstacles involving the speed of development and the difficulty in isolating sufficient numbers of cells [16]. The use of ESCs provides an alternative method of studying vascular development. Doetschman [17] showed that blastocyst-derived ESCs formed yolk sacs and blood islands similar to in vivo embryos. The use of ESCs as a model for early vascular development was further justified by research showing ESCs giving rise to hematopoietic cells in a developmental sequence similar to normal mouse development [18]. Upon aggregation, ESCs formed three-dimensional embryoid bodies (EBs) that contain representatives of the three germ layers [17, 19, 20]. Using EBs developed from ESCs and differentiated for three to three and a half days, researchers identified a blast colony-forming cell that developed into numerous hematopoietic lineages [21]. Further experiments showed that blast colony-forming cells could give rise to both hematopoietic and endothelial lineages, confirming the existence of a common precursor to endothelial and hematopoietic precursors [22].

2.2 Formation of Blood Vessels

Blood vessel formation in the developing embryo and postnatal (adult) is achieved through two distinct pathways: vasculogenesis and angiogenesis. Vasculogenesis occurs in the embryo during the de novo formation of the primitive vascular network from the mesoderm. During this sequence of events, hemangioblast cells form in the yolk sac, eventually giving rise to endothelial and hematopoietic precursors and finally forming a primary plexus [1].

Vascular development continues with angiogenesis, where new blood vessels form from preexisting vessels through sprouting, migration, and proliferation of existing ECs [1, 23]. Angiogenesis proceeds through two phases – vessel growth and subsequent vessel stabilization [23]. This process is initialized by the breakdown of the basement membrane and extracellular matrix (ECM) by proteinases, which provides a pathway for EC migration. Proliferating ECs then assemble to form a lumen [6]. Once proliferation is arrested, stabilization of the newly formed vessel proceeds with reconstruction of the basement membrane. Additionally, pericytes and smooth muscle cells (SMCs) are recruited around the immature capillary to provide further stabilization and to prevent ECs from undergoing apoptosis [24, 25].

Whereas vasculogenesis is the primary mechanism for forming the vasculature in the embryo, in adults, angiogenesis leading to development of new blood vessels occurs during the remodeling of the vascular system in response to such changes in physical homeostasis as wounds, pregnancy, or disease states, including tumor growth [1, 26, 27]. Previously, it was thought that angiogenesis in adults involved proliferation of preexisting ECs. Recent studies, however, have shown that a circulating bone-marrow-derived endothelial progenitor cell is capable of differentiating into ECs and integrating into the vasculature [28]. Furthermore, Asahara and colleagues showed that, in models of ischemia, circulating EPCs contributed to neovascularization of the ischemic tissue. This led to the idea of a potential mechanism for postnatal vasculogenesis in tissue regeneration in the adult [29].

2.3 Diseases Affecting the Vascular System

Cardiovascular disease is currently the leading cause of morbidity and mortality in the human population, with atherosclerosis, an inflammation of the arterial walls, being one of the major causes of cardiovascular disease [30]. The basic structure of a blood vessel is composed of ECs surrounded by a basement membrane and supporting pericyte cells. Capillaries consist of a single layer of ECs. Arteries, on the other hand, are composed of an inner layer of ECs and basement membrane, the intima; a middle layer of SMCs, the media, and an outer layer of collagen, the externa.

When damaged, the endothelium has the ability to repair itself through a reendothelialization process [31]. If a small section of the endothelium is removed, a healthy endothelium, sensing the lack of contact inhibition, triggers ECs at the edges to proliferate and repair the damage. This repair mechanism recruits not only local ECs but also circulating EPCs [28].

The importance of circulating EPCs has been shown in recent studies which have observed that increased EPC mobilization in the peripheral blood contributes to faster reendothelialization [32]. However, endothelia that are subject to risk factors, such as cholesterol or hypertension, have a reduced ability to self-renew, and the reendothelialization process may result in a plaque, contributing to the progression of atherosclerosis [33]. Over time, complications of the plaque will develop, including thrombosis and stenosis, leading to ischemia and myocardial infarction [34].

Additionally, the level of circulating EPCs tends to correlate with vascular health, with a decrease implicating endothelial dysfunction, while an increase implicates enhanced function. Aging, for example, has been shown to negatively affect such EPC characteristics as proliferation, migration, and survival [35]. In human clinical settings, a decrease in the level of circulating EPCs among the general population correlates with early atherosclerosis and can serve as a marker for cardiovascular risk [36]. Other disorders and diseases have been associated with endothelial dysfunction: diabetes [37], high blood pressure [38], chronic obstructive pulmonary disease [39], and high cholesterol levels [40]. In contrast, physical training helps maintain vascular homeostasis and induces mobilization of EPCs to increase circulating EPCs in both healthy subjects and subjects with coronary artery disease [41, 42].

3 Endothelial Cells and Progenitors

3.1 Isolation of Primary ECs

More than 20 years ago, attempts were made to identify, and then isolate and culture, ECs. Jaffe et al. isolated ECs, which they identified via morphologic and immunohistologic procedures, from human umbilical cord veins and was able to maintain them in long-term in vitro culture [43]. These cells, termed HUVECs (human umbilical cord vein endothelial cells) have played a major role in the development of the vascular biology field by providing a critical in vitro model for insights into cellular and molecular events, including atherogenesis [44, 45], the effects of inflammatory cytokines on ECs [46], mechanisms of atherosclerosis [47], angiogenesis in response to ischemia, and stress-mediated endothelial signaling [48]. Cultures of HUVECs have been shown to maintain nearly all features of native ECs, EC-specific markers, and VEGF signaling pathways, making them a good tool for the study of ECs. Additionally, culturing HUVECs on Matrigel in a hypoxic environment results in the formation of tube-like structures that resemble tube formation in vivo, providing evidence for blood vessel recruitment to tissues under ischemic stress [49].

The need for new EC lines arose from observations by numerous groups that ECs from microvessels differ significantly from macrovessel ECs, such as those derived from the umbilical cord vein [50–52]. Research within the past 20 years has shown that microvascular ECs differentiate into capillary-like structures more rapidly, possess a different secretory profile, and differ in their expressed cell adhesion molecules when compared to those from large vessels.

Collectively, all primary EC lines fail to retain EC characteristics after prolonged passage in culture and undergo senescence and, therefore, have limited lifespans.

3.2 Discovery of a Circulating Endothelial Progenitor Cell (EPC)

It was previously known that angiogenesis occurred through the proliferation and migration of preexisting ECs in blood vessel walls in order to carry out the repair process [1, 26]. ECs, however, are terminally differentiated cells that lack the high proliferative potential of stem cells and thus are limited in their ability to repair the endothelium. Brown et al. discovered that patients with neutropenia, a disorder characterized by abnormally low levels of white blood cells, recovered after transfusion of peripheral blood progenitor cells [53]. Studies involving the use of Dacron grafts identified ECs present in locations on the graft that could not be reached by preexisting ECs through the process of angiogenesis [54, 55]. Additionally, several studies have demonstrated the existence of circulating ECs contributing to endothelium growth on similar vascular prostheses [56–59]. It appears, then, that there may be another cell type with high proliferative characteristics of a stem cell that contributes to endothelial repair by differentiating into ECs [60].

In 1997, Isner and colleagues announced the isolation of a circulating EPC from human peripheral blood capable of differentiating into mature ECs [28]. The study focused on two antigens expressed on angioblasts and hematopoietic stem cells: CD34, a marker expressed by all hematopoietic stem cells, to isolate CD34-positive mononuclear blood cells (MB^{CD34+}) and Flk-1, a vascular endothelial growth factor receptor 2 (VEGFR2; KDR in humans), to isolate Flk-1-positive mononuclear blood cells (MB^{Flk1+}). MB^{CD34+} cells were initially plated on tissue culture plastic and collagen, whereby a subset became attached and displayed spindle-shaped morphology. Upon replating these attached cells on fibronectin, a small fraction that quickly attached were spindle shaped and were denoted as AT^{CD34+} (i.e., attached CD34+ cells). When MB^{CD34+} DiI-labeled cells were cocultured with MB^{CD34-} cells, a tenfold increase in the proliferation rates of MB^{CD34+}, compared to MB^{CD34+} plated alone, was observed. Furthermore, the coculture cells quickly formed basic cellular networks and tube-like structures, and, after 12 h, cell clusters composed mainly of MB^{CD34+} cells formed, resembling the blood islands observed by Flamme and Risau in their quail epiblast cultures [61]. MB^{Flk1+} cells displayed similar behavior, suggesting that MB^{CD34+} and MB^{Flk1+} cells from peripheral blood are of a progenitor nature and contribute to postnatal angiogenesis.

Next, Isner and colleagues evaluated the differentiation capabilities of MB^{CD34+} cells. The expression of leukocyte and EC-lineage markers – including CD34, CD31, Flk-1, Tie-2, and E-selectin – was compared among MB^{CD34+}, AT^{CD34+}, and HUVECs; HUVECs and AT^{CD34+} cells had similar expression profiles. Additionally, after 7 days of culture the expression of EC markers on MB^{CD34+} cells was similar to that on AT^{CD34+} and HUVECs. To further confirm the EC-like phenotype, a functional assay was performed. AT^{CD34+} cells produced nitric oxide in response to the EC-dependent agonist acetylcholine (Ach) and responded to VEGF. Additionally, AT^{CD34+} cells possessed the ability, characteristic of functional ECs, to take up acetylated low-density lipoprotein (acLDL).

Next, models of hindlimb ischemia were used to examine whether these cells contributed to angiogenesis in vivo. The femoral artery was excised from one mouse hindlimb, followed by injection into the bloodstream of DiI-labeled human MB^{CD34+} or MB^{CD34-} cells. Histological examination revealed proliferative DiI-labeled cells integrating with capillary walls in the ischemic hindlimb, with consistent colabeling of DiI and CD31 cells. DiI-labeled cells in MB^{CD34-} mice, on the contrary, localized near capillaries but did not integrate with the vessel wall. Furthermore, the regeneration potential of autologous EPCs was examined by isolating MB^{CD34+} cells from New Zealand white rabbit blood, labeling the cells with DiI dye, and reinjecting them into the same rabbit after induction of ischemia. Examination revealed that DiI-labeled cells localized to neovascular zones of the ischemic region. Isner and colleagues concluded that MB^{CD34+} or MB^{Flk1+} cells have the ability to differentiate into ECs. Additionally, these cells can home to areas of angiogenesis, incorporating into the microvasculature and contributing to vasculogenesis.

3.3 Bone Marrow Origin of EPCs

The discovery of circulating EPCs by Isner and colleagues prompted the question of the origin of EPCs that contributed to postnatal vasculogenesis, which remained unclear at the time. Initial speculations determined the sources of these cells to be (1) mature ECs that detached from upstream proximal vascular walls, (2) a naturally circulating pool of EPCs, and (3) bone-marrow (BM)-derived EPCs [62]. Shi et al. isolated $CD34^+$ mononuclear cells from bone marrow and evaluated whether ECs could be derived. Rapidly proliferating cells were observed after culture on fibronectin and were determined to be of endothelial lineage with the ability to incorporate acLDL and staining for von Willebrand factor (vWF).

To evaluate whether the BM-derived cells participated in endothelialization of vascular prostheses, beagle dogs were implanted with Dacron grafts at the descending thoracic aorta and injected with marrow cells. The graft was made impermeable to prevent host migration of ECs. Cells isolated from the graft surface stained positive vWF, confirming their EC phenotype. Genotyping further revealed that the isolated cells were of donor origin, strongly suggesting that the ECs derived from host BM cells.

Isner and colleagues conducted further studies with murine BM transplant models to confirm the BM origin of EPCs and the existence of a physiologically relevant postnatal vasculogenesis process [63]. BM cell propagation and localization were tracked by use of a constitutively expressed beta-galactosidase lacZ transcriptionally regulated by an EC-specific promoter, Flk-1 or Tie-2, yielding mice with BM genotypes Flk-1/lacZ/BMT and Tie-2/lacZ/BMT. Flk-1 has been shown to be involved with embryonic angioblast differentiation [64] and postnatal neovascularization [65, 66]. Tie-2 is expressed in endothelial lineage cells participating in angiogenesis [67].

Isner and colleagues' study detected physiological localization of EPCs to normal organs (peripheral blood, spleen, lung, ovary, uterus), indicating a possible role of EPCs in physiological organ maintenance. Vasculogenesis in the physiological neovascularization of the ovaries was examined by inducing an ovarian cycle, and it was shown that EPCs integrated into the uterus and corpus luteum of the ovary, suggesting EPC involvement with postnatal regenerative processes. A wound-healing model was explored, and, within 1 week, a large number of BM-derived EPCs were incorporated into the foci of neovascularization, with downregulation of lacZ expression after 4 weeks. Lastly, a hindlimb ischemia model was studied, and EPC colonies were observed in tissue stroma at sites of ischemia and were incorporated into capillaries [63].

3.4 Alternative Sources of EPCs

The discussion thus far has shown us two primary sources of EPCs. First, from the periphery blood, circulating EPCs have a high proliferative potential and the capacity to home to sites of neovascularization. Second, BM transplants showed that EPCs can derive from a precursor cell lying dormant in BM.

Following these studies, efforts were made to isolate EPCs from various sources: umbilical cord blood, the liver, and tissue or vascular wall.

Peichev et al. showed that human-derived EPCs can be isolated via anti-CD34$^+$/anti-VEGFR2$^+$ using cells isolated from fetal livers, mobilized peripheral blood, and cord blood. Such cells were shown to express endothelial-specific markers, to migrate in response to VEGF, and to home to sites of neovascularization and to differentiate into mature ECs in vivo [68]. Igreja et al. used human cord blood to isolate CD133$^+$/KDR$^+$ (kinase insert domain-containing receptor; also known as VEGFR2) and CD34$^+$/KDR$^+$ EPCs; they observed these EPCs differentiate into ECs expressing mature endothelial markers [69]. Additionally, cord blood EPCs cultured on type I collagen dishes formed a stable network of microvessels distributed uniformly throughout the collagen gel with an intact basement membrane [70]. Blood vessels formed from cord blood EPCs last longer, and the number of EPCs present in cord blood is much greater than in adult blood [68]. The fetal liver has recently been discovered to contain a population of EPCs that are CD31$^+$/Sca1$^+$ and can be isolated at high yield with high angiogenic capacity. Culturing on Matrigel led to the formation of capillary-like structures, and in vivo studies determined their capacity to effectively integrate with the host vasculature [71].

Recently, a novel class of cells termed tissue-resident endothelial precursors was isolated from small blood vessels in dermal, adipose, and skeletal muscle tissues. Evaluation of tissue-resident endothelial precursors determined that these cells were of mesodermal origin, displayed angiogenic precursors, and were capable of differentiating into ECs to form vascular networks upon transplantation [72]. Another report indicated the presence of cells within skeletal muscle that coexpressed myogenic and EC markers. When cultured in endothelial growth medium, adherent cells continued to coexpress these markers [73]. A different report has proposed that the vascular wall may be another source of EPCs and may contribute to postnatal vasculogenesis. CD34 cells isolated from a distinct zone of the vascular wall of large and middle-sized adult blood vessels were shown to have the ability to differentiate into mature ECs and form capillary-like structures in vitro [74].

3.5 Acceleration of Research Through Use of EPCs

While vasculogenesis was previously believed to be restricted to blood vessel development in the embryo, Isner and colleagues established the existence of BM-derived EPCs and their participation in neovascularization after injury. Their study introduced the prospect of a novel approach to cellular therapy. Researchers have since investigated the potential use of circulating EPCs from other sources for therapeutic revascularization of injured and ischemic tissues. A reliable model for vascular development and repair has allowed an in-depth study of EC biology and the role of EPCs and ECs in regenerative processes and the development of vascular diseases.

3.5.1 Additional EPC Markers

CD34 and VEGFR2, are well established as crucial markers of EPCs from BM and circulation [28, 62, 63, 75].

AC133 is a hematopoietic stem cell marker that typically coexpressed with VEGFR2 but is rapidly lost as ECs mature. Peichev et al. demonstrated that AC133 can be used to isolate EPCs from fetal livers, mobilized peripheral blood, and cord blood [68]. $CD34^+/VEGFR2^+/AC133^+$ cells expressed the endothelial-specific markers VE-cadherin and E-selectin, possessed the chemokine receptor CXCR-4, and migrated in response to VEGF. Nonadherent $CD34^+/VEGFR2^+$ cells cultured in medium containing VEGF, FGF2 and collagen, proliferated and differentiated to mature adherent ECs ($AC133^-/VEGFR2^+$). Taken together, these results show that $CD34^+/VEGFR2^+/AC133^+$ cells are a potential source for new ECs.

Isolated and cultured $CD34^+$ EPC cells that undergo differentiation are determined to be mature ECs by the expression of CD31 and the absence of CD45, which is a typical marker on hematopoietic cells. The gradual decrease of AC133 also indicates maturation towards endothelial lineage and can serve as an early marker for EPCs. Alternatively, EPCs have been isolated by screening for cells expressing $CD34^+/AC133^+/CD45^{low}$. EPCs derived via this method formed primitive cord-like networks, and upon injection into an ischemic hindlimb model, EPCs were shown to integrate with the host vasculature [76].

Attempts to expand EPCs ex vivo have led to the conclusion that at least two types of EPCs exist in culture from peripheral blood: an early-outgrowth population and a late-outgrowth population named outgrowth endothelial cells. Gulati et al. derived EPCs and OECs from one pool of peripheral blood mononuclear cells. The study utilized the surface endotoxin receptor CD14 to demonstrate that most EPCs with typical EPC characteristics (acLDL incorporation, CD31, VEGFR2, Tie-2) derived from a $CD14^+$ subpopulation of peripheral blood mononuclear cells, whereas outgrowth endothelial cells originate from $CD14^-$. Comparison of tube-forming capabilities revealed that outgrowth endothelial cells have greater angiogenic capacity and a more pronounced endothelial phenotype. Overall, the study shows two distinct populations of EPCs derived from peripheral blood [77].

CD14 is also a marker commonly expressed in populations of monocytes and macrophages. The discovery by Gulati and others that in certain populations of cultured EPCs, a high proportion of cells express $CD14^+$, raised the question whether $CD14^+$ cells represent EPCs or are simply monocytes and macrophages. Previous studies have shown $CD14^+$-derived monocytes display EC-like characteristics, such as expression of vWF and formation of vascular-like structures under angiogenic culture conditions [78, 79]. Similar to Gulati et al., Rehman et al. cultured mononuclear cells to isolate a population of adherent cells that were $acLDL^+$/ulex-$lectin^+$, typical of the EPC phenotype. Analysis of surface markers revealed low expression of AC133, CD31, and CD117 (c-kit), and high expression of CD14 and CD45. Additionally, two monocyte lineage-specific markers, the monocyte activation marker CD11c and the hemoglobin scavenger receptor CD163, were significantly upregulated in comparison to $CD14^+$ isolated monocytes. These EPCs also secreted various angiogenic growth factors (VEGF, hepatocyte growth factor, granulocyte colony-stimulating factor, granulocyte-macrophage colony-stimulating factor) and exhibited low proliferative rates. The lack of EPC-specific markers, such as AC133, led the authors to conclude that this population of acLDL$^+$/lectin$^+$ cells should be considered circulating angiogenic cells instead, contributing to angiogenesis as supporting cells along another route [80].

Recent studies have examined more closely the monocyte versus EPC relationship, questioning the typical EPC acLDL$^+$/lectin$^+$ phenotype. Rohde et al. isolated a highly purified CD14$^+$ population and examined surface markers and growth characteristics in angiogenic culture in comparison to genuine EPCs. Cultured monocytes expressed EPC phenotype: acLDL$^+$/lectin$^+$, CD31, CD45, CD14. When cultured in angiogenic conditions, however, many of these endothelial-specific markers are downregulated, and these cells lack proliferative potential and do not form vascular networks. The authors concluded the complement of markers once thought to be EPC-specific are actually expressed on normal monocytes, allowing these cells to mimic EPCs but not actually have true EPC function [81].

3.5.2 Neovascularization

Augmenting vasculogenesis and angiogenesis has been shown to be a therapeutic strategy to treat ischemia and vascular insufficiency [82]. Numerous studies using infusions of EPCs in various animal ischemia models and clinical trials have shown improvement in capillary density and neovascularization. When mice with hind limb ischemia received transplants of ex vivo-expanded EPCs, they showed improved neovascularization and blood flow with a reduced rate of limb loss [83]. Studies found a marked improvement in blood flow when patients with ischemic limbs (because of peripheral arterial disease) received infusions of BM mononuclear cells containing CD34, whereas patients infused with peripheral blood mononuclear cells showed much smaller improvements [84]. While CD34$^+$ cells contributed to neovascularization, injection of mature ECs is not sufficient to induce neoangiogenesis [85].

Despite a vast number of studies clearly showing the contribution of EPCs to neovascularization, the mechanisms and processes of neovascularization are still the focus of many studies. Studies show that the basal incorporation rate of EPCs is very low in stable adult tissue [86] compared to ischemic conditions, where the incorporation rate is higher but varies widely [87]. Even with detectable numbers of EPCs integrated into the vasculature, the actual number of cells is still small. Urbich hypothesized that while EPCs do incorporate into vessel walls, the angiogenesis rate may also be influenced by paracrine signals and growth factors secreted by EPCs, such as VEGF and hepatocyte growth factor, which may signal mature ECs to proliferate and migrate [88].

EPCs have also been used to investigate tumor angiogenesis in animal models. Studies demonstrated that ECs are recruited from preexisting capillaries during tumor angiogenesis [89]. Transplantation of VEGF-responsive BM-derived EPCs into radiated mice reconstituted tumor angiogenesis. On the other hand, inhibition of VEGFR1 and VEGFR2 was shown to block tumor angiogenesis [90].

3.5.3 Endothelial Regeneration

As previously discussed, the repair of injured endothelia is crucial to maintaining a healthy vasculature. A previous theory of endothelium regeneration posited that neighboring ECs proliferated and migrated. Studies have shown statins, however, to increase mobilization of circulating EPCs and to accelerate reendothelialization, similar to the actions of VEGF. In addition to increasing EPCs, statins enhance EPC adhesion by upregulating integrin receptor subunits and thus promote homing of EPCs to foci of ischemia or vascular injury [91].

Repairing the endothelium as quickly as possible is advantageous for preventing the recurrence of stenosis, the narrowing of vessels. Infusion of EPCs into mice with an endothelial injury of the carotid artery caused enhanced reendothelialization with reduced neointima formation [92]. While research has primarily shown that EPCs exhibiting the standard endothelial marker CD34 contribute to reendothelialization, infusion with $CD34^-/CD14^+$ – which display EC-specific markers and gain new EC-specific markers when exposed to vascular GFs – was also shown to contribute to endothelial regeneration and functionality [93].

While some studies have shown extensive integration of transplanted EPCs within newly formed capillaries during the repair process, other studies have indicated very low levels of integration despite significant improvements in blood flow [94, 95], implying that other mechanisms and processes besides cell incorporation contribute to reendothelialization and vascularization. Several studies have reported that EPCs secrete angiogenic cytokines [96, 97]. Rehman et al. showed that cells, which displayed many characteristics of EPCs but were actually of monocyte lineage, secreted proangiogenic cytokines that may contribute to endothelial repair [80]. Another study, demonstrated that exposure of mature ECs to conditioned medium from EPC cultures resulted in increased EC migration. Analysis of the conditioned medium revealed increased levels of proangiogenic factors including: VEGF, hepatocyte GF, insulin-like GF-1, and stromal cell-derived factor-1. This led the authors to hypothesize that EPCs which homed to sites of injury could potentially secrete GFs that provide survival signals to existing ECs, thereby helping to accelerate revascularization and new vessel formation [98].

A recent study investigated the sources of paracrine signaling upon transplantation of EPCs into a mouse model of ischemic myocardium. The researchers discovered two phases of cytokine release in the host. First, upregulation of cytokine release was observed from the transplanted EPCs. Upon determining that transplanted EPCs did not remain in the site of transplantation after 1 week, continued upregulation of angiogenic cytokines was concluded to be from the host cells [99]. Another study investigated the effects of transplantation of autologous EPCs, showing that these EPCs secreted a large array of angiogenic cytokines, and transplantation into a rabbit model of carotid artery denudation resulted in improved endothelialization and endothelial function. Closer inspection of the injured vascular wall revealed an uneven distribution of EPC colonies, leading the authors to suggest that only a small fraction of regenerated tissue is derived directly from transplanted EPCs, while the dominant pathway of regeneration is through paracrine effects of EPCs on the existent endothelium [100].

3.6 Redefining the EPCs

The first isolation of an EPC led to a flurry to research focused on its origin and its involvement with vasculogenesis and reendothelialization. Despite numerous studies using animal models that showed EPC incorporation into vessel walls, BM-derived EPC infusion into patients with cardiovascular disease provided limited improvements with no evidence of long-term incorporation [101]. It was assumed that EPCs would function like stem cells, and no studies have verified that they truly behave in ways characteristic of stem cells. The discovery that the level of circulating EPCs correlates to vascular health and serves as an indicator of cardiovascular disease was quantified by identification of EC colony-forming units (CFU-ECs) [5]. Until now, EPCs have typically been identified by EC-specific cell markers using flow cytometry (CD34, CD133, VEGFR-2) or by their ability to uptake acLDL [28, 68].

Because EPCs express markers similar to hematopoietic stem cells, isolation via these markers will result in an impure pool of cells. In contrast, Ingram et al. took a novel approach *of using single-cell clonogenic assays* to construct a hierarchy of EPCs based on their proliferative potential [102]. Using these assays, Ingram's group isolated cells that expressed EC-surface antigens, but neither expressed hematopoietic cell-specific surface antigens nor displayed hematopoietic activity. This established the endothelial colony-forming cell (ECFC) as an important assay to evaluate the proliferative state of EPCs.

The results by Ingram et al. and the identification of ECFCs led to controversy about what truly defined an EPC. Recent research by Yoder et al. demonstrated that CFU-ECs and ECFCs, both of which were once thought to fall under the umbrella term EPCs, are clonally distinct with differing functions [103]. Using mononuclear cells isolated from human subjects, and following two common methods to generate EPCs, Yoder et al. obtained two separate colony types, CFU-ECs and ECFCs, from nonadherent and adherent cells. Analysis of cell surface antigen expression showed significant differences between the two sets of colonies. CFU-ECs were discovered to be progeny of hematopoietic-derived monocytes and macrophages, which was determined by observing the ability to ingest bacteria and by their displaying macrophage molecules, despite their incorporation of acLDL, which is characteristic of ECs. Single-cell clonogenic assays showed that ECFCs replated and formed secondary colonies, but CFU-ECs never did so. Finally, infusions of CFU-ECs and EPCs into mouse models were performed to observe establishment of new blood vessels. Whereas CFU-ECs failed to form chimeric blood vessels, ECFCs formed vessels that integrated with the mouse vasculature, as evidenced by perfusion of mouse red blood cells.

Although an exact definition of what constitutes an EPC remains unclear, Yoder and colleagues have provided the field with the ECFC, which has the properties of a circulating EPC and which has clearly been shown to contribute to angiogenesis and vasculogenesis [104]. On the basis of these findings, some researchers believe that most of the current work using EPCs as a therapeutic strategy has used monocyte-derived cells that do not directly contribute to the blood vessel but instead recruit cells that help facilitate this process. Further research is ongoing to determine the true nature of EPCs and their regeneration potential.

4 Vascular Derivatives of Embryonic Stem Cells

4.1 Human Embryonic Stem Cells

Despite the proliferative potential of adult EPCs and other progenitor cells, several reports have indicated that a decline in the proliferation and differentiation abilities of these cells is associated with aging and chronic diseases [105]. For example, transplantation of young BM-derived EPCs in aging mice restored neovascularization, whereas aging BM did not [106]. Furthermore, studies using EPCs from human patients with coronary artery disease reported impaired migratory capacities [107], while diabetic patients were found to have smaller amounts of circulating EPCs than healthy patients, with lower proliferation rates and impaired functionality [108]. ESCs, in contrast, have been regarded as immortal and are capable of multiple passages while maintaining normal karyotype and function, making them an alternative source of cells for vascular regeneration and tissue engineering.

Pluripotent ESCs are characterized by their ability to proliferate in an unlimited manner while maintaining an undifferentiated state with normal karyotype. Derivation of ESCs typically occurs by extraction of totipotent cells from the blastocyst [109, 110]. Previous methods of establishing pluripotent cell lines involved forming a teratocarcinoma and culturing from this disorganized population of undifferentiated embryonic cells [111, 112]. An inherent problem with using teratoma cells for studies, however, is their abnormal karyotype. Evans and Kaufman overcame this obstacle by establishing pluripotent cells from mouse embryos with normal karyotype. Their technique involved extracting inner cell mass cells from the mouse embryonic blastocyst. In 1995, Thomson and colleagues successfully isolated a primate ESC line using similar methods [113]. While primate ESCs can be useful for in vitro developmental studies, as primates and humans are closely related, the ideal model for studying differentiation of human tissues would be human ESCs (hESCs). The establishment of an ESC line derived from human blastocysts was first achieved in 1998 by Thomson et al. In order to be considered pluripotent, human ESCs derived from the ICM need to satisfy the three-germ-layer differentiation requirement and be capable of unlimited self-renewal in culture while maintaining a normal phenotype and karyotype [110].

ICM cells isolated from blastocysts were cultured on mouse embryonic fibroblast (MEF) feeder layers and selected for colonies of uniform, undifferentiated morphology. Examination of cell surface markers found expression of stage-specific embryonic antigen (SSEA)-3, SSEA-4, TRA-1–60, TRA-1–81, and alkaline phosphatase – markers consistent with primate ESCs and human embryonal carcinoma cells.

Currently, several methods exist for the derivation of hESCs, such as extraction of the inner cell mass by immunosurgery [110], mechanical isolation [114, 115], and placing the whole blastocyst on an inactivated MEF feeder layer [116]. Recent efforts have also investigated the potential of dedifferentiating somatic cells into pluripotent ESCs [117, 118].

Recently, several papers have reported on the generation of induced pluripotent stem cell lines derived from human somatic cells, bypassing the need to obtain viable stem cells from embryonic lines. It was discovered that transduction of somatic cells with OCT4, SOX2, NANOG, and LIN28 [119] or OCT4, SOX2, Klf4, and c-Myc [120] sufficed to produce induced pluripotent stem cells with normal ESC morphology, karyotype, surface markers, and the potential to differentiate into the three germ layers.

Because hESCs are cultured on a mouse feeder layer, a need arose to develop an animal-free culture system to prevent exposure of human cells to xenogenic pathogens. A recent study, for example, reports that hESCs cultured in animal serum replacements on MEFs incorporate nonhuman sialic acid [121]. Successful animal-free culture systems, however, have been developed that remove the need for nonhuman feeder layers and serum environments. In these studies, hESCs, cultured on a number of different human feeder layers, including human embryonic fibroblasts, adult fallopian tube epithelial cells, and human foreskins, have shown the ability to maintain hESC- pluripotency, immortality, and proliferative capabilities [122–124]. An alternative is to forego feeder layers and use a serum-free environment, attained by using serum replacement in combination with a variety of cytokines. Human ESCs have been successfully derived and cultured on Matrigel with defined media, and they expressed normal hESC markers and maintained a normal karyotype [125]. Alternatively, a culture of hESCs in a defined matrix will allow strict control over what components are incorporated. For example, hESCs encapsulated in hyaluronic acid (HA)-based hydrogels could be maintained in their undifferentiated state when cultured in MEF-conditioned media [126]. HA was further found to be involved in the undifferentiated state of hESCs and, later, in the initiation of cardiovascular differentiation [126, 127].

4.2 Human ESC Differentiation

Differentiation of hESCs can be induced via three popular methods. The first involves growing the cells in suspension to form aggregates, leading to EBs. EBs induce spontaneous differentiation into the various cell types of the three germ layers in an organizational manner that parallels embryonic development [128]. Various methods exist for EB formation: suspension culture, hanging drop culture, and spinner flasks [128–130]. Alternatively, hESCs can spontaneously differentiate when cultured in small aggregate colonies on inactivated feeder layers [131]. Desired cell lineages can then be isolated and cultured in fresh medium. This method, however, is conducted in a two-dimensional (2D) culture that does not mimic the natural microenvironment. A third method involves 2D culturing the cells on ECM proteins, such as collagen, which can induce directed differentiation towards a specific lineage [132].

4.3 Directing hESCs Towards Endothelial Fate

While the murine system of vascular development during embryogenesis has been extensively studied, the same could not be said of the human system before Thomson et al. established hESC lines. Numerous studies have shown the vasculogenic potential of hESCs. Mouse ESCs can give rise to both endothelial and smooth muscle cells, reproducing the vascular developmental process [133]. These cells can be coaxed to form tube-like structures, and infusion into chick embryos can contribute to the developing vasculature. Deriving vascular progenitor cells from hESCs promises to offer potential clinical applications, tissue engineering of the vasculature, and insights into early human development.

ECs were first derived from differentiated human EBs by Levenberg et al. [134]. Human ESCs were cultured in conditions tuned for the formation of EBs. EBs were examined for endothelial-specific markers to confirm the existence of ECs, and observation of EBs over several days showed spontaneous differentiation and formation of blood vessel-like structures. Using CD31 (PECAM1), a marker for embryonic ECs, the group isolated pure ECs from EBs at day 13. Examination of these cells showed coexpression of endothelial markers (VE-cadherin, CD34, and Flk1, along with expression of vWF), while maturation of these cells after 2 weeks of differentiation led to continued expression of typical EC markers with similar morphology. Additionally, these cells were capable of organizing into functional blood vessel-like and capillary structures both in vitro (2D cultured on Matrigel) and in vivo (seeded in 3D polymeric scaffolds [134].

Since then, several studies have demonstrated success in differentiating EBs in order to isolate ECs using different cell lines, growth factors, and stages of isolation. The two main methods of purifying ECs from hESCs are (1) culturing in endothelial maturation conditions by supplementing either the feeder layer or the media, and (2) selecting cells derived from EBs displaying endothelial-specific markers.

A study by Wang et al. isolated a population that express PECAM-1, Flk-1, and VE-cadherin, but not CD45. Despite expression of these markers and functional DiI-AcLDL uptake, these cells lacked expression of more mature endothelial properties, such as vWF and eNOS. Thus, these cells were termed "primitive endothelial-like cells" [135]. Culture of these cells in either endothelial maturation medium or Hem-culture medium led to the generation of mature ECs expressing vWF and eNOS, and hematopoietic progenitors expressing CD34+. More recently, Ferreira et al. derived endothelial-like and smooth muscle-like cells from human EBs, isolating CD34+ cells and differentiating them into endothelial-like and smooth muscle-like cells by supplementation of the medium with VEGF or platelet-derived growth factor (PDGF), respectively. Endothelial-like cells displayed a high proliferation rate, expressed typical EC-markers – VE-cad, PECAM1, CD34, vWF, and Tie2 – and maturation was indicated by the loss of SSEA4. Culturing endothelial-like cells on Matrigel resulted in the formation of tube-like structures with a similar morphology to ECs. Transplantation studies with endothelial-like cells encapsulated within Matrigel revealed the formation of microvessels, with a small fraction (about 5%) containing host blood cells. Additionally, culturing in PDGF medium caused cells to differentiate towards a smooth muscle lineage, with expression of alpha-smooth muscle actin, smooth muscle-myosin heavy chain, calponin, and angiopoietin-1 [136].

Whereas previous studies used an EB-based 3D system to differentiate hESCs, we have developed an alternative 2D differentiation model, based on type-IV collagen that was optimized to facilitate a better-controlled differentiation procedure for the formation of vascular lineage cells. Isolated progenitor cells displayed specific endothelial progenitor markers (CD31, CD34, AC133, and Tie2), and these cells were subsequently exposed to VEGF to induce differentiation. These conditions produced cells displaying the uptake of Dil-AcLDL and vWF, although only 20% of cells displayed the endothelial marker CD31 (Fig. 1a). Additionally, formation of capillary-like structures expressing endothelial markers, including sprouting and branching networks, were observed within type-I collagen and Matrigel hydrogels (Fig. 1b) [132]. Another study used a similar 2D culture system to differentiate hESCs on MEF towards an endothelial lineage to bypass EB formation. Differentiated cells were isolated by CD34+ and cultured in VEGF-supplemented medium to generate adherent cells that have the ability to uptake Dil-AcLDL and to express the endothelial markers CD31, VE-cad, vWF, KDR, and Tie2. When cultured on Matrigel, formation of vascular network-like structures occurred. Transplantation of ECs with 10T1/2 mouse mesenchymal precursor cells within a fibronectin/collagen gel in mice resulted in the formation of a functional microvascular network that anastomosed with the host vasculature [137].

Several investigators have recently reported developing refined methods of isolating ECs derived from hESCs and generating viable vascular networks within hydrogels. Human ESCs were induced to differentiate into ECs by coculture with OP9 stromal cells [138]. On day seven of hESC/OP9 coculture, ECs were isolated and defined by CD31+/CD43− phenotype [138]. Era et al. described a method for deriving ECs from hESCs by culturing in a defined culture medium containing bone morphogenetic protein 4 or activin. VEGFR2+ cells were generated and, when cultured on OP9 feeder cells, gave rise to EC colonies [139].

EPCs isolated from umbilical cord blood, bone marrow, or circulating peripheral blood have the capacity to proliferate and home to sites of injury or ischemia and contribute to the neovascularization process. While this offers a potential therapeutic

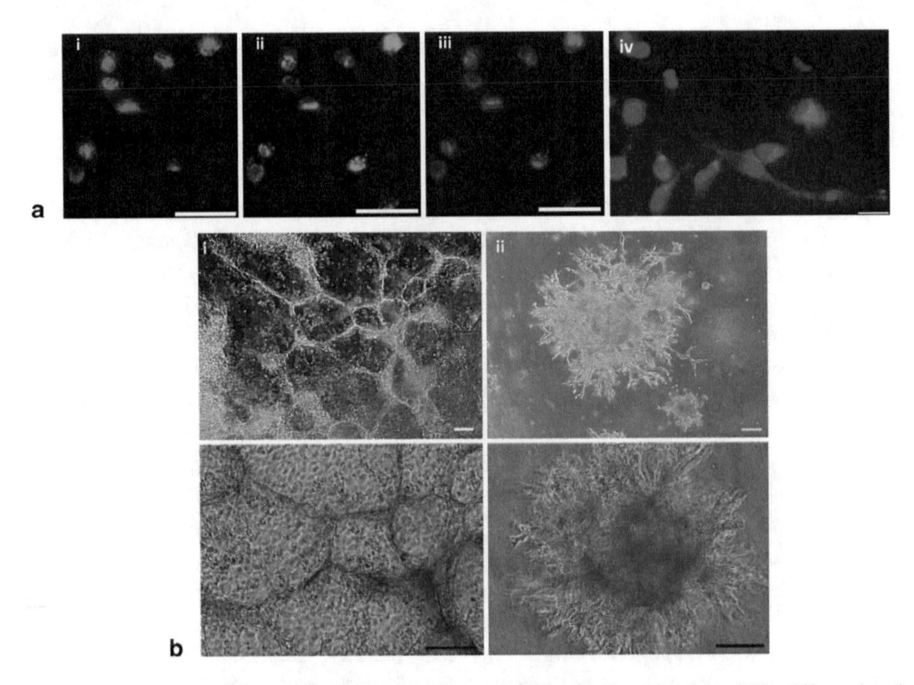

Fig. 1 Two-dimensional vascular differentiation of hESCs. (**a**) Examination of 2D differentiated hESCs supplemented with VEGF revealed Dil-AcLDL uptake (*red*) as demonstrated in (i) and (iii), and, perinuclear von Willebrand factor (*green*) as demonstrated in (ii) and (iii), in the majority of cells while only ~20% of the cells expressed CD31 (iv). (**b**) 2D differentiated cells were allowed to aggregate for 24 h in medium supplemented with human VEGF, after which they were seeded into type-I collagen or Matrigel in the same medium. Tube-like formation after 7 days in type-I collagen gel (i) and aggregate sprouting in Matrigel (ii) (both with low and high magnification). Adapted from [132]

strategy through cell transplantation and tissue engineering, a limitation is imposed upon the number of available cells, which are typically very low in adult peripheral blood. In contrast, hESCs have an unlimited proliferative potential, and their EC derivative possesses great therapeutic potential. ECs are generated from hESCs in a microenvironment similar to embryonic development. Tissue engineering offers the opportunity to study vascular development, organogenesis and, ultimately, regeneration in defined in vitro microenvironments.

5 Vascular Tissue Engineering

5.1 Introduction

While readily available protocols for differentiating hESCs to ECs will allow researchers to study the vascular developmental processes of embryos and the molecular cell-signaling pathways of ECs in greater depth, a more pressing concern lies within the realm of tissue engineering. Two areas of vascular tissue engineering research are (1) developing methods to vascularize thick tissues and (2) developing methods to engineer whole blood vessels, such as arteries. While large vessels containing SMCs provide resistance and help regulate the flow of blood, microvessels, such as capillaries, are the functional component of a tissue. Thus, for this review, we will focus on common techniques for vascularizing thick tissues.

The goal of tissue engineering is to develop biological substitutes that restore, maintain or improve tissue function. One major obstacle preventing tissue engineering from progressing rapidly is the inability to sustain a large mass of cells once transferred from in vitro to in vivo (i.e., to enable engineering of vascularized tissues and to help repair ischemic tissues in vivo). In any large living organism, the vascular system is essential for supplying oxygen and nutrients, allowing autocrine signaling, and removing wastes. Additionally, because of oxygen diffusion limitations, most cells are limited in that they cannot survive further than approximately 150 μm from the nearest capillary [140, 141]. This explains why current tissue-engineering products have been most successful with tissues that are either thin (e.g., bladder) or have very low oxygen requirements (e.g., cartilage).

While numerous studies have developed methods for vascularizing thick tissues [23, 141], successful integration, which depends on rapid oxygen and nutrient supply to the transplant, remains problematic due to the time it takes for blood vessels to develop from the surrounding tissue [142]. Even with the onset of vascular penetration, only peripheral cells are nourished, while the center cells continue to experience hypoxic conditions. Thus, the establishment and maintenance of the vascular system is required for continued growth of almost all normal tissues.

An ideal bioengineered vascular network is achieved once it can fully function as a natural vascular network in vivo. Important characteristics include a capillary-like branching network having a lumen, responsiveness to biochemical and biophysical signals, and the ability to form anastomosis with the host vasculature. Currently, the formation of an integrated microvascular system has been achieved with human dermal microvascular ECs to perfuse a network of adipocyte cells in the chorioallantoic membrane model [143]. However, vascularization in an animal model has yet to be examined. Other studies have developed techniques to achieve formation of organized capillaries in vitro, but efforts are still focusing on a universal method of vascularizing thick tissues that can be quickly integrated into the host system.

5.2 Cell Sources

Utilization of cells isolated from a patient's own body drastically minimizes the risk of immunorejection of an engineered tissue. Several types of cells are potential sources for vascular tissue engineering, including mature ECs, such as vessel-wall-derived ECs, HUVECs, and human dermal microvascular ECs. While numerous studies with mature ECs have demonstrated their ability to form capillary-like structures, their rapid in vitro senescence limits their utility for cell-based therapies. To address this, attempts have been made to immortalize EC lines by genetic manipulation resulting in an unlimited proliferation [144]. However, the insertion of foreign genes to modify the genome would not provide a reliable source of cells, due to the unforeseen risk of tumor development if proliferation could not be restrained. Adult EPCs from various sources (cord blood, peripheral blood, and BM) – which, in some cases, can be isolated from a specific patient and then expanded and matured in vitro – are another important source for vascular tissue engineering. Extensive efforts currently focus on investigating the proliferative limitations, functionality, and ultimate regenerative potential of these cells [145–150]. Various protocols for the derivation of EPCs and ECs from hESCs offer another justifiable source for cellular therapies and regeneration of the vasculature. Furthermore, recent publications demonstrating the ability to reprogram somatic cells into pluripotent hESCs offer the prospect of overcoming immunorejection following transplantation. Here as well, many studies focus on investigating the purity, maturation, and functionality of the hESC-vascular derivatives, with the goal of enabling their usage for cellular and tissue engineering-based therapies.

5.3 Scaffolds as 3D Templates

In most cases, tissue engineering attempts to recapitulate certain aspects of normal development in order to stimulate cell differentiation and functional tissue assembly. Numerous studies have shown that cells cultured on a 2D surface behave differently from those cultured in a 3D matrix. Cells typically form a monolayer on a flask, growing at the liquid/substratum interface, and they cease proliferation once confluency is reached due to contact inhibition. At this point, cells may undergo changes in morphology and function. In a developing embryo or in any tissue structure, cells migrate, proliferate, and exist in a 3D microenvironment containing a diverse array of biochemical and biophysical signals. Thus, the development of a 3D scaffold compatible with vascular and tissue-specific cells is essential to achieving the goal of tissue bioengineering. Therefore, in vitro tissue engineering generally involves the use of scaffolds that are designed to provide 3D structural and logistic templates for tissue development, to control the cellular microenvironment and to provide the necessary molecular and physical regulatory signals.

Currently, several strategies exist for vascularization of tissue engineered on biological or synthetic scaffolds. One approach is to integrate growth factors into the polymer scaffold itself. Since the scaffolds are biodegradable, degradation will slowly release growth factors that diffuse into the local environment, inducing vessels from surrounding tissue to grow into and vascularize the scaffold [151]. An alternative approach is to seed the scaffold with ECs so that these cells may form a capillary network that can integrate with ingrowing vessels from the host tissue. A problem that remains in both cases, however, is that cells in the center of the scaffold still experience a limited oxygen supply prior to blood vessels' ingrowth followed by anastomosis with the host for a viable blood supply. A third approach involves prevascularizing the tissue in vivo [152, 153]. An arteriovenous loop (AVL) is formed within an animal model, and this AVL is encapsulated by a polycarbonate chamber. A natural scaffold consisting of fibrin will initially form, and then angiogenesis will proceed with sprouting, leading to the formation of a microcirculatory network. A transient stage exists where stem or progenitor cells can be seeded to develop into a specific tissue in the presence of an active vascularization. Subsequent vascular remodeling produces a viable tissue construct with great potential. Because the use of an AVL requires a live animal subject to provide progenitor cells and cytokines, it may not be the most effective method of bioengineering tissues on a large scale. Thus, an alternative approach is to utilize stem cells and incorporate bioactive molecules within the scaffold, which can direct preferential differentiation towards a vascular lineage [154].

5.4 *Scaffold Design Specifications*

Some design criteria for a capable and vascularizable scaffold include (1) a biocompatible material with a surface chemistry permitting cell adhesion, promoting cell growth, and allowing cells to maintain a differentiated state; (2) a controllable rate of degradation and resorption to match tissue growth, with nontoxic byproducts that can be eliminated from the body via metabolic processes or kidney filtration; (3) permeability to the flow transport of chemicals, nutrients, and metabolic wastes; (4) ability to support the presence of several different cell types of the vasculature (ECs, SMCs, fibroblasts) and other cell types (organ-specific, such as hepatocytes, cardiomyocytes, etc.); (5) a continuous porous mesh structure and orientation, with a high surface-area-to-volume ratio that allows cells to migrate and physically interact with each other; (6) mechanical strength to support supercellular organization; and (7) the ability to conform to various shapes [155–157]. Biocompatible materials that have been shown to satisfy one or more of these criteria can be classified into synthetic, biodegradable polymers (e.g., polyurethanes) or natural, enzymatically degradable biopolymers (e.g., hyaluronic acid) depending on the desired application and in vivo environment [158].

5.5 *Synthetic Polymer Scaffolds*

Synthetic polymer scaffolds, both biodegradable and degradable, are often used in biomedical devices and as scaffolds in bioengineered tissues. One advantage to using synthetic polymers is the ability to tailor a scaffold's ultrastructure and mechanical properties and its degradation kinetics through chemistry and processing [159]. This review will focus on common materials on which cells have been grown successfully and that have the potential to be used for engineering a vascular network.

Important scaffold parameters that should be controlled during the design process for vasculature tissue engineering include pore size, shape, distribution and interconnectivity, and elastic modulus [23]. These parameters will affect the effectiveness of cell seeding, cell migration and blood vessel infiltration, ECM deposition, and nutrient and waste transport.

5.5.1 PGA

Polyglycolic acid (PGA) is an inelastic polyester that is degraded by water through hydrolysis, producing glycolic acid. The controlled rate of degradation allows time for ECM deposition to occur and for cells to establish themselves within a microenvironment [160, 161]. PGA scaffolds are typically used as nonwoven sheets. Cells are cultured directly on the surface, where they attach and grow to form a tissue-like structure [162]. Thus, in this instance, the scaffold serves as a developmental guide for cell growth to establish a tissue structure once the scaffold degrades. While ECs have been grown successfully on PGA, the cells are still growing on a 2D surface and do not form organized capillary-like structures [162, 163]. The inability to encapsulate cells, due to a harsh polymerization process, limits the usefulness of PGA as a scaffold material.

5.5.2 PLA

Another polymer similar to PGA in mechanical strength is polylactic acid (PLA). Unlike PGA, however, PLA is more hydrophobic and thus more resistant to hydrolysis. Additionally, growth factors (GFs), such as VEGF, can be incorporated into a PLA scaffold by the process of freeze drying followed by gas foaming, where the PLA/VEGF mixture is placed under high pressure and then released to form a porous structure [164]. HUVECs seeded onto such a scaffold quickly adhered (within 24 h) and were maintained in culture for 28 days, retaining their proliferative capacities, though no blood vessels were formed. Implantation into a chorioallantoic membrane showed increased blood vessels in proximity, demonstrating that the encapsulated GF can stimulate angiogenesis nearby [164]. A recent study has reported the successful incorporation of more than one GF into PLA [165]. Additionally, although cells seeded onto a PLA scaffold can migrate, the unequal distribution of cells on the outer surface compared to the inner surface remains a problem.

Because PLA is intrinsically unreactive towards cellular components, many studies have investigated the surface modification of poly-l-lactic acid (PLLA) scaffolds to enhance cell–polymer interactions and cytocompatibility. Surface modifications involve the immobilization of natural materials within the PLLA polymer, including gelatin, collagen, alginate, and biotin [166–169]. Modified surfaces have been shown to improve cell attachment, growth, and proliferation across a diverse range of cell types.

A study that investigated the effects of various matrix modifications to PLLA, such as amide groups, covalently attached biomacromolecules (collagen), and carboxylic groups on cell interaction [170] showed that all modified PLLA variants improved HUVEC growth and cytocompatibility, with PLLA–collagen scaffolds having the highest viability and attachment rates. A similar study by the same group introduced free amino groups onto the PLLA surface, along with immobilized gelatin and collagen, to improve the endothelium regeneration rate [170]. HUVECs cultured on this scaffold had improved attachment rates and cell morphology compared to the control PLLA polymer.

5.5.3 PLGA

Since the properties of PGA and PLA are complementary, many studies have produced scaffolds consisting of varying proportions of each polymer to produce a hybrid material, poly(lactide-co-glycolide) (PLGA). The advantage is the ability to control for factors such as degradation rates (which can range from weeks to years). GFs and cytokines can be easily incorporated into the scaffold and delivered in a sustained manner, either through encapsulation within PLGA microspheres or as part of the fabrication process [151, 171]. Studies with rat SMCs and rat aortic ECs on PLGA scaffolds showed increased cell density in response to changes in the nanotopography of the scaffold surface [172]. Additionally, differentiating hESCs have been successfully used with laminin-coated PLGA scaffolds. Upon implantation, the hESCs formed liver-specific tissues, and extensive vascularization of the scaffold was observed from host vasculature incursion and from differentiation of hESCs to EPCs [173].

PLGA/PLLA hybrid scaffolds have been developed to harness the advantages of biodegradability for cellular ingrowth and mechanical strength for 3D structure. A hybrid PLLA/PLGA scaffold, biodegradable and highly porous, was constructed and implanted into mice to study the therapeutic properties of ECs derived from hESCs [134]. Subcutaneous transplantation of PLLA/PLGA seeded with hESC-derived ECs revealed after 7 days that, amidst fibrous connective tissue, functional blood-carrying microvessels had formed and anastomosed with the host vasculature. Further research has investigated the possibility of constructing a microvascular network in vitro in conjunction with other tissue types (please see Sect. 5.7). Differentiating hESCs, derived from EBs seeded onto the PLGA/PLLA hybrid scaffold, were shown to adhere, proliferate, and maintain viability [174]. After an in vitro incubation period, primitive tissue structures and development of ECM were evident. Additionally, capillary-like networks, marked by the presence of endothelial markers, indicated their differentiation into EC lineage and the organization of vessel structures. The results showed the ability of cells grown on a PLGA/PLLA scaffold to differentiate and organize in parallel among a variety of lineages.

5.5.4 PEG

Homogeneous distribution of cells with PLA, PGA, and hybrid scaffolds is difficult to achieve, because of the low bioactivity of the material, which therefore limits the use of thick scaffolds. An alternative utilizes polyethylene glycol (PEG), a hydrophilic polymer that is intrinsically nonadhesive to cells. PEG has the advantage that its polymer backbone can be functionalized with adhesion factors, growth factors, and biochemical groups to improve bioactivity. Reactive groups on the PEG polymer can then be cross-linked by chemical or light-induced reactions to form a hydrogel that mimics the ECM, either in vitro or directly in vivo [175, 176]. Additionally, alternative forms can be synthesized from PEG molecules, such as branching arms and acrylation [177, 178]. Because some of these peptide groups are involved in the cross-linking process, migrating cells that naturally secrete proteases can break down the PEG scaffold and mediate remodeling processes [179]. Experiments with matrix-conjugated VEGF found that local VEGF release is critical for EC survival and migration throughout the scaffold [180, 181]. In vivo studies with VEGF-conjugated PEG hydrogels revealed significant cellular ingrowth and formation of blood vessels, with the hydrogel slowly being remodeled and resorbed into host tissue.

An advantage of PEG is the ability to prepare the scaffold within a cell suspension. For example, cell encapsulation can be achieved via photoinitiated polymerization with a light source such as long-wavelength ultraviolet (UV) light. This method was shown to enable a homogeneous cellular distribution of viable cells [182]. The hydrogel can also be subjected to physical deformations to study the effects of physical stresses on cells adhering within the hydrogel [183]. Success with this encapsulation technique has been achieved with osteoblasts [182], chondrocytes [184], and ECs and SMCs [185]. Experiments have also been performed with the encapsulation of human mesenchymal stem cells (hMSCs) to study their differentiation towards an osteoblastic lineage in osteogenic differentiation media [186]. Human MSCs were cultured up to 6 weeks while maintaining their undifferentiated state. These results suggest that further modification of PEG-derived scaffolds with appropriate macromolecules may provide an environment for vasculature differentiation.

Indeed, a recent study developed a dextran-based hydrogel with a synthesized combination of dextran-acrylate, acryloyl-PEG, and the Arg-Gly-Asp (RGD) attachment motif found in fibronectin [154]. Human ESC colonies were suspended within the scaffold solution along with VEGF-loaded microparticles and were photopolymerized. Encapsulated hESCs aggregated while forming EBs and preferentially differentiated towards a vascular lineage compared to suspension-borne EBs [154]. Further modifications of PEG-derived scaffolds with appropriate macromolecules may provide an environment for vasculature differentiation.

5.5.5 PGS and PGSA

Tough, biodegradable elastomers that exhibit mechanical properties similar to those of soft tissue [187–189] could prove useful in multiple medical applications, for example, in the area of small vascular grafts [190]. Wang and colleagues [191] synthesized a tough biodegradable elastomer, poly(glycerol sebacate), which degrades via surface erosion, exhibits in vitro and in vivo biocompatibility, consists of inexpensive FDA-approved building blocks (within certain products), and features functionalizable groups throughout the polymer matrix [191, 192]. A recent modification to poly(glycerol sebacate) incorporated acrylate groups into the polymer backbone, allowing for photopolymerization while preserving the elastic and biocompatible properties of PGS [193]. However, the modified elastomer, poly(glycerol-co-sebacate)-acrylate, polymerizes to form a uniform scaffold in which the density and hydrophobicity of the material, which lacks pores, does not allow for rapid diffusion of media and therefore may not be well suited for cell encapsulation. In order to form porous, photocurable poly(glycerol-co-sebacate)-acrylate scaffolds, we have added nontoxic and nonreactive glycerol during polymerization, resulting in the formation of a 3D photocurable elastomeric porous matrix (Fig. 2a; [194]). Human ESCs encapsulated in porous PGSA were found to adhere to the scaffold wall and to form protrusions and apparent interconnections between each other within 24 h. After 7 days, undifferentiated hESCs proliferated and differentiated in the poly(glycerol-co-sebacate)-acrylate scaffold, forming 3D tissue-like structures (Fig. 2b; [194]). Subcutaneous transplantations showed that porous scaffolds have biocompatibility profiles similar to nonporous PGSA, in which minimal inflammatory zones surrounding the scaffolds decrease along the 7-week experiment, but porous poly(glycerol-co-sebacate)-acrylate promotes tissue ingrowth and integration with host vasculature, unlike nonporous PGSA (Fig. 2c; [194]).

5.6 *Natural Biopolymer Scaffolds*

In contrast to synthetic scaffolds, biopolymer scaffolds are composed of components found in the living tissue, which makes them appealing for cell-based therapies as a 3D scaffold for tissue-engineering applications. Furthermore, natural biopolymer scaffolds are enzymatically degradable, while some are also recognizable by cells, resulting in beneficial cell–scaffold interactions; therefore, such scaffolds have the potential to serve as instructive 3D environments.

5.6.1 Matrigel

Matrigel, the commercial name for ECM secreted from Engelbreth-Holm-Swarm mouse sarcoma cells, is commonly used to study angiogenesis. Its main components are collagen type IV, laminin, heparin sulfate (HS) proteoglycan, and entactin. Under incubation conditions (37°C), Matrigel self-assembles into a thin film or a thick gel, depending on the application. It has been useful for studying the mechanisms of neovascularization and angiogenesis [195–198]. Typically, Matrigel is used to coat culture dishes, where ECs cultured on Matrigel will organize themselves into capillary-like structures. Additionally, encapsulation of cells within Matrigel offers

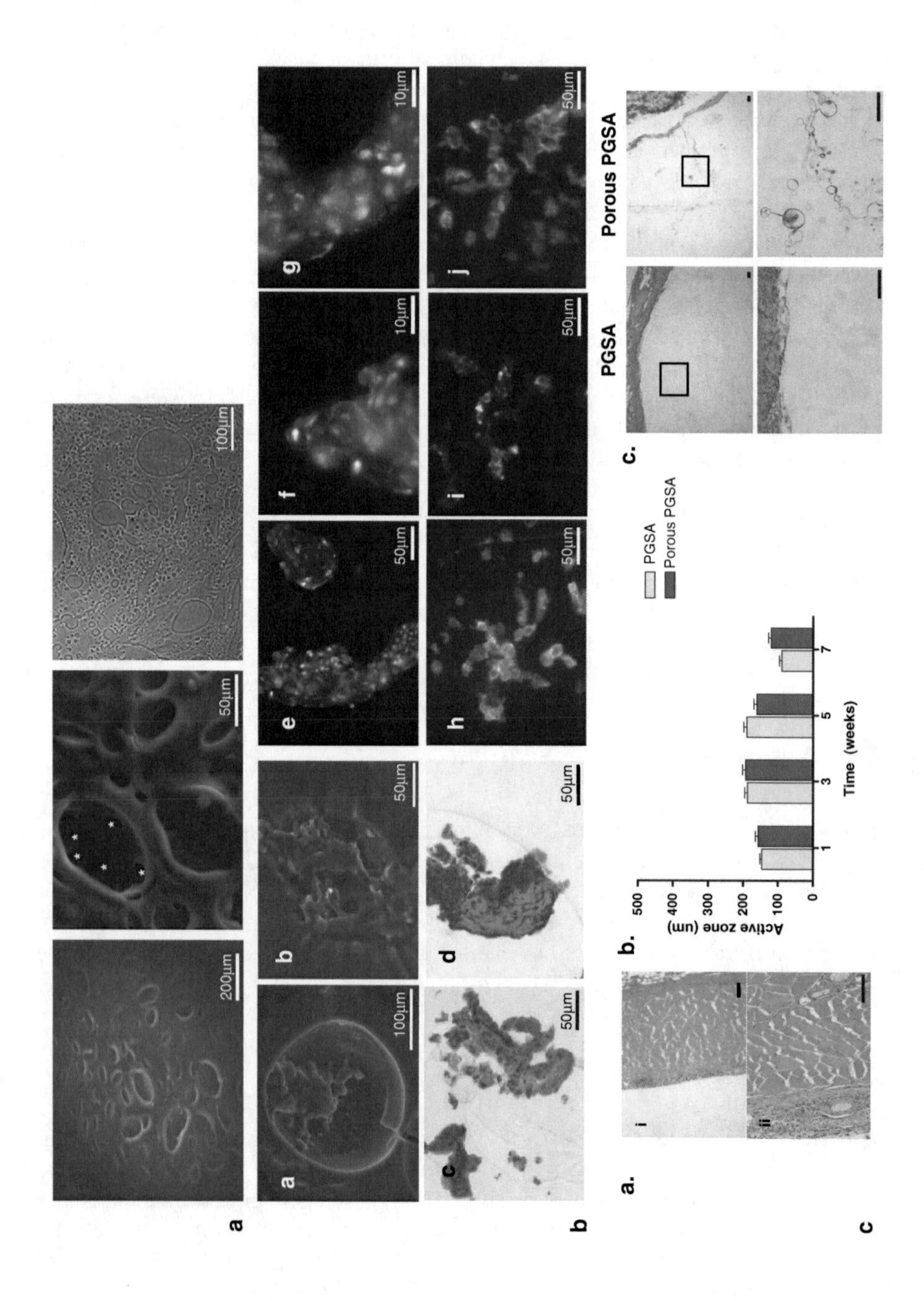

a
200µm
50µm
100µm
b
a
100µm
b
50µm
c
50µm
d
50µm
e
50µm
f
10µm
g
10µm
h
50µm
i
50µm
j
50µm
c
a.
i
ii
b.
Active zone (um)
500
400
300
200
100
0
1
3
5
7
Time (weeks)
PGSA
Porous PGSA
c.
PGSA
Porous PGSA

Fig. 2 Porous PGSA. (**a**) Characterization of porous PGSA. Environmental scanning electron microscope images at low (*left*) and high (*middle*) magnification, revealing the presence of macropores (>50 μm) and interconnecting pores (20–50 μm, *asterisks*). Light microscopy images of cryostat-sectioned scaffolds (*right*) showing the presence of micropores (<20 μm). (**b**) Cell encapsulation. Environmental scanning electron microscope and Hematoxylin and Eosin stained micrographs showing encapsulated hESC colonies after 1 day (a, b) settled primarily within macropores. After 7 days (c), they formed tissue-like structures covering most of the scaffold pores, and (d) were organized in 3D structures containing cells with various morphologies. Immunofluorescent staining of hESCs encapsulated in porous PGSA and culturing for a further 7 days revealed proliferating cells positive for Ki67 at (e) low and (f, g) high magnification. Differentiated hESCs were found to express early markers of the three germ layers: (h) brachury (mesoderm), (i) cytokeratin 18 (ectoderm), and (j) α-feto protein (endoderm). (**c**) In vivo biocompatibility. Scaffolds were transplanted subcutaneously in rats. (a) After 1 week, a minimal inflammatory zone was found surrounding the porous PGSA (i), with no detectable inflammation in the muscle (ii). (b) Inflammatory zones of both porous and nonporous scaffolds were reduced along the 7-week experiment ($p < 0.001$), while (c) Hematoxylin and Eosin-stained histological slices of explants revealed tissue ingrowth in porous scaffolds that were not observed in the nonporous scaffolds (all boxes in *upper panel* are shown at higher magnification in *lower panels*). Scale bar = 100 μm. Adapted from [194]

an alternative for 3D studies of the vasculature. For example, a recent study demonstrated that ECs that were cultured on beads and encapsulated within a suspension of Matrigel displayed cellular outgrowths with a tubular branching morphology [199]. Vascular derivatives of hESCs were mixed within cold-liquid Matrigel, followed by subcutaneous injection and polymerization. Examination of explants revealed that vascular derivatives of hESCs were functional and were capable of integrating with the host's circulatory system [136].

Though Matrigel is widely utilized as a scaffold material for in vitro studies of vasculature network formation and functionality, its sarcoma cell origin and undefined composition make it not feasible for clinical use.

5.6.2 Hyaluronic Acid

HA (also known as hyaluronan or hyaluronate), a linear glycosaminoglycan, is biocompatible and has been used as a biomaterial for tissue-engineering applications, partly for its unique viscoelastic properties and its ability to retain water [200]. It exists in the ECM of almost all tissues and is broken down by the specific protein hyaluronidase. HA has a crucial role in regulating the angiogenic process and vascular EC function. Notably, the high-molecular weight HA polymer (~107 kDa) is antiangiogenic and inhibits EC proliferation and migration, as well as capillary formation in collagen gels. In contrast, low-molecular weight degradation products (three to ten disaccharide units) stimulate EC proliferation, migration, and sprouting [201–203] and induce angiogenesis in the infarcted myocardium [204] and in the chick chorioallantoic membrane [205]. Generation of this "angiogenic" HA from naturally occurring HA is mediated by the endoglycosidase hyaluronidase (via polysaccharide degradation), by processes associated with tissue damage, inflammatory disease, and certain types of tumors [206].

Properties that make the linear polymer HA a desirable biomaterial include (1) the ability to control molecular weight through degradation; (2) the ability to remodel it by proteases secreted from specific cell types; (3) its capacity to be functionalized with desirable GFs and other biomacromolecules; (4) its ability to undergo cross-linking to form hydrogels; and (5) the modifiability of its adhesion properties through functional groups or hybrid materials [207].

An early study of HA attempted to develop an HA-based hydrogel with the following criteria: (1) have a mild cross-linking technique, (2) promote cell adhesion, and (3) be selectively degradable. Cell adhesion was introduced by adding RGD functional groups, and the biodegradable hydrogel was formed by photopolymerization [207]. A culture of human dermal fibroblasts on the scaffold surface displayed high proliferation rates.

Growth factors can also be preloaded into a HA hydrogel [208]. Upon implantation of a VEGF-HA hydrogel into mice to study its in vivo effects, an increased microvessel density and growth was observed, demonstrating that ECs are able to migrate through the hydrogel and that the VEGF contributed a strong angiogenic effect. Because the breakdown of HA contributes to angiogenesis, controlled breakdown can be used to advantage to promote tissue growth.

Additionally, because of the liquid nature of HA before polymerization, there exists the possibility of encapsulating cells to allow for a homogeneous distribution. Mouse ESCs and fibroblast cells encapsulated within HA hydrogels proved viable and were evenly distributed throughout the hydrogel [209]. Varying the concentrations and molecular weights of HA macromers led to the development of hydrogels with controllable degradation rates that were able to sustain mouse fibroblasts and chondrocytes [176]. To further validate the use of HA hydrogels for stem cell and vascularization studies, we recently developed a HA-based hydrogel that can successfully culture and differentiate hESCs [126]. Observations revealed that hESCs encapsulated within the hydrogel scaffold were uniformly distributed and maintained their undifferentiated state while continually remodeling the scaffold structure (Fig. 3a). Cells within the scaffold can be further released by degrading the scaffold with hyaluronidase (Fig. 3b). Controlled differentiation of hESCs could also have been achieved within the HA hydrogels by supplementing the media with VEGF. Replacement with angiogenic differentiation medium resulted in cell sprouting and elongation, which were associated with specific vascular markers, providing a good foundation to study vascular differentiation within a 3D environment (Fig. 3c).

Development and improvement of HA-based hydrogels for vascular tissue engineering could include incorporation of different ECM molecules and the generation of HA gradient matrices.

5.6.3 Collagen

Collagen, being the most abundant and ubiquitous protein in the body, is another integral component of the ECM matrix and is part of the structure of blood vessels [210, 211]. While many different types of collagen exist, of particular interest for vascular regeneration are types I and IV. Type I collagen is the most dominant form of this protein, and its structure is nearly identical across different animal species [212]. In vivo, collagen typically forms fibrils, aggregates of collagen molecules that are further organized into parallel arrangements. In vitro, collagen hydrogels and matrices can be formed using a variety of techniques (e.g., cross-linking or freeze-drying) [213]. Collagen cross-linking is temperature dependent, and therefore a rapid polymerization can be carried out at room temperature or higher.

Because collagen matrices typically have poor loading capacities for biomacromolecules, in order to incorporate these functional compounds into collagen scaffolds, collagen can be modified by covalently linking HS or glycosaminoglycan chains, whereby GFs can attach and affect local tissue response, such as promoting vascular ingrowth into the matrices [214, 215]. Incorporation of VEGF and fibroblast growth factor 2 (FGF2) into implanted collagen scaffolds has enhanced angiogenesis and promoted blood vessel growth [216].

Similar to HA hydrogels, encapsulation of cells within collagen scaffolds can be carried out by adding cells to a solution of collagen and then cross-linking to form a homogeneous hydrogel. Feraud et al. examined whether an ESC model system could recapitulate angiogenesis [217]. Eleven-day-old mouse EBs were subcul-

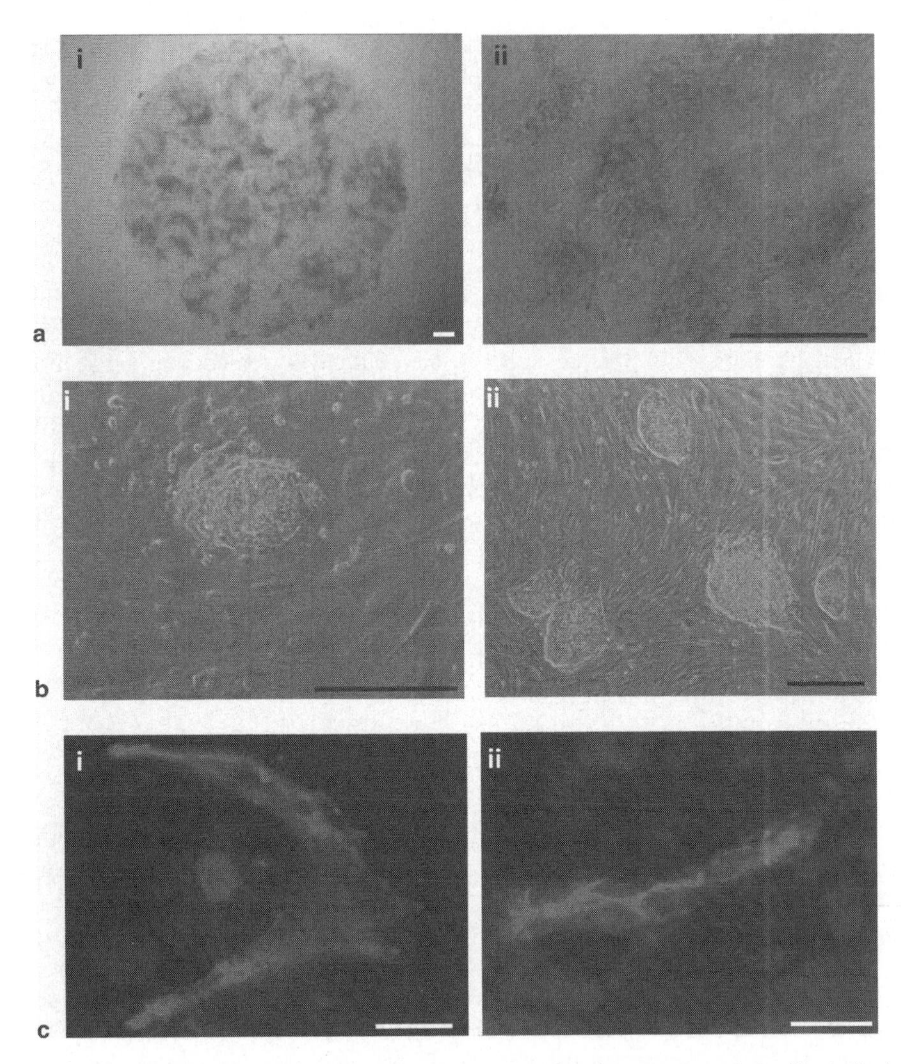

Fig. 3 HA hydrogels. (**a**) Human ESC encapsulation. (i) Light microscopy revealed uniform distribution of encapsulated hESC colonies and (ii) XTT (2,3-bis(2-methoxy-4-nitro-5-sulfophenyl)-5-[(phenylamino)carbonyl]-2H-tetrazolium hydroxide) revealed orange dye in metabolically active hESCs encapsulated in HA hydrogels. (**b**) Cell release. (i) Human ESCs, released from the hydrogel using hyaluronidase and cultured on MEFs, were found to form small colonies of undifferentiated cells after 24 h and (ii) were propagated on MEFs for three passages. (**c**) Vasculogenesis. Encapsulated hESCs were cultured in MEF-conditioned media for 1 week followed by replacement by medium containing VEGF. Cell sprouting and elongation were observed after 48 h in the gels where such cells were mainly positive for (i) α-smooth muscle actin, while some were positive for (ii) early stage endothelial marker CD34 (in situ 3D staining of gels). Scale bars = A–B – 100 μm; C – 25 μm. Adapted from [126]

tured into a type I collagen matrix. In the presence of VEGF, EBs rapidly developed endothelial sprouting, suggesting that ESC differentiation could recapitulate, in addition to vasculogenesis, the early stages of angiogenesis. On the basis of this study, we have demonstrated that vascular cells derived from hESCs, once encapsulated in collagen gels, sprout and form networks [132].

Although collagen-based, tissue-engineered blood vessels have many interesting properties and have been utilized to study aspects of vascular biology, these constructs are too weak to be implanted as bypass grafts for in vivo investigations. Attempts to improve collagen scaffolds have included the addition of PEG or poly(ethylene glycol) diacrylate (PEGDA) as a cross-linker [218] and the incorporation of elastin to form hybrid constructs that better mimic arterial physiology and exhibit improved mechanical properties [219].

5.6.4 Fibrin

Fibrin is a major component of blood clots, which form in response to vessel injury, and acts as a temporary scaffold for regeneration and new cell ingrowth [180, 181]. Upon injury, the precursor molecule fibrinogen is proteolytically cleaved to form fibrin, which then proceeds to form a dense network through physical polymerization and enzymatic cross-linking. Subsequently, invading cells naturally secrete proteases that gradually degrade the fibrin network. All of these make fibrin an attractive candidate as a scaffold for vascular regeneration. Fibrin, injected subcutaneously, was found to contribute to significant neovascularization and development of ECM [220]. It was also shown that fibrin can serve as an efficient vehicle for delivery of GFs, leading to its potential use as a scaffold with incorporated biomacromolecules. VEGF was incorporated into fibrin by mixing with fibrinogen and then initiating the clotting mechanism with thrombin. HUVECs cultured on such a surface coated with VEGF-releasing fibrin were observed to display increased proliferation, indicating that VEGF retained its active form and that fibrin contains binding sites for VEGF [221]. Additionally, the incorporation of FGF2 via equilibrium binding within a fibrin scaffold was shown to promote EC growth. HUVECs grown on fibrin-FGF2 scaffold surfaces maintained their proliferative state without any soluble growth factors for up to 96 h [222]. Because diffusion was the mode of release of GFs in the previous two studies, burst release was a common problem with GF incorporation. Therefore, two alternative methods were examined for controlled release: covalent linkage between the GF and fibrin [223], and covalent linkage of heparin to fibrin, which provides binding sites for heparin-binding GFs [224, 225]. Using these methods, the release of GFs will depend on cleavage from the matrix by proteases secreted by cells that are infiltrating and remodeling the matrix. HUVECs grown on these VEGF-incorporating hydrogels displayed enhanced proliferation, with GF release dependent upon the rate of matrix degradation, in contrast to soluble VEGF [223].

Numerous studies have investigated the use of fibrin hydrogels as a scaffold for ECs to migrate and form vascular structures. For example, ECs encapsulated within

VEGF-fibrin migrated and formed tubes up to 100 μm in diameter [180, 181]. In another 3D angiogenesis system, HUVECs were cultured as a monolayer, whereupon a fibrin solution was placed on top of the cells and polymerized with thrombin and the entire construct was immersed in media [226]. Observations revealed the formation of tube-like structures that penetrated the fibrin hydrogel, with tube formation occurring at increasing heights above the monolayer as time progressed, up to 21 days. In a more recent study, HUVECs were either encapsulated within fibrin hydrogel or seeded on beads and encapsulated within fibrin. In both cases, a fibroblast layer was cultured on the surface. HUVECs on beads sprouted from microbeads into fibrin and formed interconnected, elongated lumen-like structures. HUVECs encapsulated within fibrin also formed interconnected cord-like structures, but to a lesser extent [227].

Another study developed a new VEGF construct, binding VEGF to fibrin via a plasmin-sensitive sequence that displayed characteristics of being efficiently retained in the fibrin matrix and having a release profile dependent upon plasmin. EPCs from umbilical cords were cultured on the VEGF-fibrin hydrogel and maturation and differentiation towards EC was observed, as indicated by the loss of CD133 and the gain of CD31 and VE-cad [228].

These results, taken together, show the potential of fibrin-GF-bound hydrogels to serve as controllable environments for 3D vascular differentiation, organization, and delivery.

5.6.5 Elastin

While collagen-based scaffolds have been shown to be effective for cell encapsulation and delivery, the constructs are usually mechanically weak. Whereas collagen provides tensile support, elastin fibers contribute to vessel recovery from pulsatile deformations. Therefore, a large number of studies have attempted to engineer elastin-incorporated blood vessels. The inherent insolubility of unmodified elastin fibers, due to intermolecular cross-links, presents a problem, and researchers have searched for a way to incorporate elastin into matrices [219]. An extensive review on the use of elastin in tissue engineering has been written [229], so this review will focus on studies oriented towards vascular engineering.

Recently, a simple technique for synthesizing elastin-based materials was developed [230]. The final product retained many properties of natural elastin, such as effective SMC adhesion, formation into fibrils, and the ability to regulate vascular SMC migration and proliferation. Studies have also worked at developing purification methods to isolate pure elastin, which is useful for its ability to be molded into any shape and integrated into collagen scaffolds [231]. Whereas unmodified elastin fibers are insoluble, chemical techniques and degradation of specific peptide bonds by proteolytic enzymes can produce a solubilized form of elastin called hydrolyzed elastin [232]. It has been observed that elastin peptides, the degradation products of elastin, can modulate the physiology of many cell types, including ECs and SMCs, through the existence of an elastin receptor [233]. Thus, biomaterials incorporating elastin peptides may influence cellular function and mimic a more natural microenvironment. A recent study developed a scaffold comprising insoluble collagen and soluble elastin that, upon implantation, proved capable of inducing angiogenesis, providing motivation for using hybrid scaffolds to study vascular development [229].

Besides using elastin in its native fibril form or its peptide form, researchers have also utilized protein engineering to develop elastin-like molecules to incorporate specific sequences, such as the RGD sequence. One such study developed an elastin-mimetic copolymer capable of temperature-dependent self-assembly into hydrogels [234]. Another bioengineered elastin protein is tropoelastin, a biosynthetic precursor to elastin that does not contain any cross-links [235]. Tropoelastin, however, can self-assemble into sheets, sponges, or tubes and then can be cross-linked through chemical methods [236].

Future studies investigating the use of elastin-based scaffolds will determine its potential for stem-cell-based vascular tissue engineering.

5.6.6 Alginate

Alginate, a natural hydrophilic anionic polysaccharide derived from algae and composed of beta-D-mannuronic acid and alpha-L-glucuronic acid, can be cross-linked in a gel by complexing with calcium ions. The solubility of alginate gels depends upon their calcium level and is independent of temperature. Fabrication of alginate can produce scaffolds that are characterized by high porosity and pore interconnectivity. The porosity and pore shape can be controlled through different methods of scaffold formation, contributing to nanotopographic and morphogenic effects on embedded cells [237]. Thus, it is possible to reliably generate isotropic spherical pores. Because of its hydrophilic nature, alginate scaffolds allow efficient cell loading, and they maintain viability during culture within the scaffold.

Encapsulating proteins within alginate is advantageous, because the process occurs under very mild conditions [238]. However, initial work with alginate gels incorporating VEGF faced the problem of burst release within 4 days [239]. Because VEGF is a potent signal transduction molecule that must be released in a controlled manner, quick release is not ideal. A recent study discovered that, by manipulating the amount of calcium chloride, a sustained release of VEGF could be achieved [238]. An alternative solution to this problem involved encapsulating the growth factor within PLGA microspheres capable of controlling release and embedding these within an alginate gel [240]. Implantation into a mouse model revealed significant host tissue ingrowth into the alginate scaffold, with a high density of capillaries within and around the scaffold. A similar study, using a similar VEGF-releasing alginate scaffold consisting of biodegradable VEGF microspheres to provide for a sustained VEGF release, was implanted on liver lobes. Within 3 days, a thin layer of capillaries surrounded the scaffold, and within 2 weeks, nearly half the scaffold had host tissue and blood capillary ingrowth. The VEGF-alginate scaffold also accelerated and promoted the formation of more-mature blood vessels, indicated by larger capillary sizes (greater than 16 μm) and the presence of smooth muscle actin-positive cells that help to stabilize new capillaries [241].

Alginate scaffolds have also been used to grow cells with a homogeneous distribution within the scaffold [242]. After forming the hydrogel, cells were seeded upon the scaffold. Because of the hydrophilic nature of alginate, media and cells were pulled into the scaffold via capillary action, and moderate centrifugation distributed cells throughout the scaffold. A benefit of having a dense cell culture is the maximization of cell contact interactions. Using the same method, hESCs were homogeneously distributed within a 3D porous alginate scaffold [243]. Formation of EBs was consistently and evenly distributed across the scaffold. After 1 month of culture, EBs proliferated and differentiated to different tissue types of the three germ layers. Specifically, enhanced vasculature formation was observed in the alginate-borne EBs compared to suspension-formed EBs. This was achieved via the 3D culture and differentiation of hESCs within the scaffolds without the use of any chemical modifications of the alginate scaffold.

5.6.7 Dextran

Dextran is a bacteria-derived polysaccharide that has been proposed as a potential biomaterial for tissue engineering for its biocompatibility and hydrogel properties. Because dextran is naturally resistant to cell adhesion and protein adsorption, modification of its polymer backbone allows the development of a hydrogel with specific characteristics. Cytodex, one commercial product based on dextran, is often used to culture cells within hydrogels. Dextran is degraded by the enzyme dextranase, which has been discovered to exist in the human colon. Additionally, the controlled degradation rate of dextran offers potential for controlled protein release and drug delivery [244, 245].

Similar to PEG, dextran is naturally resistant to cell adhesion and thus has often been used as a biomaterial surface coating to limit cell adhesion. Although natural dextran has no protein-binding sites, the dextran polymer chain has multiple reactive bioactive binding sites which can be modified with functional groups, which contrasts with PEG, which has only one reactive site. Dextran modified with the RGD peptide was shown to promote EC adhesion and spreading compared to unmodified dextran [246]. While this study only focused on surface coatings, it showed dextran's potential for incorporating a variety of biomolecules to provide a range of surface signals for cells.

Dextran hydrogels can be formed by either physical or chemical cross-linking. Physical cross-linking involves the coupling of lactic acid to dextran, where, once dissolved in water, it will form physical cross-links resulting in a hydrogel [247]. Chemical cross-linking can be carried out in a variety of ways by covalently attaching polymerizable groups. Early research on dextran focused on drug delivery, as pore sizes were too small for cell migration, which requires pore sizes of between 50 and 300 μm [248]. An acrylated dextran hydrogel was developed and proved able to support cell encapsulation and growth [249]. Levasque and colleagues further developed dextran hydrogels from methacrylated dextran and PEG, which resulted in the formation of macropores ranging from 10 to 120 μm [248], as well as a dextran hydrogel modified with p-maleimidophenyl isocyanate and a peptide cross-linker that is susceptible to matrix metalloprotease 2 degradation [250]. These modifications showed promise in cell culture and adhesion [250, 251].

A recent study developed a dextran-based hydrogel incorporating either RGD molecules or soluble VEGF embedded within PLGA microparticles to preferentially differentiate hESCs towards a vascular lineage. Human ESC colonies were encapsulated within the hydrogel and allowed to differentiate. After 10 days in culture, upregulation of vascular markers and well-organized vasculature networks were observed in hESCs encapsulated in the dextran hydrogels. Furthermore, when these cells were released from the dextran hydrogels and cultured on a Petri dish in EC medium, the number of vascular cells increased, suggesting preferable proliferation along a vascular lineage [154].

Having a successful method for generating larger pore sizes and the ability to direct vascular differentiation within the hydrogel make dextran a potentially suitable material for vascular stem cell engineering. Future work will need to determine the ability of vascular cells to form organized networks and whether implantation of dextran hydrogels can result in a functional vascular network.

5.7 Coculture

Although cells cultured in flasks are typically grown as a pure cell monolayer, this is not the case in natural development. In tissues or in a developing embryo, cells are constantly exposed to a wide variety of biochemical signals, and they engage in complex interactions with other cell types that provide signals for stability and survival. In designing scaffolds, functional groups need to be available for cells to adhere and proliferate. While numerous studies have shown this to be effective, and sometimes required, for cell adhesion, growth, and proliferation, the amount of GF decreases over time and does not provide the large number of signals that neighboring cells provide. Thus, contact with other cell types through coculture would be necessary to engineer complex tissues.

One cell type involved with vascular development in vivo is the pericyte. When developing blood vessels establish a primitive vascular network through angiogenesis, becoming a functional network requires the recruitment of periendothelial cells, such as pericytes, that contribute to remodeling and maturation. The EC–pericyte relationship is complex and involves regulation of mitotic rates and responses to hypoxia. Pericytes serve multiple functions for ECs, such as releasing soluble factors that inhibit EC proliferation, acting as a SMC substitute and promoting microvessel constriction, and producing the mitogen VEGF [252]. Surrounding capillaries are pericytes that help to stabilize vessels by secreting ECM components and providing mechanical strength [253]. Lack of supporting signals for capillaries will eventually lead to regression [254]. The evidence supporting the importance of pericytes indicates the need for coculture within scaffolds to better mimic the proper microenvironment for vascular development.

An early study showed that porcine microvascular endothelial cells support the sustained expansion of primitive hematopoietic progenitor cells when in direct contact [255]. Being able to discern the interactions that are occurring between different cell types is a complex and daunting challenge, but many recent studies have begun to investigate the coculture of cells within scaffolds. Recently, a study investigated the coculture of bovine aortic ECs and bovine fibroblasts encapsulated within Matrigel. Results indicated that the presence of nonendothelial cells improved the long-term survival of ECs [256]. Similarly, utilizing the chorioallantoic membrane model to coculture preadipocytes with human dermal microvascular ECs within a fibrin hydrogel resulted in the formation of a capillary network that anastomosed with the host vasculature without needing any exogenous angiogenic GFs [143].

A study investigated the coculture of neural progenitor cells and brain ECs within a PEG-based hydrogel implanted in a mouse model. The group observed that neural progenitor cells helped to augment formation and stabilization of microvascular networks, as indicated by reduced regression of capillary-like structures. A functional vascular network was established within 2 weeks, and it had a significantly higher density of capillary-like structures relative to the control [257].

Several recent studies have concentrated on developing long-lasting vascular networks from hECs, hEPCs, and hESCs through coculture of different cell types. An early study cocultured HUVECs with mesenchymal precursor cells encapsulated within fibronectin-collagen hydrogels that were subsequently implanted in mice. Observations indicated increased capillary density and the formation of a stabilized vascular network that anastomosed to the host within 2 weeks. The vascular network was monitored and remained functional for up to 1 year [258]. A continuous study compared formation and functions of tissue-engineered blood vessels generated by peripheral-blood- and umbilical-cord-blood-derived EPCs in a model of in vivo vasculogenesis. The study found that adult peripheral blood EPCs formed blood vessels that were unstable and regressed within 3 weeks. In contrast, umbilical cord blood EPCs formed normal functioning blood vessels that lasted for more than 4 weeks [70].

Another group engineered vascularized muscle by coculturing the mouse myoblast with human embryonic ECs and embryonic fibroblasts seeded within a PLGA/PLLA scaffold. Embryonic fibroblasts are known to promote vascularization through secretion of VEGF, and this was confirmed by an increase in the number of endothelial structures. Within 2 weeks, a functional microvascular network formed and anastomosed to the host [259]. Additionally, coimplantation of hESC-derived ECs with mouse mesenchymal precursor cells into a mouse model resulted in the formation of a functional microvascular network that anastomosed with the host within 2 weeks [137]. More recently, a tri-culture system was developed using hESC-cardiomyocytes, human embryonic ECs, and embryonic fibroblasts within PLGA/PLLA scaffolds. Embryonic fibroblasts were found to be essential because they augmented vascularization, promoted vessel organization, promoted EC proliferation, and stabilized microvessels by differentiating into SMCs [260].

These results indicate the necessity of developing effective coculture systems capable of quickly vascularizing and integrating with the host vasculature. Engineered skeletal muscle and cardiac muscle have been successfully created by prevascularizing the tissue. Thus, it is likely that in order to engineer viable tissues for clinical applications, future research will need to focus on developing an effective biodegradable scaffold capable of supporting the appropriate engineered tissue along with an integrated vasculature.

6 Conclusion

In recent years, tremendous advances have been made in the field of stem cells, specifically in vascular differentiation of hESCs, and in the sources and maturation of adult hEPCs. Development of biomaterials capable of presenting specific environmental cues of choice offers a unique opportunity to generate instructive 3D environments for vascular assembly. Therefore, vascular stem cell engineering is in an advantageous position to develop new methods for designing and controlling in vitro vascular stem cell microenvironments, which will enable fundamental insight into the individual and interactive effects of factors that guide differentiation and cellular organization. Furthermore, prospective research may result in a construct engineered in vitro that has significant therapeutic implications.

References

1. Risau W (1997) Nature 386: 671
2. Yancopoulos GD, Davis S , Gale NW, Rudge JS, Wiegand SJ, Holash J (2000) Nature 407: 242
3. Ingo FlammeTFWR(1997) J Cell Physiol 173: 206
4. Gerecht-Nir S, Dazard JE, Golan-Mashiach M, Osenberg S, Botvinnik A, Amariglio N, Domany E, Rechavi G, Givol D, Itskovitz-Eldor J (2005) Dev Dyn 232: 487
5. HillJ M, Zalos G, Halcox JP, Schenke WH, Waclawiw MA, Quyyumi AA, Finkel T (2003) N Engl J Med 348: 593
6. Carmeliet P(2000) Nat Med 6: 389
7. Tabibiazar R, Rockson SG (2001) Eur Heart J 22: 903
8. Wu X, Rabkin-Aikawa E, Guleserian KJ, Perry TE, Masuda Y, Sutherland FW, Schoen FJ, Mayer JE, Jr., Bischoff J (2004) Am J Physiol Heart Circ Physiol 287: H480
9. Risau W, Flamme I(1995) Annu Rev Cell Dev Biol 11: 73
10. Ema M, Rossant J (2003) Trends Cardiovasc Med 13: 254
11. Barker JE (1968) Dev Biol 18: 14
12. Wagner RC (1980) Adv Microcirc 9: 45
13. Sabin F (1920) Contribut Embryol 9: 213
14. Murray PDF (1932) Proc R Soc Lond B Biol Sci 111: 497
15. Pardanaud L, Yassine F, Dieterlen-Lievre F (1989) Development 105: 473
16. Park C, Ma YD, Choi K (2005) Exp Hematol 33: 965
17. Doetschman TC, Eistetter H, Katz M, Schmidt W, Kemler R (1985) J Embryol Exp Morphol 87: 27

18. Keller G, Kennedy M, Papayannopoulou T, Wiles MV (1993) Mol Cell Biol 13: 473
19. Keller GM (1995) Curr Opin Cell Biol 7: 862
20. Keller G, Lacaud G, Robertson S (1999) Exp Hematol 27: 777
21. Kennedy M, Firpo M, Choi K, Wall C, Robertson S, Kabrun N, Keller G (1997) Nature 386: 488
22. Choi K, Kennedy M, Kazarov A, Papadimitriou JC, Keller G (1998) Development 125: 725
23. Ko HC, Milthorpe BK, McFarland CD (2007) Eur Cell Mater 14: 1
24. Vailhe B, Vittet D, Feige JJ (2001) Lab Invest 81: 439
25. Benjamin LE, Hemo I, Keshet E (1998) Development 125: 1591
26. Risau W (1998) Circ Res 82: 926
27. Gunsilius E, Duba HC, Petzer AL, Kahler CM, Grunewald K, Stockhammer G, Gabl C, Dirnhofer S, Clausen J, Gastl G (2000) Lancet 355: 1688
28. Asahara T, Murohara T, Sullivan A, Silver M, van der Zee R, Li T, Witzenbichler B, Schatteman G, Isner JM (1997) Science 275: 964
29. Asahara T, Kawamoto A (2004) Am J Physiol Cell Physiol 287: C572
30. Libby P, Ridker PM, Maseri A (2002) Circulation 105: 1135
31. Hirsch EZ, Chisolm GM, 3rd, White HM (1983) Atherosclerosis 46: 287
32. Fontaine V, Filipe C, Werner N, Gourdy P, Billon A, Garmy-Susini B, Brouchet L, Bayard F, Prats H, Doetschman T, Nickenig G, Arnal JF (2006) Am J Pathol 169: 1855
33. Reidy MA, Bowyer DE (1978) Atherosclerosis 29: 459
34. Fadini GP, Agostini C, Sartore S, Avogaro A (2007) Atherosclerosis 194: 46
35. Heiss C, Keymel S, Niesler U, Ziemann J, Kelm M, Kalka C (2005) J Am Coll Cardiol 45: 1441
36. Fadini GP, Avogaro A, Agostini C (2006) Am J Respir Cell Mol Biol 35: 403
37. Loomans CJ, de Koning EJ, Staal FJ, Rookmaaker MB, Verseyden C, de Boer HC, Verhaar MC, BraamB, Rabelink TJ, van Zonneveld AJ (2004) Diabetes 53: 195
38. Fadini GP, Coracina A, Baesso I, Agostini C, Tiengo A, Avogaro A, de Kreutzenberg SV (2006) Stroke 37: 2277
39. Palange P, Testa U, Huertas A, Calabro L, Antonucci R, Petrucci E, Pelosi E, Pasquini L, Satta A, Morici G, Vignola MA, Bonsignore MR (2006) Eur Respir J 27: 529
40. Chen JZ, Zhang FR, Tao QM, Wang XX, Zhu JH (2004) Clin Sci (Lond) 107: 273
41. Laufs U, Werner N, Link A, Endres M, Wassmann S, Jurgens K, Miche E, Bohm M, Nickenig G (2004) Circulation 109: 220
42. Steiner S, Niessner A, Ziegler S, Richter B, Seidinger D, Pleiner J, Penka M, Wolzt M, Huber K, Wojta J, Minar E, Kopp CW (2005) Atherosclerosis 181: 305
43. Jaffe EA, Nachman RL, Becker CG, Minick CR (1973) J Clin Invest 52: 2745
44. Yamada T, Fan J, Shimokama T, Tokunaga O, Watanabe T (1992) Am J Pathol 141: 1435
45. Libby P (2000) Clin Cardiol 23 (Suppl 6): VI
46. Bevilacqua MP, Gimbrone MA, Jr. (1987) Semin Thromb Hemost 13: 425
47. Burns MP, DePaola N (2005) Am J Physiol Heart Circ Physiol 288: H194
48. Davies PF (1997) J Vasc Res 34: 208
49. Nagata D, Mogi M, Walsh K (2003) J Biol Chem 278: 31000
50. Folkman J, Haudenschild CC, ZetterBR (1979) Proc Natl Acad Sci U S A 76: 5217
51. Fujimoto T, Singer SJ (1988) J Histochem Cytochem 36: 1309
52. Gerritsen ME (1987) Biochem Pharmacol 36: 2701
53. Brown R, Adkins D, Goodnough M, Wehde M, Hendricks D, Ehlenbeck C, Laub LDiPersio J (1995) Blood 86: 293a
54. Shi Q, Wu MH, Hayashida N, Wechezak AR, Clowes AW, Sauvage LR (1994) J Vasc Surg 20: 546
55. Wu MH, Shi Q, Wechezak AR, Clowes AW, Gordon IL, Sauvage LR (1995) J Vasc Surg 21: 862
56. Stump MM, Jordan GL, Jr, Debakey ME, Halpert B (1963) Am J Pathol 43: 361
57. Gonzalez IE, Vermeulen F, Ehrenfeld WK (1969) Isr J Med Sci 5: 648
58. Frazier OH, Baldwin RT, Eskin SG, Duncan JM (1993) Tex Heart Inst J 20: 78
59. Scott SM, Barth MG, Gaddy LR, Ahl ET, Jr (1994) J Vasc Surg 19: 585
60. Hristov M, Erl W, Weber PC (2003) Trends Cardiovasc Med 13: 201

61. Flamme I, Risau W (1992) Development 116: 435
62. Shi Q, Rafii S, Wu MH, Wijelath ES, Yu C, Ishida A, Fujita Y, Kothari S, Mohle R, Sauvage LR, Moore MA, Storb RF, Hammond WP (1998) Blood 92: 362
63. Asahara T, Masuda H, Takahashi T, Kalka C, Pastore C, Silver M, Kearne M, Magner M, Isner JM (1999) Circ Res 85: 221
64. Fong GH, Rossant J, Gertsenstein M, Breitman ML (1995) Nature 376: 66
65. Li J, Brown LF, Hibberd MG, Grossman JD, Morgan JP, Simons M (1996) Am J Physiol 270: H1803
66. Banai S, Shweiki D, Pinson A, Chandra M, Lazarovici G, Keshet E (1994) Cardiovasc Res 28: 1176
67. Schnurch H, Risau W (1993) Development 119: 957
68. Peichev M, Naiyer AJ, Pereira D, Zhu Z, Lane WJ, Williams M, Oz MC, Hicklin DJ, Witte L, Moore MA, Rafii S (2000) Blood 95: 952
69. Igreja C, Fragoso R, Caiado F, Clode N, Henriques A, Camargo L, Reis EM, Dias S (2008) Exp Hematol 36: 193
70. Au P, Daheron LM, Duda DG, Cohen KS, Tyrrell JA, Lanning RM, Fukumura D, Scadden DT, Jain RK (2007) Blood
71. Cherqui S, Kurian SM, Schussler O, Hewel JA, Yates JR, 3rd, Salomon DR (2006) Stem Cells 24: 44
72. Grenier G, Scime A, Le GrandF, AsakuraA, Perez-IratxetaC, Andrade-NavarroMA, LaboskyPA, Rudnicki MA(2007) Stem Cells 25: 3101
73. Tavian M, Zheng B, Oberlin E, Crisan M, Sun B, Huard J, Peault B (2005) Ann N Y Acad Sci 1044: 41
74. Zengin E, Chalajour F, Gehling UM, Ito WD, Treede H, Lauke H, Weil J, Reichenspurner H, Kilic N, Ergun S (2006) Development 133: 1543
75. Kawamoto A, Asahara T (2007) Catheter Cardiovasc Interv 70: 477
76. Aoki M, Yasutake M, Murohara T (2004) Stem Cells 22: 994
77. Gulati R, Jevremovic D, Peterson TE, Chatterjee S, Shah V, Vile RG, Simari RD (2003) Circ Res 93: 1023
78. Pujol BF, Lucibello FC, Gehling UM, Lindemann K, Weidner N, Zuzarte ML, Adamkiewicz J, Elsasser HP, Muller R, Havemann K (2000) Differentiation 65: 287
79. Schmeisser A, Garlichs CD, Zhang H, Eskafi S, Graffy C, Ludwig J, Strasser RH, Daniel WG (2001) Cardiovasc Res 49: 671
80. Rehman J, Li J, Orschell CM, March KL (2003) Circulation 107: 1164
81. Rohde E, Malischnik C, Thaler D, Maierhofer T, Linkesch W, Lanzer G, Guelly C, Strunk D (2006) Stem Cells 24: 357
82. Isner JM, Asahara T (1999) J Clin Invest 103: 1231
83. Kalka C, Masuda H, Takahashi T, Kalka-Moll WM, Silver M, Kearney M, Li T, Isner JM, Asahara T (2000) Proc Natl Acad Sci U S A 97: 3422
84. Tateishi-Yuyama E, Matsubara H, Murohara T, Ikeda U, Shintani S, Masaki H, Amano K, Kishimoto Y, Yoshimoto K, Akashi H, Shimada K, Iwasaka T, Imaizumi T (2002) Lancet 360: 427
85. Kocher AA, Schuster MD, Szabolcs MJ, Takuma S, Burkhoff D, Wang J, Homma S, Edwards NM, Itescu S (2001) Nat Med 7: 430
86. Crosby JR, Kaminski WE, Schatteman G, Martin PJ, Raines EW, Seifert RA, Bowen-Pope DF (2000) Circ Res 87: 728
87. Hess DC, Hill WD, Martin-Studdard A, Carroll J, Brailer J, Carothers J (2002) Stroke 33: 1362
88. Urbich C, Dimmeler S (2004) Circ Res 95: 343
89. Holash J, Maisonpierre PC, Compton D, Boland P, Alexander CR, Zagzag D, Yancopoulos GD, Wiegand SJ (1999) Science 284: 1994
90. Lyden D, Hattori K, Dias S, Costa C, Blaikie P, Butros L, Chadburn A, Heissig B, Marks W, Witte L, Wu Y, Hicklin D, Zhu Z, Hackett NR, Crystal RG, Moore MA, Hajjar KA, Manova K, Benezra R, Rafii S (2001) Nat Med 7: 1194

91. Walter DH, Rittig K, Bahlmann FH, Kirchmair R, Silver M, Murayama T, Nishimura H, Losordo DW, Asahara T, Isner JM (2002) Circulation 105: 3017
92. Werner N, Junk S, Laufs U, Link A, Walenta K, Bohm M, Nickenig G (2003) Circ Res 93: e17
93. Fujiyama S, Amano K, Uehira K, Yoshida M, Nishiwaki Y, Nozawa Y, Jin D, Takai S, Miyazaki M, Egashira K, Imada T, Iwasaka T, Matsubara H (2003) Circ Res 93: 980
94. Tomita S, LiR K, Weisel RD, Mickle DA, Kim EJ, Sakai T, Jia ZQ (1999) Circulation 100: II247
95. Iba O, Matsubara H, Nozawa Y, Fujiyama S, Amano K, Mori Y, Kojima H, Iwasaka T (2002) Circulation 106: 2019
96. He T, Peterson TE, Katusic ZS (2005) Am J Physiol Heart Circ Physiol 289: H968
97. Hur J, Yoon CH, Kim HS, Choi JH, Kang HJ, Hwang KK, Oh BH, Lee MM, Park YB (2004) Arterioscler Thromb Vasc Biol 24: 288
98. Urbich C, Aicher A, Heeschen C, Dernbach E, Hofmann WK, Zeiher AM, Dimmeler S (2005) J Mol Cell Cardiol 39: 733
99. Cho HJ, Lee N, Lee JY, Choi YJ, Ii M, Wecker A, Jeong JO, Curry C, Qin G, Yoon YS (2007) J Exp Med 204: 3257
100. He T, Smith LA, Harrington S, Nath KA, Caplice NM, Katusic ZS (2004) Stroke 35: 2378
101. Badorff C, Dimeler S (2006) Handb Exp Pharmacol 283
102. Ingram DA, Mead LE, Tanaka H, Meade V, Fenoglio A, Mortell K, Pollok K, Ferkowicz MJ, Gilley D, Yoder MC (2004) Blood 104: 2752
103. Yoder MC, Mead LE, Prater D, Krier TR, Mroueh KN, Li F, Krasich R, Temm CJ, Prchal JT, Ingram DA (2007) Blood 109: 1801
104. Prater DN, Case J, Ingram DA, Yoder MC (2007) Leukemia 21: 1141
105. Dimmeler S, Vasa-Nicotera M (2003) J Am Coll Cardiol 42: 2081
106. Edelberg JM, Tang L, Hattori K, Lyden D, Rafii S (2002) Circ Res 90: E89
107. Vasa M, Fichtlscherer S, Aicher A, Adler K, Urbich C, Martin H, Zeiher AM, Dimmeler S (2001) Circ Res 89: E1
108. Capla JM, Grogan RH, Callaghan MJ, Galiano RD, Tepper OM, Ceradini DJ, Gurtner GC (2007) Plast Reconstr Surg 119: 59
109. Evans MJ, Kaufman MH (1981) Nature 292: 154
110. Thomson JA, Itskovitz-Eldor J, Shapiro SS, Waknitz MA, Swiergiel JJ, Marshall VS, Jones JM (1998) Science 282: 1145
111. Stevens LC (1960) Dev Biol 2: 285
112. Pierce GB, DixonF J, Jr (1959) Cancer 12: 573
113. Thomson JA, Kalishman J, Golos TG, Durning M, Harris CP, Becker RA, Hearn JP (1995) Proc Natl Acad Sci U S A 92: 7844
114. Amit M, Itskovitz-Eldor J (2002) J Anat 200: 225
115. Skottman H, Hovatta O (2006) Reproduction 132: 691
116. Suss-Toby E, Gerecht-Nir S, Amit M, Manor D, Itskovitz-Eldor J (2004) Hum Reprod 19: 670
117. Cibelli J (2007) Science 318: 1879
118. Oliveri RS (2007) Regen Med 2: 795
119. Yu J, Vodyanik MA, Smuga-Otto K, Antosiewicz-Bourget J, Frane JL, Tian S, Nie J, Jonsdottir GA, Ruotti V, Stewart R, Slukvin II, Thomson JA (2007) Science 318: 1917
120. Takahashi K, Yamanaka S (2006) Cell 126: 663
121. Martin MJ, Muotri A, Gage F, Varki A (2005) Nat Med 11: 228
122. Amit M, Margulets V, Segev H, Shariki K, Laevsky I, Coleman R, Itskovitz-Eldor J (2003) Biol Reprod 68: 2150
123. Lee JB, Song JM, Lee JE, Park JH, Kim SJ, Kang SM, Kwon JN, Kim MK, Roh SI, Yoon HS (2004) Reproduction 128: 727
124. Lee JB, Lee JE, Park JH, Kim SJ, Kim MK, Roh SI, Yoon HS (2005) Biol Reprod 72: 42
125. Ludwig TE, Levenstein ME, Jones JM, Berggren WT, Mitchen ER, Frane JL, Crandall LJ, Daigh CA, Conard KR, Piekarczyk MS, Llanas RA, Thomson JA (2006) Nat Biotechnol 24: 185

126. Gerecht S, Burdick JA, Ferreira LS, Townsend SA, Langer R, Vunjak-Novakovic G (2007) Proc Natl Acad Sci U S A 104: 11298
127. Choudhary M, Zhang X, Stojkovic P, Hyslop L, Anyfantis G, Herbert M, Murdoch AP, Stojkovic M, Lako M (2007) Stem Cells 25: 3045
128. Itskovitz-Eldor J, Schuldiner M, Karsenti D, Eden A, Yanuka O, Amit M, Soreq H, Benvenisty N (2000) Mol Med 6: 88
129. Wartenberg M, Gunther J, Hescheler J, Sauer H (1998) Lab Invest 78: 1301
130. Hopfl G, Gassmann M, Desbaillets I (2004) Methods Mol Biol 254: 79
131. Reubinoff BE, Pera MF, Fong CY, Trounson A, Bongso A (2000) Nat Biotechnol 18: 399
132. Gerecht-Nir S, Ziskind A, Cohen S, Itskovitz-Eldor J (2003) Lab Invest 83: 1811
133. Yamashita J, Itoh H, Hirashima M, Ogawa M, Nishikawa S, Yurugi T, Naito M, Nakao K (2000) Nature 408: 92
134. Levenberg S, Golub JS, Amit M, Itskovitz-Eldor J, Langer R (2002) Proc Natl Acad Sci U S A 99: 4391
135. Wang L, Li L, Shojaei F, Levac K, Cerdan C, Menendez P, Martin T, Rouleau A, Bhatia M (2004) Immunity 21: 31
136. Ferreira LS, Gerecht S, Shieh HF, Watson N, Rupnick MA, Dallabrida SM, Vunjak-Novakovic G, Langer R (2007) Circ Res 101: 286
137. Wang ZZ, Au P, Chen T, Shao Y, Daheron LM, Bai H, Arzigian M, Fukumura D, Jain RK, Scadden DT (2007) Nat Biotechnol 25: 317
138. Vodyanik MA, Thomson JA, Slukvin II (2006) Blood 108: 2095
139. Era T, Izumi N, Hayashi M, Tada S, Nishikawa S, Nishikawa SI (2007) Stem Cells
140. Awwad HK, el Naggar M, Mocktar N, Barsoum M (1986) Int J Radiat Oncol Biol Phys 12: 1329
141. Griffith CK, Miller C, Sainson RC, Calvert JW, Jeon NL, Hughes CC, George SC (2005) Tissue Eng 11: 257
142. Bartynski J, Marion MS, Wang TD (1990) Otolaryngol Head Neck Surg 102: 314
143. Borges J, Mueller MC, Padron NT, Tegtmeier F, Lang EM, Stark GB (2003) Tissue Eng 9: 1263
144. O'Hare MJ, Bond J, Clarke C, Takeuchi Y, Atherton AJ, Berry C, Moody J, Silver AR, Davies DC, Alsop AE, Neville AM, Jat PS (2001) Proc Natl Acad Sci U S A 98: 646
145. Eggermann J, Kliche S, Jarmy G, Hoffmann K, Mayr-Beyrle U, Debatin KM, Waltenberger J, Beltinger C (2003) Cardiovasc Res 58: 478
146. Nagano M, Yamashita T, Hamada H, Ohneda K, Kimura K, Nakagawa T, Shibuya M, Yoshikawa H, Ohneda O (2007) Blood 110: 151
147. Finney MR, Greco NJ, Haynesworth SE, Martin JM, Hedrick DP, Swan JZ, Winter DG, Kadereit S, Joseph ME, Fu P, Pompili VJ, Laughlin MJ (2006) Biol Blood Marrow Transplant 12: 585
148. Duan HX, Cheng LM, Wang J, Hu LS, Lu GX (2006) Cell Biol Int 30: 1018
149. Melero-Martin JM, Khan ZA, Picard A, Wu X, Paruchuri S, Bischoff J (2007) Blood 109: 4761
150. Droetto S, Viale A, Primo L, Jordaney N, Bruno S, Pagano M, Piacibello W, Bussolino F, Aglietta M (2004) FASEB J 18: 1273
151. Sheridan MH, Shea LD, Peters MC, Mooney DJ (2000) J Control Release 64: 91
152. Lokmic Z, Stillaert F, Morrison WA, Thompson EW, Mitchell GM (2007) FASEB J 21: 511
153. Polykandriotis E, Arkudas A, Beier JP, Hess A, Greil P, Papadopoulos T, Kopp J, Bach AD, Horch RE, Kneser U (2007) Plast Reconstr Surg 120: 855
154. Ferreira LS, Gerecht S, Fuller J, Shieh HF, Vunjak-Novakovic G, Langer R (2007) Biomaterials 28: 2706
155. Seal BL, Otero TC, Panitch A (2001) Mater Sci Eng R Rep 34: 147
156. Hutmacher DW (2001) J Biomater Sci Polym Ed 12: 107
157. Chen GP, Ushida T, Tateishi T (2001) Mat Sci Eng C Biomimetic Supramol Syst 17: 63
158. Ifkovits JL, Burdick JA (2007) Tissue Eng 13: 2369
159. Gunatillake PA, Martin DJ, Meijs GF, McCarthy SJ, Adhikari R (2003) Aust J Chem 56: 545

160. Miller RA, Brady JM, Cutright DE (1977) J Biomed Mater Res 11: 711
161. Mooney DJ, Mazzoni CL, Breuer C, McNamara K, Hern D, Vacanti JP, Langer R (1996) Biomaterials 17: 115
162. Zund G, Hoerstrup SP, Schoeberlein A, Lachat M, Uhlschmid G, Vogt PR, Turina M (1998) Eur J Cardiothorac Surg 13: 160
163. Shen G, Tsung HC, Wu CF, Liu XY, Wang XY, Liu W, Cui L, Cao YL (2003) Cell Res 13: 335
164. Kanczler JM, Barry J, Ginty P, Howdle SM, Shakesheff KM, Oreffo RO (2007) Biochem Biophys Res Commun 352: 135
165. Richardson TP, Peters MC, Ennett AB, Mooney DJ (2001) Nat Biotechnol 19: 1029
166. Zhu H, Ji J, Lin R, Gao C, Feng L, Shen J (2002) Biomaterials 23: 3141
167. Suh H, Hwang YS, Lee JE, Han CD, Park JC (2001) Biomaterials 22: 219
168. Black FE, Hartshorne M, Davies MC, Roberts CJ, Tendler SJB, Williams PM, Shakesheff KM (1999) Langmuir 15: 3157
169. Tetsuji Y, Yoshiyuki T, Yoshiharu K (1998) Jpn J Polym Sci Technol 55: 328
170. Zhu Y, Gao C, Liu X, He T, Shen J (2004) Tissue Eng 10: 53
171. Linn T, Erb D, Schneider D, Kidszun A, Elcin AE, Bretzel RG, Elcin YM (2003) Cell Transplant 12: 769
172. Miller DC, Thapa A, Haberstroh KM, Webster TJ (2004) Biomaterials 25: 53
173. Lees JG, Lim SA, Croll T, Williams G, Lui S, Cooper-White J, McQuade L R, Mathiyalagan B, Tuch BE (2007) Regen Med 2: 289
174. Levenberg S, Huang NF, Lavik E, Rogers AB, Itskovitz-Eldor J, Langer R (2003) Proc Natl Acad Sci U S A 100: 12741
175. Kannan RY, Salacinski HJ, Sales K, Butler P, Seifalian AM (2005) Biomaterials 26: 1857
176. Burdick JA, Chung C, Jia X, Randolph MA, Langer R (2005) Biomacromolecules 6: 386
177. Wacker BK, Alford SK, Scott EA, Das Thakur M, Longmore GD, Elbert DL (2008) Biophys J 94: 273
178. Xin A X, Gaydos C, Mao JJ (2006) Conf Proc IEEE Eng Med Biol Soc 1: 2091
179. Lutolf MP, Weber FE, Schmoekel HG, Schense JC, Kohler T, Muller R, Hubbell JA (2003) Nat Biotechnol 21: 513
180. Zisch AH, Lutolf MP, Hubbell JA (2003) Cardiovasc Pathol 12: 295
181. Zisch AH, Lutolf MP, Ehrbar M, Raeber GP, Rizzi SC, Davies N, Schmokel H, Bezuidenhout D, Djonov V, Zilla P, Hubbell JA (2003) FASEB J 17: 2260
182. Burdick JA, Anseth KS (2002) Biomaterials 23: 4315
183. Schmidt O, Mizrahi J, Elisseeff J, Seliktar D (2006) Biotechnol Bioeng 95: 1061
184. Nicodemus GD, Villanueva I, Bryant SJ (2007) J Biomed Mater Res A 83: 323
185. Almany L, Seliktar D (2005) Biomaterials 26: 2467
186. Nuttelman CR, Tripodi MC, Anseth KS (2004) J Biomed Mater Res A 68: 773
187. Gu F, Younes HM, El-Kadi AO, Neufeld RJ, Amsden BG (2005) J Control Release 102: 607
188. Yang J, Webb AR, Ameer GA (2004) Adv Mater 16: 511
189. Pego AP, Poot AA, Grijpma DW, Feijen J (2002) Macromol Biosci 2: 411
190. Nugent HM, Edelman ER (2003) Circ Res 92: 1068
191. Wang Y, Ameer GA, Sheppard BJ, Langer R (2002) Nat Biotechnol 20: 602
192. Wang Y, Kim YM, Langer R (2003) J Biomed Mater Res A 66: 192
193. Nijst CL, Bruggeman JP, Karp JM, Ferreira L, Zumbuehl A, Bettinger CJ, Langer R (2007) Biomacromolecules 8: 3067
194. Gerecht S, Townsend SA, Pressler H, Zhu H, Nijst CL, Bruggeman JP, Nichol JW, Langer R (2007) Biomaterials 28: 4826
195. Kragh M, Hjarnaa PJ, Bramm E, Kristjansen PE, Rygaard J, Binderup L (2003) Int J Oncol 22: 305
196. Stieger SM, Bloch SH, Foreman O, Wisner ER, Ferrara KW, Dayton PA (2006) Ultrasound Med Biol 32: 673
197. Yoon CH, Hur J, Park KW, Kim JH, Lee CS, Oh IY, Kim TY, Cho HJ, Kang HJ, Chae IH, Yang HK, Oh BH, Park YB, Kim HS (2005) Circulation 112: 1618
198. Albig AR, Roy TG, Becenti DJ, Schiemann WP (2007) Angiogenesis 10: 197

199. Crabtree B, Subramanian V (2007) In Vitro Cell Dev Biol Anim 43: 87
200. Prestwich GD, Marecak DM, Marecek JF, Vercruysse KP, Ziebell MR (1998) J Control Release 53: 93
201. Slevin M, Kumar S, Gaffney J (2002) J Biol Chem 277: 41046
202. Slevin M, Krupinski J, Kumar S, Gaffney J (1998) Lab Invest 78: 987
203. Rooney P, Kumar S, Ponting J, Wang M (1995) Int J Cancer 60: 632
204. Kumar S, Ponting J, Rooney P, Kumar P, Pye D, Wang M (1994) Angiogenesis: molecular biology and clinical applications. Plenum Press, New York.
205. West DC, Hampson IN, Arnold F, Kumar S (1985) Science 228: 1324
206. Stern R (2003) Glycobiology 13: 105R
207. Park YD, Tirelli N, Hubbell JA (2003) Biomaterials 24: 893
208. Peattie RA, Nayate AP, Firpo MA, Shelby J, Fisher RJ, Prestwich GD (2004) Biomaterials 25: 2789
209. Khademhosseini A, Eng G, Yeh J, Fukuda J, Blumling J, 3rd, Langer R, Burdick JA (2006) J Biomed Mater Res A 79: 522
210. Olsen BR (2007) In: Lanza R, Langer R, Vacanti JP (eds.) Principles of tissue engineering., Burlington, MA, p. 101
211. Brewster L, Brey EM, Greisler HP (2007) In: Lanza R, Langer R, Vacanti JP (eds.) Principles of tissue engineering. Elsevier, Burlington, MA, p. 569
212. Rabkin E, Schoen FJ (2002) Cardiovasc Pathol 11: 305
213. Li S-T (2003) In: JB P, JD B (eds.) Biomaterials: principles and applications. CRC Press, Boca Raton, FL, p. 117
214. Pieper JS, Hafmans T, van Wachem PB, van Luyn MJ, Brouwer LA, Veerkamp JH, van Kuppevelt TH (2002) J Biomed Mater Res 62: 185
215. Geutjes PJ, Daamen WF, Buma P, Feitz WF, Faraj KA, van Kuppevelt TH (2006) Adv Exp Med Biol 585: 279
216. Nillesen ST, Geutjes PJ, Wismans R, Schalkwijk J, Daamen WF, van Kuppevelt TH (2007) Biomaterials 28: 1123
217. Feraud O, Cao Y, Vittet D (2001) Lab Invest 81: 1669
218. Sun GM, Chu CC (2006) Carbohydrate Polymers 65: 273
219. Berglund JD, Nerem RM, Sambanis A (2004) Tissue Eng 10: 1526
220. Chekanov VS, Tchekanov GV, Rieder MA, Eisenstein R, Wankowski DM, Schmidt DH, Nikolaychik VV, Lelkes PI (1996) ASAIO J 42: M480
221. Sahni A, Francis CW (2000) Blood 96: 3772
222. Sahni A, Altland OD, Francis CW (2003) J Thromb Haemost 1: 1304
223. Zisch AH, Schenk U, Schense JC, Sakiyama-Elbert SE, Hubbell JA (2001) J Control Release 72: 101
224. Sakiyama-Elbert SE, Hubbell JA (2000) J Control Release 69: 149
225. Sakiyama-Elbert SE, Hubbell JA (2000) J Control Release 65: 389
226. Gagnon E, Cattaruzzi P, Griffith M, Muzakare L, LeFlao K, Faure R, Beliveau R, Hussain SN, Koutsilieris M, Doillon CJ (2002) Angiogenesis 5: 21
227. Martineau L, Doillon CJ (2007) Angiogenesis 10: 269
228. Ehrbar M, Metters A, Zammaretti P, Hubbell JA, Zisch AH (2005) J Control Release 101: 93
229. Daamen WF, Veerkamp JH, van Hest JC, van Kuppevelt TH (2007) Biomaterials 28: 4378
230. Leach JB, Wolinsky JB, Stone PJ, Wong JY (2005) Acta Biomater 1: 155
231. Daamen WF, Hafmans T, Veerkamp JH, van Kuppevelt TH (2005) Tissue Eng 11: 1168
232. Wei SM, Katona E, Fachet J, Fulop T, Jr., Robert L, Jacob MP (1998) Int Arch Allergy Immunol 115: 33
233. Duca L, Floquet N, Alix AJ, Haye B, Debelle L (2004) Crit Rev Oncol Hematol 49: 235
234. Wright ER, Conticello VP (2002) Adv Drug Deliv Rev 54: 1057
235. Indik Z, Abrams WR, Kucich U, Gibson CW, Mecham RP, Rosenbloom J (1990) Arch Biochem Biophys 280: 80
236. Mithieux SM, Rasko JE, Weiss AS (2004) Biomaterials 25: 4921
237. Zmora S, GlicklisR, Cohen S (2002) Biomaterials 23: 4087

238. Gu F, Amsden B, Neufeld R (2004) J Control Release 96: 463
239. Elcin YM, Dixit V, Gitnick G (2001) Artif Organs 25: 558
240. Perets A, Baruch Y, Weisbuch F, Shoshany G, Neufeld G, Cohen S (2003) J Biomed Mater Res A 65: 489
241. Kedem A, Perets A, Gamlieli-Bonshtein I, Dvir-Ginzberg M, Mizrahi S, Cohen S (2005) Tissue Eng 11: 715
242. Dar A, Shachar M, Leor J, Cohen S (2002) Biotechnol Bioeng 80: 305
243. Gerecht-Nir S, Cohen S, Ziskind A, Itskovitz-Eldor J (2004) Biotechnol Bioeng 88: 313
244. Hovgaard L, Brondsted H (1995) J Controlled Release 36: 159
245. Hennink WE, Cadee JA, JognS Jd, Franssen O, Stenekes RJH, Talsma H, Kijk-Wolthuis WNEv (2001) In: Chiellini E (ed.) Biomedical polymers and polymer therapeutics. Springer, London
246. Massia SP, Stark J (2001) J Biomed Mater Res 56: 390
247. de Jong SJ, van Eerdenbrugh B, van Nostrum CF, Kettenes-van den Bosch JJ, Hennink WE (2001) J Control Release 71: 261
248. Levesque SG, Lim RM, Shoichet MS (2005) Biomaterials 26: 7436
249. Ferreira L, Gil MH, Dordick JS (2002) Biomaterials 23: 3957
250. Levesque SG, Shoichet MS (2007) Bioconjug Chem 18: 874
251. Levesque SG, Shoichet MS (2006) Biomaterials 27: 5277
252. Sims DE (2000) Clin Exp Pharmacol Physiol 27: 842
253. Thomas WE (1999) Brain Res Brain Res Rev 31: 42
254. Armulik A, Abramsson A, Betsholtz C (2005) Circ Res 97: 512
255. Davis TA, Robinson DH, Lee KP, Kessler SW (1995) Blood 85: 1751
256. Sieminski AL, Padera RF, Blunk T, Gooch KJ (2002) Tissue Eng 8: 1057
257. Ford MC, Bertram JP, Hynes SR, Michaud M, Li Q, Young M, Segal SS, Madri JA, Lavik EB (2006) Proc Natl Acad Sci U S A 103: 2512
258. Koike N, Fukumura D, Gralla O, Au P, Schechner JS, Jain RK (2004) Nature 428: 138
259. Levenberg S, Rouwkema J, Macdonald M, Garfein ES, Kohane DS, Darland DC, Marini R, van Blitterswijk CA, Mulligan RC, D';Amore PA, Langer R (2005) Nat Biotechnol 23: 879
260. Caspi O, Lesman A, Basevitch Y, Gepstein A, Arbel G, Habib IH, Gepstein L, Levenberg S (2007) Circ Res 100: 263

Adv Biochem Engin/Biotechnol (2009) 114: 173-184
DOI: 10.1007/10_2008_20

Published online: 03 June 2009

Embryonic Stem Cells: Isolation, Characterization and Culture

Michal Amit and Joseph Itskovitz-Eldor

Abstract Embryonic stem cells are pluripotent cells isolated from the mammalian blastocyst. Traditionally, these cells have been derived and cultured with mouse embryonic fibroblast (MEF) supportive layers, which allow their continuous growth in an undifferentiated state. However, for any future industrial or clinical application hESCs should be cultured in reproducible, defined, and xeno-free culture system, where exposure to animal pathogens is prevented. From their derivation in 1998 the methods for culturing hESCs were significantly improved. This chapter wills discuss hESC characterization and the basic methods for their derivation and maintenance.

Keywords Blastocyst, Embryoid body, Embryonic stem cell, Immunosurgery, Teratoma

Contents

M. Amit and J. Itskovitz-Eldor (✉)
Department of Obstetrics and Gynecology, Rambam Medical Center, Haifa, and the
Stem Cell Center, Bruce Rappaport Faculty of Medicine, Technion-Israel Institute of Technology, Haifa, Israel
e-mail: Itskovitz@rambam.health.gov.il

1 Introduction

Embryonic stem cells (ESCs) constitute a unique type of stem cells derived from the inner cell mass (ICM) of the mammalian blastocyst. ESCs differ from their adult counterpart by their distinctive potential to differentiate into every cell type of the adult body. Several items of evidence were presented for ESCs pluripotency: (1) when transferred into suspension culture in vitro, ESCs form cell aggregates known as embryoid bodies (EBs), with regions differentiate into embryonically-distinct cell types [17, 30]; (2) injection of ESCs into the hind limb of severe combined immunodeficient (SCID) mice induces the formation of teratomas which may include tissues representative for all three germ layers [70, 66]; (3) mouse ESCs were shown to contribute to chimeras and particularly to the germ cell line [7]; and finally, (4) several murine ESC lines were demonstrated to form entire viable fetuses [46]. Since the first derivation of ESCs in 1981 from mouse blastocysts [21, 40], mouse ESCs were induced to differentiate in vitro into haematopoietic stem cell-like cells [47, 31], neural precursors [8, 9], cardiomyocytes [33], endothelial cells [26, 71] and insulin-secreting cells [55, 38]. Thus the ability of these cells to differentiate into representative cell types of the three embryonic germ layers was proven.

Since their initial derivation from mice [21, 40], ESC lines or ESC-like lines have been derived from other rodents [18, 22, 23], domestic animal species [49, 53, 43], and from three non-human primates [64, 65, 56, 42]. However, only mouse ESCs demonstrate the entire set of features typical of ESCs, rendering them the most potent research model amongst other existing ESC lines.

The first step toward isolating human ESCs (hESCs) was achieved by Bongso and colleagues who described for the first time the ability to isolate ICM cells from human blastocysts and to culture them with inactivated mouse embryonic fibroblasts (MEFs) for two passages while expressing alkaline phosphate activity and demonstrating ESC-like morphology [10]. In 1998, the first hESC lines were derived by Thomson and colleagues [66]. Accumulating knowledge shows that hESCs meet most of the criteria described for mouse ESCs.

The exceptional differentiation potential of ESCs underlines them as one of the best models to study early human development, lineage commitment and differentiation processes; hopefully, in future they could also be used for cell-based therapy. Recently, a new source for pluripotent cells was proposed by Yamanaka et al., who succeeded in reprogramming mouse somatic cells and, later on, human somatic cells, to ESC-like cells [62, 63]. As their report states, an overexpression of four transcription factors, c-Myc, Oct4, Flf4 and Sox2, caused by retroviral infection, was sufficient to reprogram somatic cells [62, 41, 75]. These induced pluripotent stem (iPS) cells expressed typical ESC markers, formed the same colony morphology and were able to differentiate into representative tissues of the three embryonic germ layers both in vitro and in vivo. Later on it was shown that reprogramming of somatic cells could be obtained, albeit with lower efficiency, when oncogene C-Myc was replaced and Oct4, Sox2, Nanog and Lin28 were used [76, 45].

iPS cells were already derived from embryonic fibroblasts [62, 41], hESC-derived fibroblasts [59], fetal fibroblasts [59, 76], foreskin fibroblasts [59, 76], adult skin [25,37] and adult liver and stomach cells [4]. Future studies will reveal which culture and differentiation protocols developed for hESCs will suit these cells as well.

2 Methods for Isolating Escs

2.1 Source for Embryos

For the derivation of hESC lines, human embryos from in vitro fertilization (IVF) programs and embryos produced for research purposes [34] were used. These include surplus, apparently normal, embryos [2, 14, 57, 66], or low-grade or abnormally fertilized oocytes that were disqualified for clinical uses [35, 61, 77]. Some of the embryos are genetically-abnormal embryos after pre-implantation genetic diagnosis (PGD) that would otherwise have been discarded [68, 39]. In these studies, hESC lines harboring specific genetic diseases were derived, demonstrating all hESC characteristics.

Alongside the traditional sources of embryos for the isolation of ESC lines, other optional sources were also suggested; parthenogenetic embryos resulting from activated oocytes, or single blastomers isolated from developing embryos using similar methods to those used for PGD (allowing using the donor embryo for reproductive purposes). Vrana and colleagues demonstrated that an activated oocyte of a non-human primate can be used successfully for the derivation of ESC lines that exhibit all ESC features [67], though the extent of their differentiability is unknown. Mouse ESC lines were successfully derived from a single blastomer [13], using a technique in which a single blastomer is mixed with an already-established cell line, expansion of the newly derived line takes place and isolation is carried out by a selective tag. Both techniques have not yet been applied to human embryos.

Due to the progress in assisted reproductive medicine techniques, more embryos are currently available for hESC line derivation. It is estimated that over 500 hESC lines are obtainable for research worldwide [60]. This number indicates that the derivation of these lines is a reproducible procedure. The use of embryos for research, however, raised ethical concerns that were addressed by the publication of specific guidelines for the use of embryos for hESC studies [15].

2.2 Extraction of ICM

hESC lines are derived using the techniques developed in the 1970s for embryonal carcinoma (EC) cell lines and in the 1980s for mouse ESC line derivation.

Two principle methods can be used to isolate ICM cells from the blastocyst, namely immunosurgical and mechanical isolation.

Immunosurgical isolation is a simple method developed by Solter and Knowles [54], which aims to remove selectively the trophoectoderm layer of the blastocyst, leaving an isolated and intact ICM. A potential drawback of this method is the exposure of the embryo to anti-human whole serum antibodies, which normally attach to any human cell.However, penetration of the antibodies into the blastocyst is prevented due to cell–cell connections within the outer layer of the trophoblast, thus leaving the ICM cells unharmed. This is followed by incubation with guinea pig complement-containing medium which lysises all antibody-marked cells. The intact ICM is further rinsed and cultured with mitotically inactivated MEFs or an alternative feeder-layer that is known to support hESCs culture.

Alternatively, ICM cells can be isolated by selective and mechanical removal of the trophoectoderm layer under a stereoscope. After the embryo is released from the zona pellucida, the trophoblast layer is gently removed using 27 G needles or pulled Pasteur pipettes. Similarly to using the immunosurgery method, the isolated ICM cells should be further expanded using a suitable supportive layer.

2.3 Plating Intact Embryos Whole

ESCs lines can be derived simply by plating a whole zona-free embryo with mitotically inactivated MEFs or another suitable feeder-layer. The exposed embryo attaches to the feeder layer which, in return, permits the continuous growth of the ICM with the surrounding trophoblasts as monolayer. When the ICM reaches sufficient size it is selectively removed using mechanical methods and further propagated. Although simple, this method bears the risk of ICM differentiation, and the success rates tend to be lower as compares to the initial selective removal of the ICM.

2.4 Esc Characterization

Because of their uniqueness, much effort was invested in characterizing ESC cells. The first to be derived, i.e., mouse ESCs, are the most characterized ESCs, and therefore their list of features is used as a golden standard for other types of ESCs. The complete list of features is listed in Table 1.

When cultured in suitable conditions, ESCs are capable of prolonged undifferentiated proliferation. During culture, the cells create uniform colonies exhibiting high nucleus-to-cytoplasm ratio, two or more nucleoli, and typical spaces between the cells.

ESCs exhibit and maintain normal diploid karyotype even after prolonged culture [1]. Incidences of karyotypic instability are uncommon [1, 20, 19, 14], suggesting that those observed represent random changes which often occur in cell culture.

Table 1 List of ESC characteristics

Derived from the ICM of pre-implantation embryo, at the blastocyst stage
Capable of prolonged undifferentiated proliferation in culture
Exhibit and maintain normal diploid karyotype
Pluripotent
Able to integrate into all fetal tissues during embryonic development following injection into the blastocyst, including the germ layer (For obvious ethical reasons, the ability to examine how hESCs integrate into fetal tissues during embryonic development is restricted)
Clonogenic, i.e., each single ESC possesses all other features
Express high levels of *OCT 4* and Nanog, transcription factors known to be involved in the process of ESCs self maintenance
Can be induced to differentiate after continuous culture in an undifferentiated state
Remain in the S phase of the cell cycle for the majority of their lifespan
Do not show X chromosome inactivation

ESCs had been shown to be pluripotent, both in vitro and in vivo by EB formation [17, 30] and teratoma formation [70, 66], respectively.

ESCs express surface markers specific to the undifferentiation stage. While mouse ESCs strongly express surface marker stage-specific embryonic antigen-1 (SSEA-1), and do not express SSEA3, SSEA4, tumor recognition antigen-60 (TRA-1-60) and TRA-1-81, non-human primate ESCs and hESCs strongly express SSEA-4, TRA-1-60, and TRA-1-81, weakly express SSEA-3 and do not express SSEA1 at all [66, 57]. ESCs also express some specific genes, the most recognized is Oct 4, a transcription factor known to be involved in the process of ESC self maintenance [48]. Another transcription factor, Nanog, was recognized as having a role in the cells' renewal and is often used to define undifferentiated ESCs [11, 44]. Additional genes were found to be strongly expressed in hESCs and mESCs and were collected into a set of markers that identify undifferentiated ESCs [6].

Mouse ESCs remain in the S phase of the cell cycle for the majority of their lifespan; HESCs, like mouse ESCs, do not exhibit X inactivation. While maintained at the undifferentiated stage, both X chromosomes are active and, upon differentiation, one chromosome undergoes inactivation [16]. Recently, additional support to this finding was reported; however, it was also found that some hESC lines vary in their X-inactivation status [27, 24, 51, 52]. This may be indicative of a different and later source for some of the lines rather than the ICM, such as the epiblast stage.

As with other cell lines, single human ES cells possess all other features of the tested line, and their clonallity was demonstrated [1].

3 Methods for Hesc Culture

3.1 Defined Culture System

Any future exploitation of hESCs for clinical and industrial purposes will require a reproducible, well-defined, and animal-free culture system for their routine culture.

The traditional culture and isolation methods for hESCs, however, include inactivated MEFs as feeder layers and medium supplemented with high percentage of fetal bovine serum (FBS) [66]. The feeder layer plays a dual role of supporting ESC proliferation and preventing their spontaneous differentiation. In order to prevent any exposure of the cells to animal photogenes, hESCs must be cultured with medium supplemented with serum replacement, with no animal product, and the MEFs should be replaced by human feeder or with a cellular matrix, such as fibronectin, or laminin. A few steps toward meeting these requirements have already been achieved.

The simplest alternative to the culture method based on the use of MEF and FBS is the use of human supportive layer and medium supplemented with either human serum or serum replacement.Several cell types were found suitable to support undifferentiated hESCs, including human fetal-derived fibroblasts [58], foreskin fibroblasts [3, 28], and adult marrow cells [12]. Human fetal-derived fibroblasts and foreskin fibroblasts were also found to support the isolation of new hESC lines in animal-free or serum-free conditions [58, 28, 29].

Although these culture systems move us closer to the desired goal of animal-free conditions, they cannot be regarded as well-defined. The need to culture the feeder lines themselves, which will limit the large-scale culture of hESCs, the differences between batches of feeder-layer cells and the use of human serum rule this system out as defined. The ideal culture method would therefore be a combination of an animal-free matrix and both serum and animal-free medium. In 2001, Xu and colleagues made a significant advance in this respect: their newly culture method relied on Matrigel, laminin or fibronectin as matrix and 100% MEF-conditioned medium, supplemented with serum replacement [72].When cultured in these conditions, hESCs can be stably maintained for over a year and still exhibit their ESC characteristics. However, this method still holds the disadvantages of exposure to animal pathogens through the MEF-conditioned medium or Matrigel matrix, possible variations between batches of MEFs used for the production of the conditioned medium and the needs for simultaneous culture of both the feeders and the hESCs.

Indeed, the same group proposed an improvement to this culture system, where the MEF-conditioned medium was removed by supplementing the medium with 40 ng mL^{-1} basic fibroblast growth factor (bFGF) and 75 ng mL^{-1} Flt-3 ligand [73].

Extensive work has been carried out to improve further the feeder-layer free culture system of hESCs. As a result, several agents were reported to support undifferentiated hESC cultures in feeder layer- free conditions. Amongst them the combination of TGF_{b1} and bFGF [78], activin [5], high concentration of Noggin [74], high concentration of bFGF [69, 74], Bio [50], and a blend of five factors used in defined culture media [36]. It is therefore reasonable to assume that more than one pathway is involved in maintaining hESC potency. Further study is required in order to clarify the mechanism underlying these factors' involvement in hESC self-maintenance.

The majority of the existing hESC lines were derived with feeder layers [66, 57, 2, 14].The first report of a feeder layer-free derivation of a hESC line was reported by Klimanskaya and colleagues, in which MEF-produced matrix and medium supplemented with a high dose of bFGF (16 ng mL^{-1}), LIF, serum replacement and plasmanate were used [32]. In this study, six new hESC lines were successfully derived, exhibiting ESC features after prolonged culture of over 30 passages. This pioneering work proves the feasibility of a supportive feeder layer-less derivation of hESCs, although the culture system includes some non-defined materials. A recent publication by Ludwig and colleagues reported the derivation of two new hESC lines using a defined serum- and animal-free medium, and feeder layer-free culture conditions [36].The matrix consisted of a mixture of human collagen, fibronectin and laminin, and the medium was supplemented with five growth factors, including TGF_{b1} and bFGF. The newly derived cells sustained most hESC features after several months of continuous culture. Thus, for the first time, defined, animal-, serum- and feeder-free culture conditions for hESCs are presented. However, the two new hESC lines were reported to harbor karyotype abnormalities; one 47, XXY after 4 months of continuous culture and the second exhibited trisomy 12 after 7 months of continuous culture. It is unknown whether the embryos were originally defected or whether these events of karyotype abnormalities occurred during prolonged culture.

3.2 Suspended Culture System

Culture of hESCs requires meticulous care which includes daily medium change, routine passaging every 4–6 days, and occasionally mechanical removal of differentiated colonies from the culture. Although hESCs can be cultured in these conditions in large quantities, the use of hESCs for therapy and for industrial applications requires a scalable and controlled culture system for both differentiated and undifferentiated hESCs. To this end we recently developed a novel suspension culture system for undifferentiated hESCs. The new three dimensional (3D) culture system is based on medium supplemented with 15% serum replacement, cytokines and bFGF. Four cell lines, H9.2, I3, I4 and I6, were cultured in suspension in Petri dishes where they spontaneously formed spheroid clumps. Cells cultured in this system for over a year, maintained all ESC features, including expression of specific markers, stable karyotype, and the developmental potential to differentiate into representative tissues of the three embryonic germ layers in vitro and in vivo. The calculated cell doubling time was 35.2 ± 1.3 h, similarly to a previous report on hESCs in 2D cultures [1]. Correspondingly, the cultures were split every 5–7 days – the same splitting interval of cells cultured with MEFs.

One month after being transferred into a stirred dynamic culture using either shaking Erlenmeyer's or spinner flasks, the spheroid clumps formed by the cells remained similar to those observed within cells cultured statically using Petri dishes. hESCs cultured for 3 months in the dynamic system maintained stable karyotype, were strongly positive for undifferentiation markers, and remained pluripotent. During 10 days of culture in the dynamic culture cell number increased 25-fold. Thus the novel culture system reported here makes it possible to expand undifferentiated hESCs in suspension cultures which will facilitates the large-scale culture of hESCs needed in the clinic and industry.

Acknowledgments The authors thank Mrs. Hadas O'Neill for editing the manuscript. The research conducted by the authors was partly supported by NIH grant R24RR18405.

References

1. Amit M, Carpenter MK, Inokuma MS, Chiu CP, Harris CP, Waknitz MA, Itskovitz-Eldor J, Thomson JA (2000) Clonally derived human embryonic stem cell lines maintain pluripotency and proliferative potential for prolonged periods of culture. Dev Biol 227:271–278 [not in library; checked in PubMed]
2. Amit M, Itskovitz-Eldor J (2002) Derivation and spontaneous differentiation of human embryonic stem cells. J Anat 225–232
3. Amit M, Margulets V, Segev H, Shariki C, Laevsky I, Coleman R, and Itskovitz-Eldor J (2003) Human feeder layers for human embryonic stem cells. Biol Reprod 68:2150–2156
4. Aoi T, Yae K, Nakagawa M, Ichisaka T, Okita K, Takahashi K, Chiba T, Yamanaka S (2008) Generation of pluripotent stem cells from adult mouse liver and stomach cells. Science 321:699–702
5. Beattie GM, Lopez AD, Bucay N, Hinton A, Firpo MT, King CC, Hayek A (2005) Activin A maintains pluripotency of human embryonic stem cells in the absence of feeder layers. Stem Cells 23:489–495
6. Bhattacharya B, Miura T, Brandenberger R, Mejido J, Luo Y, Yang AX, Joshi BH, Ginis I, Thies RS, Amit M, Lyons I, Condie BG, Itskovitz-Eldor J, Rao MS, Puri RK (2004) Gene expression in human embryonic stem cell lines: unique molecular signature. Blood 103(8):2956–2964
7. Bradley A, Evans M, Kaufman MH, Robertson E. (1984) Formation of germ-line chimaeras from embryo-derived teratocarcinoma cell lines. Nature 309:255–256
8. Brüstle O, Spiro AC, Karram K, Choudhary K, Okabe S, McKay RDG (1997) In vitro-generated neural precursors participate in mammalian brain development. Proc Natl Acad Sci U S A 94:14809–14814
9. Brüstle O, Jones KN, Learish RD, Karram K, Choudhary K, Wiestler OD, Duncan ID, McKay RDG (1999) Embryonic stem cell-derived glial precursors: a source of myelinating transplants. Science 285:754–756 [checked original]
10. Bongso A, Fong CY, Ng SC, Ratnam S (1994) Isolation and culture of inner cell mass cells from human blastocysts. Hum Reprod 9(11):2110–2117
11. Chambers I, Colby D, Robertson M, Nichols J, Lee S, Tweedie S, Smith A (2003) Functional expression cloning of Nanog, a pluripotency sustaining factor in embryonic stem cells. Cell 113(5):643–655
12. Cheng L, Hammond H, Ye Z, Zhan X, Dravid G (2003) Human adult marrow cells support prolonged expansion of human embryonic stem cells in culture. Stem Cells 21:131–142
13. Chung Y, Klimanskaya I, Becker S, Marh J, Lu SJ, Johnson J, Meisner L, Lanza R (2006) Embryonic and extraembryonic stem cell lines derived from single mouse blastomeres. Nature 439:216–219
14. Cowan CA, Klimanskaya I, McMahon J, Atienza J, Witmyer J, Zucker JP, Wang S, Morton CC, McMahon AP, Powers D, Melton DA (2004) Derivation of embryonic stem-cell lines from human blastocysts. N Engl J Med 350(13):1353–1356
15. Daley GQ, Ahrlund-Richter L, Auerbach JM, Benvenisty N, Charo RA, Chen G, Deng H, Goldstein LS, Hudson KL, Hyun I, Junn SC, Love J, Lee EH, McLaren A, Mummery CL, Nakatsuji N, Racowsky C, Rooke H, Rossant J, Schöler HR, Solbakk JHH, Taylor P, Trounson AO, Weissman IR, Wilmut I, Yu J, Zoloth R (2007) Science 315:603–604
16. Dhara SK, Benvenisty N (2004) Gene trap as a tool for genome annotation and analysis of X chromosome inactivation in human embryonic stem cells. Nucleic Acids Res 32(13):3995–4002

17. Doetschman TC, Eistetter H, Katz M, Schmidt W, Kemler R (1985) The in vitro development of blastocyst-derived embryonic stem cell lines: formation of visceral yolk sac, blood islands and myocardium. J Embryol Exp Morphol 87:27–45 [checked PubMed]
18. Doetschman T, Williams P, Maeda N (1988) Establishment of hamster blastocyst-derived embryonic stem (ES) cells. Dev Biol 127:224–227 [too early for library; checked in PubMed]
19. Draper JS, Smith K, Gokhale P, Moore HD, Maltby E, Johnson J, Meisner L, Zwaka TP, Thomson JA, Andrews PW (2004) Recurrent gain of chromosomes 17q and 12 in cultured human embryonic stem cells. Nat Biotechnol 22(1):53–54
20. Eiges R, Schuldiner M, Drukker M, Yanuka O, Itskovitz-Eldor J, Benvenisty N (2001) Establishment of human embryonic stem cell-transfected clones carrying a marker for undifferentiated cells. Curr Biol 11:514–518[journal not in library; checked in PubMed.]
21. Evans MJ, Kaufman MH (1981) Establishment in culture of pluripotential cells from mouse embryos. Nature 292:154–156 [checked original]
22. Giles JR, Yang X, Mark W, Foote RH (1993) Pluripotency of cultured rabbit inner cell mass cells detected by isozyme analysis and eye pigmentation of fetuses following injection into blastocysts or morulae. Mol Reprod Dev 36:130–138 [journal not in library; checked PubMed]
23. Graves KH, Moreadith RW (1993) Derivation and characterization of putative pluripotential embryonic stem cells from preimplantation rabbit embryos. Mol Reprod Dev 36:424–433 [journal not in library; checked PubMed]
24. Hall LL, Byron M, Butler J, Becker KA, Nelson A, Amit M, Itskovitz-Eldor J, Stein J, Stein G, Ware C, Lawrence JB (2008) X-inactivation reveals epigenetic anomalies in most hESC but identifies sublines that initiate as expected. J Cell Physiol 216(2):445–452
25. Hanna J, Wernig M, Markoulaki S, Sun CW, Meissner A, Cassady JP, Beard C, Brambrink T, Wu LC, Townes TM, Jaenisch R (2007) Treatment of sickle cell anemia mouse model with iPS cells generated from autologous skin. Science 318(5858):1920–1923
26. Hirashima M, Kataoka H, Nishikawa S, Matsuyoshi N, Nishikawa S (1999) Maturation of embryonic stem cells into endothelial cells in an in vitro model of vasculogenesis. Blood 93(4):1253–1263
27. Hoffman LM, Hall L, Batten JL, Young H, Pardasani D, Baetge EE, Lawrence J, Carpenter MK (2005) X-inactivation status varies in human embryonic stem cell lines. Stem Cells 23(10):1468–1478
28. Hovatta O, Mikkola M, Gertow K, Stromberg AM, Inzunza J, Hreinsson J, Rozell B, Blennow E, Andang M, Ahrlund-Richter L (2003) A culture system using human foreskin fibroblasts as feeder cells allows production of human embryonic stem cells. Hum Reprod 18:1404–1409
29. Inzunza J, Gertow K, Stromberg MA, Matilainen E, Blennow E, Skottman H, Wolbank S, Ahrlund-Richter L, Hovatta O (2005) Derivation of human embryonic stem cell lines in serum replacement medium using postnatal human fibroblasts as feeder cells. Stem Cells 23:544–549
30. Itskovitz-Eldor J, Schuldiner M, Karsenti D, Eden A, Yanuka O, Amit M, Soreq H, Benvenisty N (2000) Differentiation of human embryonic stem cells into embryoid bodies comprising the three embryonic germ layers. Mol Med 6:88–95 [journal not in library; checked e-journal]
31. Kennedy M, Firpo M, Choi K, Wall C, Robertson S, Kabrun N, Keller G (1997) A common precursor for primitive erythropoiesis and definitive haematopoiesis. Nature 386:488–93 [checked original]
32. Klimanskaya I, Chung Y, Meisner L, Johnson J, West MD, Lanza R (2005) Human embryonic stem cells derived without feeder cells. Lancet 365:1636–1641
33. Klug MG, Soonpaa MH, Koh GY, Field LJ (1996) Genetically selected cardiomyocytes from differentiating embryonic stem cells form stable intracardiac grafts. J Clin Invest 98:216–224
34. Lanzendorf SE, Boyd CA, Wright DL, Muasher S, Oehninger S, Hodgen GD (2001) Use of human gametes obtained from anonymous donors for the production of human embryonic stem cell lines. Fertil Steril 76(1):132–137
35. Lerou PH, Yabuuchi A, Huo H, Miller JD, Boyer LF, Schlaeger TM, Daley GQ. (2008) Derivation and maintenance of human embryonic stem cells from poor-quality in vitro fertilization embryos. Nat Protoc 3:923–933
36. Ludwig TE, Levenstein ME, Jones JM, Berggren WT, Mitchen ER, Frane JL, Crandall LJ, Daigh CA, Conard KR, Piekarczyk MS, Llanas RA, Thomson JA (2006) Derivation of human embryonic stem cells in defined conditions. Nat Biotechnol 24:185–187

37. Lowry WE, Richter L, Yachechko R, Pyle AD, Tchieu J, Sridharan R, Clark AT, Plath K (2008) Generation of human induced pluripotent stem cells from dermal fibroblasts. Proc Natl Acad Sci U S A 105(8):2883–2888
38. Lumelsky N, Blondel O, Laeng P, Velasco I, Ravin R, McKay R (2001) Differentiation of embryonic stem cells to insulin-secreting structures similar to pancreatic islets. Science 292:1389–1394
39. Mateizel I, De Temmerman N, Ullmann U, Cauffman G, Sermon K, Van de Velde H, De Rycke M, Degreef E, Devroey P, Liebaers I, Van Steirteghem A (2006) Derivation of human embryonic stem cell lines from embryos obtained after IVF and after PGD for monogenic disorders. Hum Reprod 21:503–511
40. Martin GR (1981) Isolation of a pluripotent cell line from early mouse embryos cultured in medium conditioned by teratocarcinoma stem cells. Proc Natl Acad Sci U S A 78:7634–7638 [checked original]
41. Meissner A, Wernig M, Jaenisch R (2007) Direct reprogramming of genetically unmodified fibroblasts into pluripotent stem cells. Nat Biotechnol 25(10):1177–1181
42. Mitalipov S, Kuo HC, Byrne J, Clepper L, Meisner L, Johnson J, Zeier R, Wolf D (2006) Isolation and characterization of novel rhesus monkey embryonic stem cell lines. Stem Cells 24(10):2177–2186
43. Mitalipova M, Beyhan Z, First NL (2001) Pluripotency of bovine embryonic stem cell line derived from precompacting embryos. Cloning 3:59–67 [journal not in library; checked e-journal]
44. Mitsui K, Tokuzawa Y, Itoh H, Segawa K, Murakami M, Takahashi K, Maruyama M, Maeda M, Yamanaka S (2003) The homeoprotein Nanog is required for maintenance of pluripotency in mouse epiblast and ES cells. Cell 113(5):631–642
45. Nakagawa M, Koyanagi M, Tanabe K, Takahashi K, Ichisaka T, Aoi T, Okita K, Mochiduki Y, Takizawa N, Yamanaka S (2008) Generation of induced pluripotent stem cells without Myc from mouse and human fibroblasts. Nat Biotechnol 26(1):101–106
46. Nagy A, Rossant J, Nagy R, Abramow-Newerly W, Roder JC (1993) Derivation of completely cell culture-derived mice from early-passage embryonic stem cells. Proc Natl Acad Sci U S A 90:8424–8428 [checked original]
47. Nakano T, Kodama H, Honjo T (1994) Generation of lymphohematopoietic cells from embryonic stem cells in culture. Science 265:1098–101 [checked original]
48. Nichols J, Zevnik B, Anastassiadis K, Niwa H, Klewe-Nebenius D, Chambers I, Schöler H, Smith A (1998) Formation of pluripotent stem cells in the mammalian embryo depends on the POU transcription factor Oct4. Cell 95(3):379–391
49. Notarianni E, Galli C, Laurie S, Moor RM, Evans MJ (1991) Derivation of pluripotent, embryonic cell lines from the pig and sheep. J Reprod Fertil Suppl 43:255–260 [too early for library; checked PubMed]
50. Sato N, Meijer L, Skaltsounis L, Greengard P, Brivanlou AH (2004) Maintenance of pluripotency in human and mouse embryonic stem cells through activation of Wnt signaling by a pharmacological GSK-3-specific inhibitor. Nat Med 10:55–63
51. Shen Y, Matsuno Y, Fouse SD, Rao N, Root S, Xu R, Pellegrini M, Riggs AD, Fan G (2008) X-inactivation in female human embryonic stem cells is in a nonrandom pattern and prone to epigenetic alterations. Proc Natl Acad Sci U S A 105(12):4709–4714
52. Silva SS, Rowntree RK, Mekhoubad S, Lee JT (2008) X-chromosome inactivation and epigenetic fluidity in human embryonic stem cells. Proc Natl Acad Sci U S A 105(12):4820–4825
53. Sims M, First NL (1994) Production of calves by transfer of nuclei from cultured inner cell mass cells. Proc Natl Acad Sci U S A 91:6143–6147 [checked original (note: volume and year citation printed incorrectly at top of journal page)]
54. Solter D, Knowles BB (1975) Immunosurgery of mouse blastocyst. Proc Natl Acad Sci U S A 72:5099–5102 [PubMed]
55. Soria B, Roche E, Berná G, León-Quinto T, Reig JA, Martín F (2000) Insulin-secreting cells derived from embryonic stem cells normalize glycemia in streptozotocin-induced diabetic mice. Diabetes 49:157–62 [checked original]

56. Suemori H, Tada T, Torii R, Hosoi Y, Kobayashi K, Imahie H, Kondo Y, Iritani A, Nakatsuji N (2001) Establishment of embryonic stem cell lines from cynomolgus monkey blastocysts produced by IVF or ICSI. Dev Dyn 222:273–279
57. Reubinoff BE, Pera MF, Fong C, Trounson A, Bongso A (2000) Embryonic stem cell lines from human blastocysts: somatic differentiation *in vitro*. Nat Biotechnol 18:399–404 [checked original]
58. Richards M, Fong CY, Chan WK, Wong PC, Bongso A (2002) Human feeders support prolonged undifferentiated growth of human inner cell masses and embryonic stem cells. Nat Biotechnol 20:933–936
59. Park IH, Zhao R, West JA, Yabuuchi A, Huo H, Ince TA, Lerou PH, Lensch MW, Daley GQ (2008) Reprogramming of human somatic cells to pluripotency with defined factors. Nature 451(7175):141–146
60. Strulovici Y, Leopold PL, O'Connor TP, Pergolizzi RG, Crystal RG (2007) Human embryonic stem cells and gene therapy. Mol Ther 15(5):850–66
61. Suss-Toby E, Gerecht S, Amit M, Manor D, Itskovitz-Eldor J (2004) Derivation of a diploid human embryonic stem cell line from a mononuclear zygote. Hum Reprod 19(3):670–675
62. Takahashi K, Yamanaka S (2006) Induction of pluripotent stem cells from mouse embryonic and adult fibroblast cultures by defined factors. Cell 126(4):663–676
63. Takahashi K, Tanabe K, Ohnuki M, Narita M, Ichisaka T, Tomoda K, Yamanaka S (2007) Induction of pluripotent stem cells from adult human fibroblasts by defined factors. Cell 131(5):861–872
64. Thomson JA, Kalishman J, Golos TG, Durning M, Harris CP, Becker RA, Hearn JP (1995) Isolation of a primate embryonic stem cell line. Proc Natl Acad Sci U S A 92:7844–7848 [checked original]
65. Thomson JA, Kalishman J, Golos TG, Durning M, Harris CP, Hearn JP (1996) Pluripotent cell lines derived from common marmoset (*Callithrix jacchus*) blastocysts. Biol Reprod 55:254–259 [checked original]
66. Thomson JA, Itskovitz-Eldor J, Shapiro SS, Waknitz MA, Swiergiel JJ, Marshall VS, Jones JM (1998) Embryonic stem cell lines derived from human blastocysts. Science 282:1145–1147 [erratum in Science (1998) 282, 1827] [checked original]
67. Vrana KE, Hipp JD, Goss AM, McCool BA, Riddle DR, Walker SJ, Wettstein PJ, Studer LP, Tabar V, Cunniff K, Chapman K, Vilner L, West MD, Grant KA, Cibelli JB (2003) Nonhuman primate parthenogenetic stem cells. Proc Natl Acad Sci U S A 100:11911–11916 (erratum in: *Proc Natl Acad Sci U S A*. 2004, 101, 693)
68. Verlinsky Y, Strelchenko N, Kukharenko V, Rechitsky S, Verlinsky O, Galat V, Kuliev A (2005) Human embryonic stem cell lines with genetic disorders. Reprod Biomed Online 10:105–110
69. Wang L, Li L, Menendez P, Cerdan C, Bhatia M (2005) Human embryonic stem cells maintained in the absence of mouse embryonic fibroblasts or conditioned media are capable of hematopoietic development. Blood 105(12):4598–603
70. Wobus AM, Holzhausen H, Jakel P, Schoneich J (1984) Characterization of pluripotent stem cell line derived from mouse embryo. Exp Cell Res 152:212–219
71. Yamashita J, Itoh H, Hirashima M, Ogawa M, Nishikawa S, Yurugi T, Naito M, Nakao K, Nishikawa S (2000) Flk1-positive cells derived from embryonic stem cells serve as vascular progenitors. Nature 408(6808):92–96
72. Xu C, Inokuma MS, Denham J, Golds K, Kundu P, Gold JD, Carpenter MK (2001) Feeder-free growth of undifferentiated human embryonic stem cells. Nat Biotechnol 19:971–974
73. Xu C, Rosler E, Jiang J, Lebkowski JS, Gold JD, O'Sullivan C, Delavan-Boorsma K, Mok M, Bronstein A, Carpenter MK (2005) Basic fibroblast growth factor supports undifferentiated human embryonic stem cell growth without conditioned medium. Stem Cells 23:315–323
74. Xu RH, Peck RM, Li DS, Feng X, Ludwig T, Thomson JA (2005) Basic FGF and suppression of BMP signaling sustain undifferentiated proliferation of human ES cells. Nat Methods 2:185–190
75. Okita K, Ichisaka T, Yamanaka S (2007) Generation of germline-competent induced pluripotent stem cells. Nature 448(19):313–317

76. Yu J, Vodyanik MA, Smuga-Otto K, Antosiewicz-Bourget J, Frane JL, Tian S, Nie J, Jonsdottir GA, Ruotti V, Stewart R, Slukvin II, Thomson JA (2007) Induced pluripotent stem cell lines derived from human somatic cells. Science 318:1917–1920
77. Mitalipova M, Calhoun J, Shin S, Wininger D, Schulz T, Noggle S, Venable A, Lyons I, Robins A, Stice S (2003) Human embryonic stem cell lines derived from discarded embryos. Stem Cells 21:521–526
78. Amit M, Shariki C, Margulets V, Itskovitz-Eldor J (2004) Feeder layer- and serum-free culture of human embryonic stem cells. Biol Reprod 70:837–845

Adv Biochem Engin/Biotechnol (2009) 114: 185-199
DOI: 10.1007/10_2008_45

Published online: 27 March 2009

Totipotency, Pluripotency and Nuclear Reprogramming

Shoukhrat Mitalipov and Don Wolf

Abstract Mammalian development commences with the totipotent zygote which is capable of developing into all the specialized cells that make up the adult animal. As development unfolds, cells of the early embryo proliferate and differentiate into the first two lineages, the pluripotent inner cell mass and the trophectoderm. Pluripotent cells can be isolated, adapted and propagated indefinitely in vitro in an undifferentiated state as embryonic stem cells (ESCs). ESCs retain their ability to differentiate into cells representing the three major germ layers: endoderm, mesoderm or ectoderm or any of the 200+ cell types present in the adult body. Since many human diseases result from defects in a single cell type, pluripotent human ESCs represent an unlimited source of any cell or tissue type for replacement therapy thus providing a possible cure for many devastating conditions. Pluripotent cells resembling ESCs can also be derived experimentally by the nuclear reprogramming of somatic cells. Reprogrammed somatic cells may have an even more important role in cell replacement therapies since the patient's own somatic cells can be used for reprogramming thereby eliminating immune based rejection of transplanted cells.

S. Mitalipov (✉)
Division of Reproductive Sciences, Oregon National Primate Research Center, Oregon Health and Science University, 505 N.W. 185th Avenue, Beaverton, Oregon 97006, USA; Oregon Stem Cell Center, Oregon Health and Science University, 505 N.W. 185th Avenue, Beaverton, Oregon 97006, USA; and Department of Obstetrics and Gynecology, School of Medicine, Oregon Health and Science University, 505 N.W. 185th Avenue, Beaverton, Oregon 97006, USA
e-mail: mitalipo@ohsu.edu

D. Wolf
Division of Reproductive Sciences, Oregon National Primate Research Center, Oregon Health and Science University, 505 N.W. 185th Avenue, Beaverton, Oregon 97006, USA

In this review, we summarize two major approaches to reprogramming: (1) somatic cell nuclear transfer and (2) direct reprogramming using genetic manipulations.

Keywords Embryonic stem cells, iPS cells, Pluripotent, Somatic cell nuclear transfer, Totipotent,

Contents

1 Totipotency

Totipotency is defined in Wikipedia as the ability of a single cell to divide and produce all the differentiated cells in an organism, including extraembryonic tissues. Totipotent cells formed during sexual and asexual reproduction include spores and zygotes. In some organisms, cells can dedifferentiate and regain totipotency. For example, a plant cutting or callus can be used to grow an entire plant. Mammalian development commences when an oocyte is fertilized by a sperm forming a single celled embryo, the zygote. Consistent with the definition, the zygote is totipotent, meaning that this single cell has the potential to develop into an embryo with all the specialized cells that make up a living being, as well as into the placental support structure necessary for fetal development. Thus, each totipotent cell is a self-contained entity that can give rise to the whole organism. This is said to be true for the zygote and for early embryonic blastomeres up to at least the 4-cell stage embryo (see Fig. 1). Experimentally, totipotency can be demonstrated by the isolation of a single blastomere from a preimplantation embryo and subsequently monitoring its ability to support a term birth following transfer into a suitable recipient. This approach was pioneered in rats and has been realized in several mammalian species including nonhuman primates [1–4]. In the latter case, we confirmed the ability of isolated blastomeres from 2- and 4-cell stage, IVF produced embryos of the rhesus monkey to support term pregnancies and to produce live animals [5]. As embryo development progresses to the 8-cell stage and beyond depending on the species, the individual blastomeres that comprise the embryo gradually lose their totipotency. It is generally believed that this restriction in developmental potential indicates irreversible differentiation and specialization of early embryonic cells into the first two lineages, the inner cell mass (ICM) that includes cells that will give rise to the fetus and the trophectoderm (TE), and an outer layer of cells that is destined to an extraembryonic fate (Fig. 1).

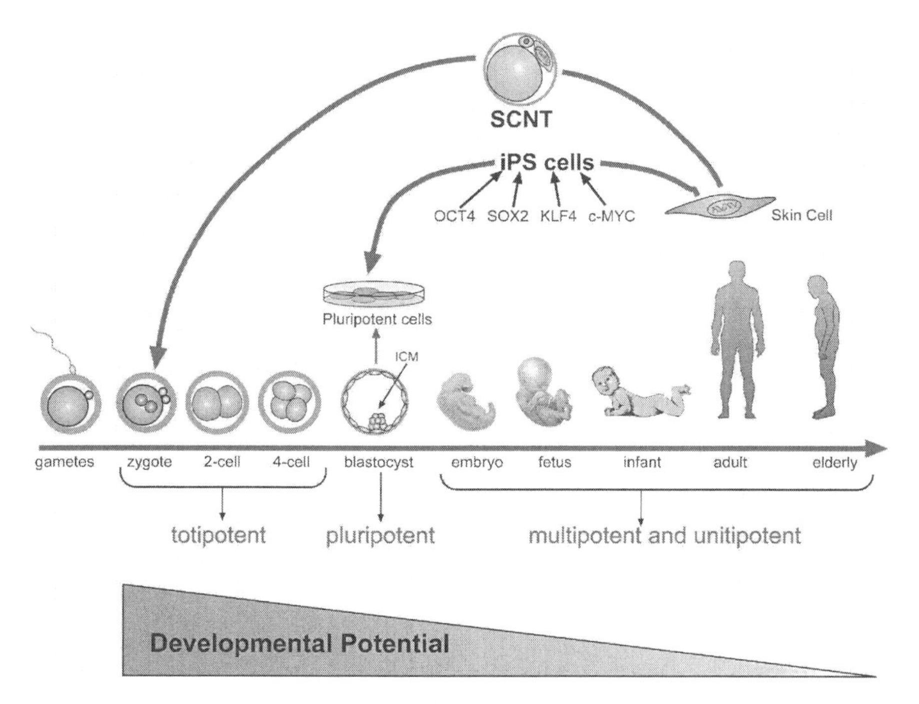

Fig. 1 Development and reprogramming. Ontogeny begins from a single cell, the zygote. The zygote and each blastomere of the early embryo are totipotent with the potential to develop into the whole organism. As development unfolds, the developmental potential of individual blastomeres gradually declines resulting subsequently in pluripotent, multipotent, unipotent and terminally differentiated somatic cells. However, developmental potential of somatic cells can be reinstated to the totipotent stage by SCNT or to the pluripotent state by direct reprogramming

A complication in assessing the state of potency of blastomeres isolated from more advanced stages of development is insufficient cytoplasmic volume. Thus, although the blastomeres may in fact be totipotent, embryonic development of relatively small isolated blastomeres arrests at or near the time of blastulation. Recall that the zygote and early blastomeres undergo several unusual mitotic or cleavage divisions that are not accompanied by a corresponding growth of cytoplasm, that is, there is no change in embryo size despite the presence of more cells or blastomeres and each individual blastomere becomes smaller. The embryonic genome at these early stages is transcriptionally quiescent and development is regulated by maternally inherited factors present at the time of fertilization in the oocyte [6]. The transition in developmental regulation with activation of the embryonic genome and a complete loss of dependence on oocyte factors occurs before the blastocyst stage in a species-specific manner. Additionally, by the late morula or early blastocyst stage the embryo ceases cleavage divisions and resumes normal mitotic divisions with concomitant increases in cell volume during the S-phase. The likelihood that early blastomeres retain totipotency for a major part of preimplantation development but experimentally we cannot prove it is directly supported by the fact that the addition of oocyte cytoplasm to a blastomere of the 8- to 16-cell stage embryo can restore, or perhaps more appropriately allow expression of, its full developmental potential. This approach, embryonic cell nuclear transfer, has been employed in the monkey to demonstrate the totipotency of 8- to 16-cell stage blastomeres whereby reconstructed embryos when transferred to a recipient resulted in a term birth [7].

It is also known that conglomerates of embryonic cells at a later stage of development can develop into an organism. An experimental manipulation that supports this concept involves blastocyst splitting. Cutting the embryo into halves with an approximately equal distribution of TE and ICM cells can lead to the production of viable infants [5, 8]. Obviously, embryo splitting that creates demi embryos with highly distorted ratios of ICM to TE cells is inconsistent with the production of live births.

2 Pluripotency

The Wikipedia definition in the broad sense means "having more than one potential outcome." In cell biology, the definition of pluripotency has come to refer to a stem cell that has the potential to differentiate into any of the three germ layers: endoderm, mesoderm or ectoderm. Pluripotent stem cells can give rise to any fetal or adult cell type. However, a single cell or a conglomerate of pluripotent cells cannot develop into a fetal or adult animal because they lack the potential to organize into an embryo. In contrast, many progenitor cells that are capable of differentiating into a limited number of cell fates are described as multipotent. Somatic stem cells such as neural, bone marrow-derived, or hematopoietic cells would fit into this latter category.

At least some of the embryo's ICM cells are pluripotent, meaning that they can form virtually every somatic and germ cell type in the body. These ICM cells are self sustained and their pluripotency is maintained by endogenously expressed factors. In vivo, pluripotent cells within the ICM exist transiently; as the developmental program unfolds they differentiate into cells of the next embryonic or fetal stage. However, they can be isolated, adapted and propagated in vitro in an undifferentiated state as embryonic stem cells (ESCs) [9, 10]. ESCs were first derived in 1981 from the ICM of the inbred mouse by Martin [10] and Evans and Kaufman [9]. In 1998, ESCs were successfully isolated from surplus, IVF-produced human embryos [11].

ESCs express specific markers or characteristics similar but not identical to the transient pluripotent cells of an embryo. This includes stage specific embryonic antigens, enzymatic activities such as alkaline phosphatase and telomerase, and "stemness" genes that are rapidly down-regulated upon differentiation, including *OCT4* and *NANOG*. Under specific conditions, ESCs can proliferate indefinitely in an undifferentiated state, suggesting that the transcriptional activity and epigenetic regulators capable of supporting pluripotency can be maintained in vitro in ESCs. However, when released from the influence of these culture conditions or following their introduction back into a host embryo, ESCs retain their ability to differentiate into any cell-type, just like ICM cells. Alternatively, they can differentiate in vivo in teratomas into cells representing the three major germ layers: endoderm, mesoderm and ectoderm or they can be directed to differentiate in vitro into any of the 200+ cell types present in the adult body. Since many human diseases result from defects in a single cell type, pluripotent human ESCs may become an unlimited source of any cell or tissue type for replacement therapy thus providing a possible cure for many devastating diseases.

Parenthetically, one of the challenges before clinical transplantation studies involving hESCs can begin concerns the immune response anticipated after transplantation [12, 13]. Human ESCs are routinely derived from IVF embryos and transplantation of such cells into genetically unrelated patients will incite an immune response and result in rejection. Histocompatibility is one of major unsolved problems in transplant medicine. Rejection of unmatched transplanted tissues is provoked by alloantigens present on graft tissues by the recipient's immune system. The alloantigens or antigenic proteins on the surface of transplant tissues that mostly cause immune rejection are the blood group antigens (ABO) and the major histocompatibility complex (MHC) proteins, also designated in humans as human leukocyte antigens (HLA). Matching donor and recipient HLA types is important to reduce a cytotoxic T-cell response in the recipient, and subsequently improve the chances of survival of the transplant. However, tissue or organ transplantation from one individual to another is a daunting task due to the existence of two classes of HLA molecules (Class I, and II), each encoded by multiple genes and most importantly, each of these genes represented by multiple alleles. For example, there are 22 different alleles identified so far for the class I HLA-A gene and 42 alleles for HLA-B. Thus, due to HLA polymorphism, the chances of finding a donor–recipient match based on just a few HLA genes (HLA-A, -B, and -DR) could be one in several million [14]. Therefore, the need for developing approaches for deriving histocompatible pluripotent cells is commonly recognized.

3 Nuclear Reprogramming

Hochedlinger and Jaenisch define nuclear reprogramming as the reversal of the differentiation state of a mature cell to one that is characteristic of the undifferentiated embryonic state [15]. Let us first look at the forward process of development and differentiation. It is now generally recognized that genetic material is usually not lost during development and differentiation. Consequently, the process of differentiation must reflect the expression at each stage of a unique cohort of specific genes, its transcriptome, and it now appears that such differential expression is determined or regulated by reversible epigenetic changes gradually imposed on the genome during development.

Epigenetic mechanisms that have been implicated in the regulation of differential gene activity include modifications to the histones (such as acetylation, methylation, phosphorylation, ubiquitination, and ADP-ribosylation) and methylation of DNA at CpG dinucleotides (see reviews by [16, 17]). These specific epigenetic modifications regulate expression or silencing of genes at the level of transcription, mediated by the level of packaging DNA into chromatin. For example, acetylation of histones H3 and H4 and methylation of H3 at the lysine 4 position (H3 Lys-4) unfolds and loosens up the DNA template and makes it accessible to transcription factors. Thus, these epigenetic mechanisms are generally associated with active gene transcription. Conversely, methylation of H3 Lys-9 and H3 Lys-27 induce DNA compaction and subsequently gene silencing.

In this review, we will summarize two major approaches to nuclear reprogramming or reversing the developmental process: (1) somatic cell nuclear transfer (SCNT) and (2) direct reprogramming using genetic manipulations. It should be noted that interest in both of these strategies derives, in large measure, from the potential production and use of histocompatible human ESCs in regenerative medicine.

4 Epigenetic Reprogramming by SCNT

The concept of reprogramming of a patient's somatic cells into pluripotent ESCs was conceived based on two independent breakthroughs in the field of developmental biology in the late 1990s: success with cloning of animals by SCNT [18, 19] and derivation of human ESCs [11]. SCNT, or cloning, dates back to 1962 when John Gurdon first demonstrated that somatic cells from *Xenopus laevis* could be reprogrammed back into an early embryonic state by factors present in an egg cytoplasm and support development of an adult frog [20]. Thus, it became clear that the cytoplasm of the oocyte has the ability to reprogram gene expression and that a single somatic cell nucleus has the capacity to yield a whole new organism [21].

Research in SCNT involving other vertebrates including mammals continued for several decades and culminated in groundbreaking announcements, first in 1996 [18] and then in 1997 [19] that sheep could be produced by SCNT using fetal and adult somatic cells. This accomplishment was quickly reproduced in other mammals including mice [22], cattle [23, 24], pigs [25], goats [26], rabbits [27], cats [28], mules [29], horses [30], rats [31], and dogs [32].

As mentioned above, ESCs were first derived in 1981 in the mouse [9, 10]. Exploiting the ability of mouse ESCs to contribute to germ-line chimeras and homologous recombination technology for the creation of knock-out mice and mammalian gene function analysis revolutionized the field of experimental biology [33]. To date, an estimated 10,000 mutated mice have been generated worldwide using the gene targeting technique. In recognizing their enormous contribution to the advances in every field of biology and medicine, the 2007 Nobel Prize in Physiology and Medicine was awarded to three scientists who pioneered the derivation of mouse ESCs and gene targeting [34].

The establishment of mouse ESCs has instigated similar studies in other mammals. Working with nonhuman primates, James Thomson of the Wisconsin National Primate Research Center reported in 1995 the successful isolation of ESCs from rhesus macaque, in vivo flushed blastocysts [35]. Unlike mouse ESCs, monkey ESCs grew as flat colonies and expressed slightly different surface markers than did mouse cells. Primate ESCs are relatively cumbersome to maintain and manipulate, requiring considerable technical expertise and attention confounded by a requirement for manual passaging and their slow growth rate. Nevertheless, these cells were indeed pluripotent and capable of differentiating into cell types of all three germ layers. ESCs were also successfully isolated in other nonhuman primate species including marmosets and cynomolgus macaques [36, 37]. In the rhesus macaque, an additional 25 cell lines were produced from in vitro produced embryos at the Oregon National Primate Research Center [38]. In 1998, following protocols and markers developed in the monkey, the isolation of ESC lines from surplus IVF-produced human embryos was reported [11]. Subsequently, approximately 65 human ESC lines were approved in the United States for Federal research support in August of 2001; however only a few of those lines are currently available and under study (http://stemcells.nih.gov/research/registry/).

Despite the remarkable strides that have been made to date with mouse and primate ESCs, success in other species has been limited. ESC-like cells have been described in several species including sheep [39], cattle [40], pigs [41], rabbits [42] and rats [43]. However, the pluripotency of these cells and their ability to maintain an undifferentiated phenotype over long term culture remains questionable.

The conceptual unification of SCNT and embryonic stem cell derivation technology suggested that it might be possible to produce preimplantation human embryos by SCNT and then derive isogenic embryonic stem cells from the resulting SCNT embryos [44, 45]. Human ESCs produced by this approach called "therapeutic cloning" would subsequently be differentiated into therapeutically useful cells and transplanted back into a patient suffering from a degenerative disease. The proof of the concept was first demonstrated in the mouse in 2000 with the isolation of pluripotent ESCs from adult somatic cell nuclei [46]. These SCNT-derived ESCs expressed canonic pluripotent markers and were able to differentiate readily into various somatic cell types in vitro or in vivo in teratomas and chimeras. Despite multiple abnormalities observed in cloned offspring, mouse ESCs derived by SCNT were transcriptionally indistinguishable from their counterparts derived from fertilized embryos [47, 48], consistent with the notion that ESCs derived from reprogrammed somatic cells have an identical therapeutic potential with "wild type" IVF-derived ESCs. This exciting scientific advance indicated that it may soon be possible to provide patients with pluripotent cells tailored for a given therapeutic purpose.

However, despite this remarkable progress, the feasibility of therapeutic cloning in primates remained questionable. Early attempts demonstrated that human and nonhuman primate SCNT embryos were unable to develop efficiently into blastocysts and typically arrested at early cleavage stages [49, 50]. This indicated an inability of primate SCNT embryos to activate embryonic genes and sustain the developmental program, possibly due to lacking or incomplete nuclear reprogramming. These challenges along with retraction of two high profile papers that contained fabricated data on human SCNT [51] significantly dampened scientific enthusiasm. The ability to derive primate ESCs by SCNT until recently was uncertain.

We initially reported incomplete nuclear remodeling following standard SCNT in the monkey, including nuclear envelope breakdown (NEBD) and premature chromosome condensation (PCC), and correlated this observation with a decline in maturation promoting factor (MPF) activity [52]. Although, a direct link between NEBD, PCC and successful reprogramming was not clear, we presumed that remodeling could be particularly beneficial for efficient nuclear reprogramming by allowing access of reprogramming factors to the somatic cell's chromatin. We introduced several modifications to SCNT protocols that prevented MPF decline and induced robust NEBD and PCC. Importantly, these modifications resulted in improved SCNT embryo development and significantly increased blastocyst rates, suggesting that MPF activity is essential for efficient nuclear reprogramming. The modified protocols allowed routine production of SCNT blastocysts from various donor somatic cells providing the foundation for rapid advances in the derivation of ESCs. More recently, we succeeded in the derivation of two ESC lines from rhesus macaque SCNT blastocysts using adult male skin fibroblasts as nuclear donors (Fig. 2) [53]. DNA analysis confirmed that nuclear DNA was identical to donor somatic cells and that mitochondrial DNA originated from oocytes. Both cell lines exhibited normal ESC morphology, expressed key stemness markers, were transcriptionally similar to control ESCs and differentiated into multiple cell types in vitro and in vivo. These results represent a significant advancement in understanding the role of nuclear remodeling events in reprogramming following SCNT and demonstrate the first successful reprogramming of adult primate somatic cells into pluripotent ESCs. Currently, we are focused on further improvements in reprogramming by SCNT and efficient derivation of ESCs in the nonhuman primate model. In our initial report, the efficiency of this approach was quite low, requiring approximately 150 oocytes to produce a single ESC line [53]. However, based on our current SCNT outcomes yielding nearly threefold higher blastocyst development and ESC derivation rates over our previously reported efficiency, as few as ten or less monkey oocytes are required to produce one ESC line (Mitalipov, unpublished results). These results suggest that systematic optimization of SCNT approaches to define critical reprogramming factors will likely succeed in the efficient generation of patient-specific ESCs for therapeutic applications.

Our recent data also strongly support the notion that oocyte-induced reprogramming of primate somatic cells results in complete erasure of somatic memory and the resetting of a new ESC-specific epigenetic state. Imprinted gene expression, methylation, telomere length and X-inactivation analyses of SCNT-derived primate ESCs were consistent with accurate and extensive epigenetic reprogramming of somatic cells by oocyte-specific factors ((Mitalipov, unpublished results).

A variation on the SCNT theme that has received recent attention is called altered nuclear transfer (ANT). Reprogramming by oocyte-specific factors after SCNT employs endogenous epigenetic pathways/programs. Thus SCNT provides a paradigm for identification of natural epigenetic factors in an egg that accompany nuclear reprogramming and promotes utilization of these factors for direct reprogramming. However, utilization of an SCNT approach for reprogramming of human somatic cells into pluripotent ESCs poses ethical concerns since it involves the creation and subsequent destruction of preimplantation stage embryos with potential for full-term development. Thus, ANT proposes the creation of pluripotent stem cells by preemptive

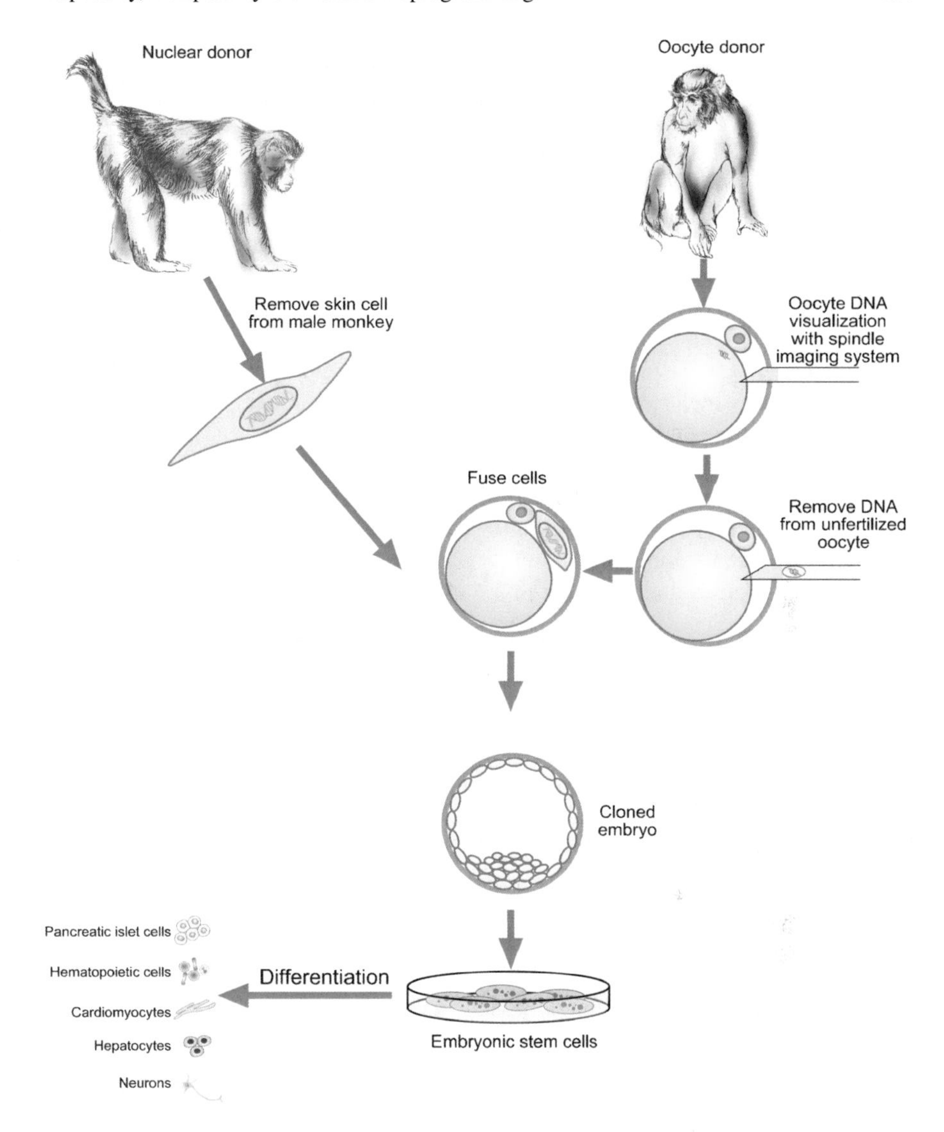

Fig. 2 A schematic diagram showing experimental steps in reprogramming of adult primate somatic cells into pluripotent embryonic stem cells via SCNT. A donor nucleus from a skin cell was introduced into an enucleated oocyte and the resulting embryo gave rise to embryonic stem cells (copied from [53], supplementary information)

alterations prior to SCNT insuring that no totipotent embryo is involved [54, 55]. These alterations should preclude the integrated organic unity and developmental potential that are the defining characteristics of a living organism, while still allowing the formation of the ICM cell lineage from which pluripotent stem cells can be derived. ANT proposes to alter the nucleus of a somatic cell and/or the cytoplasm of an enucleated oocyte prior to SCNT that would prevent formation of a totipotent zygote. However, residual oocyte factors should be capable of reprogramming an introduced nucleus with subsequent development to a stage that would support pluripotent stem cell isolation in the absence of a trophectodermal lineage.

Mature metaphase II (MII) oocytes are one of the largest cells produced by the human body. They contain key maternally inherited transcriptional and epigenetic factors essential for "natural" reprogramming of highly specialized gametic genomes into totipotent and pluripotent cells. Therefore, it is not surprising that oocyte-specific factors are able to reprogram transplanted somatic nuclei, although with less efficiency than that which occurs in the embryo after fertilization. As indicated above, maternally inherited factors in the oocyte convert a transcriptionally-quiescent embryonic genome into an active one during early embryonic development and following embryonic genome activation, control of the developmental program is gradually shifted to embryonic factors. Among these maternal transcription factors, whose functions have been well defined, are Oct4 and Sox2, which are both essential for formation of the ICM in mouse preimplantation embryos. Cdx2 plays a similar role in the development of TE. In early cleavage-stage mouse embryos these transcription factors are expressed in all blastomeres. At the blastocyst stage, Oct4 and Sox2 are detected exclusively in the ICM cells, while Cdx2 is confined to the TE [56]. The role and expression pattern of these factors is poorly studied in other species including primates. However, we have shown a similar expression profile for OCT4 in monkey preimplantation embryos [57]. The homeodomain protein, Nanog, is also detected exclusively in the ICM of mouse embryos and cooperates with Oct4 and Sox2 to control a set of target genes that have important functions in maintaining pluripotency and ICM formation. Currently, little is known about maternal epigenetic factors that induce histone modifications and DNA methylation. Recent work suggests that expression of Nanog in embryos may be regulated by the histone arginine methyltransferase Carm1 [58]. Interestingly, overexpression of Carm1 upregulates both Nanog and Sox2 and was able to direct development of individual blastomeres into an ICM fate.

As stated above, Cdx2 is one of the earliest known transcription factors that is essential for formation and function of the TE lineage [56]. Cdx2-deficient mouse embryos fail to maintain a blastocoel and cannot form the TE, but nonetheless, development of the pluripotent lineage of the ICM is relatively unaffected [59, 60]. Additional evidence for a key role of Cdx2 comes from Tead4 knockout embryos which are devoid of both the TE lineage and Cdx2 expression [61, 62]. Interestingly, Cdx2-deficient ICMs can generate functional ESCs. Recent evidence also suggests that somatic cells lacking Cdx2 can be used for SCNT, resulting in formation of the single ICM lineage suitable for isolation of ESC lines [63]. This demonstrates that inhibition of TE specific factors during SCNT can significantly alter the developmental program and prevent formation of a totipotent embryo without compromising reprogramming to the pluripotent state, thus providing a scientific basis for the ANT concept.

On the other hand, Cdx2-deficient nuclei in this study were complemented by maternal factors including Cdx2 before the onset of embryonic genome activation. Therefore, SCNT embryos were not obviously abnormal until the maternal-to-embryonic transition point. To solve this ethical dilemma, maternal Cdx2 transcripts must be inactivated as well. Moreover, it remains to be determined whether this approach will work in other species including primates.

5 Direct Reprogramming

Possibly one of the greatest developments in the stem cell research field in the past 2 years is the discovery that introduction and ectopic expression of several genes can induce pluripotency in somatic cells. A research group led by Shinya Yamanaka of Kyoto University found that murine somatic cells transduced with retroviral vectors carrying only four transcription factors, namely Oct4, Sox2, c-Myc and Klf4 can revert their epigenetic state to become ESC-like [64]. These cells termed induced pluripotent stem (iPS) cells were similar in their properties to ESCs in terms of marker expression, transcriptional activity and the ability to differentiate into a variety of cell types in chimeras. The relative simplicity with which iPS cells can be generated compared with SCNT makes this technique an attractive approach for studying the principles of nuclear reprogramming and also to evaluate their potential for clinical applications. Indeed, mouse iPS cells were quickly developed in several laboratories and have recently been used to successfully treat sickle cell anemia in mice [65].

In November of 2007, two independent groups led by Shinya Yamanaka and James Thomson reported that using a similar transduction approach they were able to generate iPS cells from human somatic cells [66, 67]. These human ESC-like cells also expressed markers of ESCs and were capable of differentiating into cell types of all three germ layers. Yamanaka's group used the same quartet of four factors that worked in the mouse, while the Thomson lab demonstrated that a slightly new combination, OCT4, SOX2, NANOG and LIN28 can also generate human iPS cells. It is interesting to note, that the efficiency of reprogramming was lower with adult somatic cells than with cells of fetal or embryonic origin. Moreover, some adult somatic cell-derived iPS cells did not contribute to all cell types following differentiation in teratomas. It is likely that there could be additional factors that may enhance production of iPS cells from adult somatic cells. Indeed, a recent report suggests that hTERT and SV40 large T can enhance the reprogramming efficiency of Yamanaka's factors on human adult somatic cells [68].

Recent reports also suggest that the kinetics of reprogramming significantly differs between iPS and SCNT approaches. Direct reprogramming of somatic cells to iPS cells appears to be a much slower process with activation of the endogenous Oct4 or Nanog in the mouse observed on day 16 post-transduction [69]. In contrast, Oct4 expression in mouse SCNT embryos can be detected after the 4-cell stage or on day 2 after SCNT [70].

The direct genetic manipulation of somatic cells into iPS cells carries an advantage over SCNT since it does not produce totipotent cells and does not require human eggs. From a bioethical viewpoint this approach would resolve concerns about producing and destroying human embryos. However, this approach currently has serious limitations as a source of cells for regenerative medicine. Reprogramming using c-Myc results in tumor development in approximately 20% of chimeric mice derived by injection of iPS cells [71]. Recent findings suggest that c-Myc is not absolutely necessary for iPS cell induction, although it appears that reprogramming efficiency is much lower when the oncogene is omitted [72]. Another concern is that introduction of multiple copies of transgenes may cause insertional mutations and disrupt the function of many endogenous genes. Continuous overexpression of transgenes is also problematic due to the possibility of incomplete silencing of these transgenes during differentiation. The residual incidence of even a few pluripotent cells in transplanted tissues may cause tumors. Although the retroviral-delivered genes are silenced in most iPS cells, there is the likelihood of reactivation of these transgenes in differentiated cells and the possibility of spontaneous reversion of transplanted cells back to the pluripotent state, leading to the risk for malignant progression.

To avoid these pitfalls each patient-specific iPS cell line must be rigorously tested in animal models before therapeutic applications. These concerns suggest that further advances in the derivation of iPS cells without gene transfer will be required to overcome these problems. In the near future, novel reprogramming approaches that involve transient gene delivery system or small molecules may prove to be a safer way of generating iPS cells suitable for clinical applications. It will be necessary to carry out a detailed analysis of iPS cells to understand fully the mechanisms of reprogramming and their role in regenerative medicine. It is also essential to continue to study SCNT-induced reprogramming and to compare carefully the properties of iPS cell lines to those derived by SCNT.

References

1. Nicholas J, Hall B (1942) Experiments on developing rats: II. The development of isolated blastomeres and fused eggs. J Exp Zool 90:441–459
2. Johnson WH et al. (1995) Production of four identical calves by the separation of blastomeres from an in vitro derived four-cell embryo. Vet Rec 137(1):15–16
3. Willadsen SM, Polge C (1981) Attempts to produce monozygotic quadruplets in cattle by blastomere separation. Vet Rec 108(10):211–213
4. Tarkowski AK (1959) Experiments on the development of isolated blastomers of mouse eggs. Nature 184:1286–1287
5. Mitalipov SM et al. (2002) Monozygotic twinning in rhesus monkeys by manipulation of in vitro-derived embryos. Biol Reprod 66(5):1449–1455
6. Minami N, Suzuki T, Tsukamoto S (2007) Zygotic gene activation and maternal factors in mammals. J Reprod Dev 53(4):707–715
7. Mitalipov SM et al. (2002) Rhesus monkey embryos produced by nuclear transfer from embryonic blastomeres or somatic cells. Biol Reprod 66(5):1367–1373
8. Ozil JP (1983) Production of identical twins by bisection of blastocysts in the cow. J Reprod Fertil 69(2):463–468
9. Evans MJ, Kaufman MH (1981) Establishment in culture of pluripotential cells from mouse embryos. Nature 292:154–156
10. Martin GR (1981) Isolation of a pluripotent cell line from early mouse embryos cultured in medium conditioned by terato-carcinoma stem cells. Proc Natl Acad Sci USA 78:7634–7638

11. Thomson JA et al. (1998) Embryonic stem cell lines derived from human blastocysts. Science 282(5391):1145–1147
12. Ginis I, Rao MS (2003) Toward cell replacement therapy: promises and caveats. Exp Neurol 184(1):61–77
13. Dawson L et al. (2003) Safety issues in cell-based intervention trials. Fertil Steril 80(5):1077–1085
14. Taylor CJ et al. (2005) Banking on human embryonic stem cells: estimating the number of donor cell lines needed for HLA matching. Lancet 366(9502):2019–2025
15. Hochedlinger K, Jaenisch R (2006) Nuclear reprogramming and pluripotency. Nature 441(7097):1061–1067
16. Gan Q et al. (2007) Concise review: epigenetic mechanisms contribute to pluripotency and cell lineage determination of embryonic stem cells. Stem Cells 25(1):2–9
17. Jenuwein T, Allis CD (2001) Translating the histone code. Science 293(5532):1074–1080
18. Campbell KH et al. (1996) Sheep cloned by nuclear transfer from a cultured cell line. Nature 380(6569):64–66
19. Wilmut I et al. (1997) Viable offspring derived from fetal and adult mammalian cells. Nature (London) 385(6619):810–813
20. Gurdon JB (1962) The developmental capacity of nuclei taken from intestinal epithelium cells of feeding tadpoles. J Embryol Exp Morph 10:622–640
21. Pomerantz J, Blau HM (2004) Nuclear reprogramming: a key to stem cell function in regenerative medicine. Nat Cell Biol 6(9):810–816
22. Wakayama T et al. (1998) Full-term development of mice from enucleated oocytes injected with cumulus cell nuclei. Nature 394(6691):369–374
23. Kato Y et al. (1998) Eight calves cloned from somatic cells of a single adult. Science 282(5396):2095–2098
24. Cibelli JB et al. (1998) Cloned transgenic calves produced from nonquiescent fetal fibroblasts. Science 280(5367):1256–1258
25. Polejaeva IA et al. (2000) Cloned pigs produced by nuclear transfer from adult somatic cells. Nature (London) 407(6800):86–90
26. Baguisi A et al. (1999) Production of goats by somatic cell nuclear transfer. Nat Biotechnol 17(5):456–461
27. Chesne P et al. (2002) Cloned rabbits produced by nuclear transfer from adult somatic cells. Nat Biotechnol 20(4):366–369
28. Shin T et al. (2002) A cat cloned by nuclear transplantation. Nature 415(6874):859
29. Woods GL et al. (2003) A mule cloned from fetal cells by nuclear transfer. Science 301(5636):1063
30. Galli C et al. (2003) Pregnancy: a cloned horse born to its dam twin. Nature 424(6949):635
31. Zhou Q et al. (2003) Generation of fertile cloned rats by regulating oocyte activation. Science 302(5648):1179
32. Lee BC et al. (2005) Dogs cloned from adult somatic cells. Nature 436(7051):641
33. Capecchi MR. (1989) Altering the genome by homologous recombination. Science 244(4910):1288–1292
34. Mak TW (2007) Gene targeting in embryonic stem cells scores a knockout in Stockholm. Cell 131(6):1027–1031
35. Thomson JA et al. (1995) Isolation of a primate embryonic stem cell line. Proc Natl Acad Sci USA 92(17):7844–7848
36. Thomson JA et al. (1996) Pluripotent cell lines derived from common marmoset (Callithrix jacchus) blastocysts. Biol Reprod 55(2):254–259
37. Suemori H et al. (2001) Establishment of embryonic stem cell lines from cynomolgus monkey blastocysts produced by IVF or ICSI. Dev Dyn 222(2):273–279
38. Mitalipov S et al. (2006) Isolation and characterization of novel rhesus monkey embryonic stem cell lines. Stem Cells 24(10):2177–2186
39. Handyside AH et al. (1987) Towards the isolation of embryonal stem cells from the sheep. Rouxs Arch Dev Biol 196:185–190

40. Evans MJ et al. (1990) Derivation and preliminary characterization of pluripotent cell lines from porcine and bovine blastocysts. Theriogenology 33:125–128
41. Notarianni E et al. (1990) Maintenance and differentiation in culture of pluripotential embryonic cell lines from pig blastocysts. J Reprod Fertil 41(Suppl):51–56
42. Giles JR et al. (1993) Pluripotency of cultured rabbit inner cell mass cells detected by isozyme analysis and eye pigmentation of fetuses following injection into blastocysts or morulae. Mol Reprod Dev 36(2):130–138
43. Iannaccone PM et al. (1994) Pluripotent embryonic stem cells from the rat are capable of producing chimeras. Dev Biol 163(1):288–292
44. Gurdon JB, Colman A (1999) The future of cloning. Nature 402(6763):743–746
45. Lanza RP, Cibelli JB, West MD (1999) Human therapeutic cloning. Nat Med 5(9):975–977
46. Munsie MJ et al. (2000) Isolation of pluripotent embryonic stem cells from reprogrammed adult mouse somatic cell nuclei. Curr Biol 10(16):989–992
47. Brambrink T et al. (2006) ES cells derived from cloned and fertilized blastocysts are transcriptionally and functionally indistinguishable. Proc Natl Acad Sci USA 103:933–938
48. Wakayama S et al. (2006) Equivalency of nuclear transfer-derived embryonic stem cells to those derived from fertilized mouse blastocysts. Stem Cells 24(9):2023–2033
49. Stojkovic M et al. (2005) Derivation of a human blastocyst after heterologous nuclear transfer to donated oocytes. Reprod Biomed Online 11(2):226–231
50. Mitalipov SM et al. (2002) Rhesus monkey embryos produced by nuclear transfer from embryonic blastomeres or somatic cells. Biol Reprod 66(5):1367–1373
51. Kennedy D (2006) Editorial retraction. Science 311:336
52. Mitalipov SM et al. (2007) Reprogramming following somatic cell nuclear transfer in primates is dependent upon nuclear remodeling. Hum Reprod 22(8):2232–2242
53. Byrne JA et al. (2007) Producing primate embryonic stem cells by somatic cell nuclear transfer. Nature 450(7169):497–502
54. Condic ML (2008) Alternative sources of pluripotent stem cells: altered nuclear transfer. Cell Prolif 41(Suppl 1):7–19
55. Hurlbut WB (2005) Altered nuclear transfer: a way forward for embryonic stem cell research. Stem Cell Rev 1(4):293–300
56. Niwa H et al. (2005) Interaction between Oct3/4 and Cdx2 determines trophectoderm differentiation. Cell 123(5):917–929
57. Mitalipov SM et al. (2003) Oct-4 expression in pluripotent cells of the rhesus monkey. Biol Reprod 69(6):1785–1792
58. Torres-Padilla ME et al. (2007) Histone arginine methylation regulates pluripotency in the early mouse embryo. Nature 445(7124):214–218
59. Chawengsaksophak K et al. (2004) Cdx2 is essential for axial elongation in mouse development. Proc Natl Acad Sci USA 101(20):7641–7645
60. Strumpf D et al. (2005) Cdx2 is required for correct cell fate specification and differentiation of trophectoderm in the mouse blastocyst. Development 132(9):2093–2102
61. Nishioka N et al. (2008) Tead4 is required for specification of trophectoderm in pre-implantation mouse embryos. Mech Dev 125:270–283
62. Yagi R et al. (2007) Transcription factor TEAD4 specifies the trophectoderm lineage at the beginning of mammalian development. Development 134(21):3827–3836
63. Meissner A, Jaenisch R (2006) Generation of nuclear transfer-derived pluripotent ES cells from cloned Cdx2-deficient blastocysts. Nature 439(7073):212–215
64. Takahashi K, Yamanaka S (2006) Induction of pluripotent stem cells from mouse embryonic and adult fibroblast cultures by defined factors. Cell 126(4):663–676
65. Hanna J et al. (2007) Treatment of sickle cell anemia mouse model with iPS cells generated from autologous skin. Science 318(5858):1920–1923
66. Takahashi K et al. (2007) Induction of pluripotent stem cells from adult human fibroblasts by defined factors. Cell 131(5):861–872
67. Yu J et al. (2007) Induced pluripotent stem cell lines derived from human somatic cells. Science 318:1917–1920

68. Park IH et al. (2008) Reprogramming of human somatic cells to pluripotency with defined factors. Nature 451(7175):141–146
69. Brambrink T et al. (2008) Sequential expression of pluripotency markers during direct reprogramming of mouse somatic cells. Cell Stem Cell 2(2):151–159
70. Boiani M et al. (2002) Oct4 distribution and level in mouse clones: consequences for pluripotency. Genes Dev 16(10):1209–1219
71. Okita K, Ichisaka T, Yamanaka S (2007) Generation of germline-competent induced pluripotent stem cells. Nature 448(7151):313–317
72. Nakagawa M et al. (2008) Generation of induced pluripotent stem cells without Myc from mouse and human fibroblasts. Nat Biotechnol 26(1):101–106

Adv Biochem Engin/Biotechnol (2009) 114: 201-235
DOI: 10.1007/10_2008_27

Published online: 10 June 2009

Large Scale Production of Stem Cells and Their Derivatives

Robert Zweigerdt

Abstract Stem cells have been envisioned to become an unlimited cell source for regenerative medicine. Notably, the interest in stem cells lies beyond direct therapeutic applications. They might also provide a previously unavailable source of valuable human cell types for screening platforms, which might facilitate the development of more efficient and safer drugs. The heterogeneity of stem cell types as well as the numerous areas of application suggests that differential processes are mandatory for their in vitro culture. Many of the envisioned applications would require the production of a high number of stem cells and their derivatives in scalable, well-defined and potentially clinical compliant manner under current good manufacturing practice (cGMP). In this review we provide an overview on recent strategies to develop bioprocesses for the expansion, differentiation and enrichment of stem cells and their progenies, presenting examples for adult and embryonic stem cells alike.

Keywords Bioreactor, Cell therapy, Differentiation, Process development, Stem cells, Teratoma

Contents

R. Zweigerdt (✉)
Institute of Medical Biology (IMB), 8A Biomedical Grove, # 06-06 Immunos, Level 5, Room # 5.04, Singapore 138648
e-mail: robert.zweigerdt@imb.a-star.edu.sg

Abbreviations

(NOD/SCID) mice	Nonobese diabetic/severe combined immunodeficient
(RWV) bioreactor	Rotating wall vessel
bFGF, FGF-2	Basic fibroblast growth factor
BM	Bone marrow
BMP	Bone morphogenetic protein
BMP	Morphogenetic protein
cGMP	Current good manufacturing practice
CHO	Hamster ovary cells
EBs	Embryoid bodies
ESC	Embryonic stem cells
GM-CSF	Granulocyte macrophage colony stimulating factor
hESC	Human embryonic stem cell
hNPC	Neural precursor cells
HSC	Hematopoietic stem and progenitor cells
LIF	Leukemia inhibitory factor bone
MASC	Magnetic activated cell sorting
MI	Myocardial infarction
MSC	Mesenchymal stem cells
NSC	Neural stem cells
PB	(Mobilized) Peripheral blood
SCF	Stem cell factor
SNM`	Spherical neural masses
TGF-beta	Transforming growth factor beta
UCB	Umbilical cord blood

1 Introduction

The title of this review is a bold claim. It implies that large scale production of stem cells is, to some extent, an established practice. Process scale-up of common mammalian cell lines such as Chinese hamster ovary cells (CHO), human tumor cells lines (such as HEK 293 and HeLa), and myelomas, which have been extensively used to produce large quantities of biopharmaceutical products (e.g., antibodies and cytokines), has indeed resulted in fermentation volumes of >1,000 or even >10,000 L in recent years [1–4]. In contrast, stem cell production and differentiation in vitro is in its infant stage. Process optimization experiments are often performed in 0.1–10 mL medium in tissue culture dishes. Spinner flask and other bioreactor volumes of 50–250 mL are considered a substantial up-scaling and lab-scale processes exceeding 1 L reactor volume are an exception.

One major underlying reason is the still limited knowledge of stem cell biology hampering the development of efficient and commercially viable processes. Not surprisingly, a recent leading edge analysis by Ann B. Parson [5] underscores that ramping up the process for stem cells products is currently one of the key success hurdles for biotech companies in the field.

1.1 Cells for Therapies: Estimating Cell Number Requirements

How many cells are actually necessary for future therapies? Obviously, this will depend on the respective application but some of the presently utilized cell therapy applications serve to highlight the dimensions. In the field of heart repair, for example, one can assume that the left ventricle of a human heart contains about 4–6 billion cardiomyocytes [6–8]. Individuals can survive myocardial infarction (MI) that affects about one-third of the left ventricle. Cardiac regeneration would thus require the replacement of as many as 1–2 billion cardiomyocytes that are irreversibly lost through hypoxia-reperfusion injury.

Similar numbers apply to beta-cell replacement in type 1 diabetic patients. The Edmonton protocol, a pancreatic islet transplantation procedure, typically utilizes a transplant of approximately 600,000 islet equivalents comprising abut 1,000 beta cells each [9] derived from cadaveric donor pancreata. This would mean that about 1 billion stem cell-derived functional beta-cell equivalents would be required per patient [10].

Another example documents the dimension of donor cell requirement to reconstitute stably blood formation in patients after chemotherapy or irradiation treatment. Using umbilical cord blood (UCB) as a cell source, cell doses of 15 million mononucleated cells containing about 1% CD34+ hematopoietic stem and progenitor cells per kg patient weight appears to be the threshold for safe transplants [11]. An adult of 80 kg receiving an unrelated UCB transplantation will thus need about 1.2 billion (1.2×10^9) nucleated cells including 12 million CD34+ cells. Supposing that UCB samples can contain about 1×10^8 mononucleated cells comprising 1% CD34+ cells, in vitro expansion would require a 12-fold increase of the cell population, thereby keeping the proportion of CD34+ cells intact, which is a key factor for successful transplant products as discussed in more detail below.

These examples suggest that 1–2 billion stem and/or differentiated progenitor cells per patient is a useful ballpark number to estimate production requirements in bioprocess development.

1.2 Cell Sources for Therapies: Adult vs Embryonic Stem Cells

Stem cells are defined as being self-renewing, pluri- or multipotent, and clonogenic. Clonogenic cells are single stem cells that are able to generate a line of genetically identical cells thereby maintaining their self-renewal and differentiation potential. Stem cells exist at different hierarchical levels throughout the development of an organism and persist in adult tissues. At one end of the spectrum, pluripotent embryonic stem cells (ESC) can give rise to all cell types in the body whereas tissue specific, multipotent stem cells only retain the ability to differentiate into a restricted subset of cell types.

With the exception of hematopoietic stem and progenitor cells (HSC), which have been used in the clinic for more than 50 years [12], the routine therapeutic application of stem cells is limited to date. Ten years after the first derivation of stable human embryonic stem cell (hESC) lines by Thomson and coworkers [13], no clinical trial based on this cell source has yet been initiated. Although trials have been announced for spinal cord repair and ophthalmic disorders by biotech companies', initiation was repeatedly delayed due to profound safety and ethical concerns [5].

Present experimental trials aimed at cell-based tissue repair have thus focused on cells isolated from patients own tissue. Autologous approaches avoid donor cell rejection and the risk of teratoma formation (benign tumors containing cells from various differentiated tissues) imposed by ESC. These personalized cell treatments require no or limited small-scale expansion of harvested cells. Examples are (1) calf biopsy-derived in vitro expanded skeletal myoblasts and (2) nonexpanded, bone marrow-derived mononucleated cells. Both of these cell types are currently being tested for heart repair in patients post MI [14]. However, poorly defined mixtures of autologous cells are often used in experimental trials simply because the (stem-) cell type(s) with a supposed therapeutic potential is not known [14]. Crude bone marrow biopsies or fractions thereof are being tested for heart repair whilst the discussion on the adequate cell type, the optimal modus of application, and the expected clinical outcome is in full swing [14–16]. Considering the controversial observations from animal models, the distrust of numerous investigators towards ongoing clinical trials is not surprising [16–18]. Results observed in rodent hearts range from efficient cardiomyogenic differentiation of bone marrow derived cells [19] and mesenchymal stem cells (MSC [20, 21]) to negligible heart muscle cell differentiation of these cell types [22, 23] and even deleterious effects like the calcification of MSC injected into heart muscle [24].

This debate not only concerns the question of which cell type is most suitable to repair a particular organ. It also relates to the underlying question of whether primitive, undifferentiated stem or progenitor cells could be delivered to regenerate damaged tissue (where the differentiation will be guided in vivo by signals in recipients damaged organ) or whether stem cells must be directed to differentiate into mature, tissue specific progenies in vitro and then transplanted. Apparently, these considerations define the goals and strategies for bioprocess development.

Notably, the interest in stem cells lies beyond direct therapeutic applications. Stem cells, or differentiated progenitors thereof, provide a promising source of valuable human cell types that have not been available for in vitro assays before. This will allow the development of novel, scalable screening platforms for compound discovery and toxicity testing which might help to develop more efficient and safer drugs [25]. Another area of stem cell research is the study of developmental and differentiation processes as well as stem cell malignancy and genetic disorders in vitro.

The heterogeneity of stem cell types as well as the numerous areas of application suggests that differential processes are mandatory for their in vitro culture. Many of the envisioned applications would require the production of a high number of stem cells and their derivatives in scalable, well-defined and potentially clinical compliant manner under current good manufacturing practice (cGMP). In this review we will provide an overview on recent strategies to develop scalable bioprocess for the expansion and differentiation of stem cells, providing examples for adult and embryonic stem cells alike.

2 Strategies in Stem Cell Scale-Up

Development of clinical/industrial scale process for cell production requires a focus on key questions of process efficiency and eventually commercial viability of an envisioned strategy. This includes estimating the process dimension defined by the (1) number of cells to be transplanted per treatment, (2) bioreactor dimensions needed to generate multiple cell doses, (3) required total medium throughput, and (4) process duration; subsequently process costs can be calculated.

Using cardiomyocytes and pancreatic cells as examples, we have calculated above that 1–2 billion cells per patient will theoretically be needed to replace the loss of functional tissue. Notably, true cell numbers for successful organ repair might be extensively higher. Recent animal models suggest that only a single-digit percentage of transplanted donor cardiomyocytes eventually survive and integrate in the heart [26, 27]. Also, the physiological potency of surrogate cells generated in vitro might require higher donor cell doses. For example, the insulin release in response to a defined glucose challenge, a potency assay used to assess beta-cell functionality in vitro, is much lower in ESC-derived beta-like cells compared to cadaveric donor-derived beta cells embedded in functional islets. The latter comprise the gold standard in the field [10, 28].

In addition, differentiation of stem cell in vitro usually results in a mixed culture with the desired cell type being a minority even if protocols for directed differentiation are applied. Let's assume that a target cell type such as beta-cells or cardiomyocytes can be generated from ESC with a relative high efficiency of 20%. Subsequently a bioprocess must generate the total amount of 5 billion differentiated progenies to produce 1 billion target cells, which would thus impact on process dimension. The resulting cell mixture might be subjected to subsequent purification steps to achieve lineage purity.

Equipped with such estimations, process development is concerned with experimental-scale approaches to provide initial real-world figures on process efficiencies, dimensions, and costs, which are subsequently subjected to up-scaling and optimization.

2.1 Culture Media and Cell Attachment Matrices: Critical, Expensive, and yet Poorly Defined

One of the most essential and costly components in stem cell production is the culture medium. Development of media that either support stem cell self renewal and proliferation or, in contrast, direct differentiation into desired lineages is at the heart of current research. Experimental reports often utilize media comprising relative high amounts of serum. Unfortunately, serum is subjected to batch-to-batch variations and represents a xenogeneic component that might conflict with the generation of clinically-compliant stem cell products. In mouse and human ESC research, the need for defined media has resulted in broad usage of commercially available serum replacement (e.g., Invitrogen, Carlsbad, CA, USA) but the formulations still generally contain bovine serum albumin.

However, studies have begun to unravel signal transduction pathways controlling self renewal and differentiation in more detail resulting in chemically-defined, xeno-free media as outlined below. In this context, synthetically manufactured compounds that can control signaling pathways and subsequently stem cell behavior are progressively tested in the field [29]. Ultimately, this strategy will not only facilitate generation of chemically defined media. Applying small molecules might also support commercial viability of bioprocesses by replacing recombinant, costly growth factors and cytokines that are currently obligatory components of many media formulations. Prominent examples are fibroblast growth factor-2 (FGF-2) supplemented to culture media for hESC expansion or numerous hematopoietic growth factors including interleukins, granulocyte macrophage colony stimulating factor (GM-CSF), stem cell factor (SCF) and others that are currently indispensable for the in vitro cultivation of HSC [12]. Other examples include the transforming growth factor beta (TGF-beta) family members activin and bone morphogenetic protein (BMP)-4, arguably among the most expensive molecules on the planet, which have recently been suggested in a sequential protocol to direct cardiomyocyte differentiation from hESC [26].

In conjunction with the culture medium, another key component controlling stem cell characteristics in vitro is the matrix provided for cell attachment. With the exception of HSC, which have historically been grown on stromal feeder cells but are now generally expanded in suspension culture, most other stem cell types have been isolated under conditions depending on surface adherence. Mouse, primate, and human ESC were all derived on a layer of embryonic fibroblast. Much effort is currently being applied to replace this coculture system, which strongly interferes with up-scaling strategies, by defined matrices. MSC, per definition, are tissue culture plastic adherent cells. Thus, it's easier to comply with their demands regarding the surface matrix used for expansion. However, culture surface enlargement to ensure efficient and reasonable mass expansion of anchorage-dependent cells is a central challenge in bioreactor design.

2.2 *Bioreactors and Microcarriers: Providing Stem Cells with a Home and a Bed*

A bioreactor may be defined as a system that simulates physiological environments for the creation, physical conditioning, and testing of cells, tissues, precursors, support structures, and organs in vitro. It thus provides for a regulated and controlled environment. At first glance, bioreactors look like highly complicated and sophisticated equipment, and indeed, very heterogeneous designs and setups exist. However, exempting some exotic models, they can be divided into a few simple categories.

The simplest and among the most extensively used reactor types in mammalian cell culture are stirred tank reactors (usually a cylinder-shaped vessel). Spinner flasks represent a simple lab-scale format of this reactor type (typical working volume of 50–250 mL), and are placed in tissue culture incubators to provide the basic growth environment, which is controlled temperature and aeration gas mixture. Spinner flask aeration is usually limited to the gas exchange at the headspace. Homogeneous mixing of the culture solution is ensured via impeller(s), turbines, or bulb-shaped stirring devices. Design of these impellers and vessel geometry as well as the stirring speed define the medium flow (direction, velocity) and thus homogeneity of culture mixing, efficiency of gas exchange, and, importantly, shear forces acting on the cells.

Compared to spinner flasks, instrumented stirred tanks allow online measurement and adaptation of parameters like the pH and oxygen tension (pO_2). Installed ports enable the simple and regular collection of culture samples. This facilitates offline (or even online) measurement of additional parameters such as cell density, cell vitality, glucose consumption, accumulation of potentially toxic metabolites such as ammonia, medium osmolarity and others. Instrumented tanks also enable additional culture aeration through a so-called sparger, a device that generates gas bubbles at the bottom of a vessel thereby adding to the gas diffusion from the headspace, to keep the pO_2 constant even in dense cultures demanding high oxygen supply. This is particularly important for stem cell cultures, as the pO_2 has been shown to impact on stem cell differentiation into specific lineages as outlined below. Another feature is the possibility for continuous feeding. Fresh medium is constantly added at a defined speed and an equivalent medium amount is constantly removed from the culture, usually without cell removal (cell retention techniques). Continuous medium perfusion, in contrast to batch feeding which is the standard feeding technique in tissue culture, results in more homogeneous culture conditions which can have profound consequences in stem cell bioprocessing. It is well established that stem cells and their differentiated progenies release inhibiting and stimulating factors that can strongly feedback on cell pluripotentiality, proliferation and differentiation. Perfusion feeding of hESC, for example, enabled growth to much higher cell densities without inducing differentiation compared to batch fed controls ([30]; perfused stationary culture). Other examples of this topic are presented for HSC expansion and cardiomyogenic differentiation of ESC below.

Stirred tanks are favored in process scale-up because established culture conditions in lab-scale can often be transferred to much higher volumes with relative ease by keeping both physical (vessel and stirrer geometry, medium flow features/shear forces, medium throughput, feeding strategy, etc.) and physiological (pO_2, pH, glucose conc., metabolic waste conc., etc.) parameters constant [1–4]. However, cells often do not immediately 'take' to culture in stirred suspension systems. Consequently, in the biopharmaceutical industry, a critical scale-up step is the adaptation of initially anchorage-dependent production cell lines to (usually serum-free) suspension culture growth without interfering with the quality and quantity of the desired, cell-derived product [1]. Such adaptation steps, however, might strongly interfere with stem cell characteristics limiting translation of this strategy to stem cell research.

Another technique to enable the growth of attachment-dependent cells in suspension is the use of microcarriers. In 1976 Van Wezel describes the use of small particles (0.2 mm), microcarriers, for the growth of anchorage-dependent cells [31]. These spherical particles are kept in suspension by stirring or other mixing techniques and provide a massively enlarged attachment surface in a relative small reactor volume due to their high surface-area-to-volume ratio. Carriers have been previously used in conventional cell culture, e.g., for vaccine production.

As with bioreactors, a "plethora" of microcarriers exists; they come in all shapes and sizes. Aiming to provide optimal cell attachment properties for diverse cell type microcarriers made from numerous materials are available. One main category of microcarriers comprises solid, spherical or disc-shaped particles made of cross-linked dextran, cellulose or polystyrene [32]. The other category is termed micro- or macroporous carrier [33]. Macroporous carriers have a sponge-like structure. They are typically made of soft materials such as gelatin or collagen and allow cells to grow in their internal pores. Due to their rough external surface, macroporous carriers generate more microeddies, resulting in higher fluid shear that acts on surface attached cells compared to solid, spherical carriers [34]. However, solid carriers also impose high mechanical stress on cells in stirred culture whereas cells grown in the interior of porous particles might be well protected. Also, the microenvironment that might develop in the vicinity of cells grown in micropores might be different from the bulk of the culture vessel and either support stem cell maintenance or differentiation. Macroporous scaffolds have therefore been used for heterogeneous hematopoietic cell cultures which entail a mixed population of adherent and suspension cells [35, 36].

More recently, porous as well as solid type microcarriers have also been tested for mouse ESC cultivation and differentiation in spinner flasks [37, 38].

To prevent the potentially detrimental shear stress on cells in stirred microcarriers culture surface enlargement for anchorage-dependent cells can also be achieved in fixed/packed bed reactors. These are fully controlled bioreactors in which macroporous microcarriers or other substrates (e.g., glass or plastic beads of various sizes, discs made of porous material etc.) are embedded in a column-shaped vessel (cell compartment). To supply cells that have been seeded into the substrate, aerated culture medium is continuously circulated through the cell compartment; in most configurations bubble-free medium aeration is established through a semipermeable membrane. Fresh medium is added according the metabolic needs of the cells and metabolic waste products are removed. Configurations of this reactor type have been applied to engineer murine and human bone marrow models to mimic ex vivo hematopoiesis [36, 39]. Hollow fiber reactors, which have also been used for HSC culturing [40], utilizes a capillary-like fiber structure for surface-enlargement; again, oxygenated medium is circulated through these fibers for cell supply.

Finally, an even lower mechanical and hydrodynamic shear but still efficient mixing and agitation of cells in suspension is enabled by rotating wall vessel (RWV) bioreactors; in contrast to the reactor types described above the incubator vessel itself is rotated to mimic gravity-free culture conditions [41]. Improved RWV systems enabling parallel bi-axial vessel rotation were recently developed and applied for efficient three-dimensional tissue engineering [42, 43]. Examples applying RWV reactors for HSC expansion and hESC differentiation are further presented below.

In the following sections we will review the status of bioprocessing with respect to several stem cell types that have an established or an envisioned role in regenerative medicine.

3 Hematopoietic Stem and Progenitor Cells: Long Medical History but Limited Ex-Vivo Expansion of a Complex Cell Mixture

Hematopoietic stem cells reside as rare cells in the bone marrow in adult mammals and sit atop a hierarchy of progenitors that become progressively differentiated to mature blood cells, including erythrocytes, megakaryocytes, myeloid cells, and lymphocytes [44]. Hematopoietic stem and progenitor cell (HSC) transplants are used as part of the treatment of a variety of genetic disorders, blood cancers, some solid tumors and when the bone marrow is damaged or diseased. Since the main forms of cancer treatments, chemotherapy and radiotherapy, are nonspecific, healthy cells including bone marrow cells are also damaged. If the intensity of the therapy destroys the bone marrow function for blood regeneration, a transplant is necessary to prevent live-threatening complications such as infections and bleeding. Full long term reconstitution of blood formation in patients receiving HSC transplants is a paradigm for successful stem cell therapies.

3.1 Stem Cell Sources and Clinical Application

Bone marrow (BM) was the first source of HSC used for transplantation but in the meantime other sources including (mobilized) peripheral blood (PB; isolated via apheresis) and umbilical cord blood (UCB) have also been utilized [45–47]. A major limitation to the clinical application of HSC has been the absolute number of stem- and progenitor- as well as mature hematopoietic cells available in stem cell products. On the other hand it has been proposed that a single stem cell is capable of more than 50 cell divisions and has the principal capacity to generate up to 10^{15} progenies, or sufficient cells for up to 60 years blood formation in an adult human [48, 49]. This potential level of expansion, if realizable ex-vivo, may have an impact on the cellular genetic stability and the differentiation potential of HSC due to the loss of telomere length [50] and oxidative stress [51]. However, even a modest in vitro expansion would have a significant effect on HSC availability and investigators have evaluated this possibility to achieve the following clinical needs:

- Generating a sufficient number of stem cells from a single bone marrow aspirate or apheresis procedure to reduce the need for large marrow harvests or multiple leukaphereses
- Generating sufficient cells from a single umbilical cord blood harvest to reconstitute an adult following high-dose chemotherapy
- Supplementing stem cell grafts with more mature precursors to limit pancytopenia (shortage of all types of blood cells)
- Increasing the number of primitive progenitors in stem cell grafts to ensure hematopoietic support for multiple cycles of high-dose chemotherapy

Adapted from Ian McNiece [12].

3.2 In Vitro Expansion and Scale-Up

Challenges associated with in vitro HSC propagation are generally applicable to most other adult stem cells as well. Particular considerations include (1) the heterogeneity of the cell source(s) available for process inoculation, (2) absence of definitive stem cell surface markers, and (3) absence of fast and reliable assays to test stem cell function. However, within the population of donor-harvested mononucleated cells expression of the CD34 antigen (CD34+), a cell surface glycoprotein, paralleled with the absence of lineage markers and CD38 expression (lin-, CD38-) has become the distinguishing feature used for the enumeration and isolation of HSC. CD34 is down regulated as cells differentiate towards hematopoietic lineages [52, 53]. Transplantation studies in several species have also shown that long-term marrow repopulation can be provided by CD34+ cells. Therefore, relevant clinical and experimental protocols aimed at the in vitro expansion of HSC are often quantifying the rare fraction of CD34+ cells pre- and postexpansion to determine success of the bioprocess. Additionally, in vitro colony forming (e.g., methylcellulose assay) and differentiation assays are combined with the in vivo ability to reconstitute multilineage hematopoieses in a xeno-transplantation model using nonobese diabetic/severe combined immunodeficient (NOD/SCID) mice to ensure the quality of the expanded cell population [54].

Aiming to reconstitute the so-called HSC niche [55], many in vitro cultures have been designed to regulate the HSC microenvironment by coculture systems utilizing supportive feeder cell lines [56–59]. However, cocultures are not only challenging in process scale-up. Most of the systems have only demonstrated maintenance of HSC numbers without achieving the desired expansion, potentially because they model steady-state hematopoietic homeostasis in vivo to some extend. Finally, feeder-based cultures may not require direct cell–cell contact but rather the secretion of HSC-supporting factors by feeder cells. This observation has driven the development of serum- and feeder-free suspension culture protocols and extensive research has been devoted in identifying optimal cocktails of hematopoietic growth factors that simultaneously inhibit apoptosis, induce mitosis, and prevent differentiation. For reviews on this topic, please see Heike and Nakahata [60], Noll et al. [61] and McNiece [12], with the later being focused on clinical studies of in vitro expanded HSC.

Besides the medium composition, HSC cultures are influenced by many other factors. Considering the general donor-to-donor variability on the expansion potential of HSC [62] it has been shown that cell production is improved by using lower seeding densities, preenrichment of stem and progenitor cells for process inoculation, increased medium exchange via culture perfusion, and applying high concentrations of early-acting growth factors [63, 64]. Adding to the complexity, it was observed that human CD34+ cells as well as differentiated hematopoietic cell types secrete numerous growth factors acting as autocrine and paracrine factors in normal hematopoiesis [65]. Exploiting this observation recent studies have improved HSC expansion by removal of lineage marker expressing differentiated progenies from the culture to avoid feedback inhibition [66, 67]. Modifying this approach in future by selectively removing cell types that secrete inhibitory factors but leaving other progenies that produce stimulatory cytokines behind might create a self-stimulating environment, thereby limiting the need for adding costly cytokines [67].

Using bone marrow, peripheral blood or umbilical cord blood for culture inoculation, permeable blood bags and conventional T-flasks are still most widely used for the expansion of human HSC in the clinic. Although they are simple to handle, these systems have the typical limitations of static cultures such as the development of gradients (e.g., dissolved oxygen, pH, cytokines and metabolites), lack of online control for environmental conditions, and a limited surface area. Due to their long medical history, HSC were among the first stem cell types to be cultured in bioreactors; numerous types of reactors have been applied including hollow fiber-, perfusion chamber-, fixed bed-, and stirred vessel bioreactors reviewed elsewhere [68, 69]. Recent studies, however, are progressing towards the long term HSC expansion in increased culture volumes. For example, UCB- and PB-derived mononuclear cells were expanded in a stirred bioreactor equipped with dissolved oxygen and pH control, whereby the process efficiency was greatly enhanced by using a cell-dilution feeding protocol [70]. Another stirred suspension approach in 250-mL scale was published by Kim and coworkers [71] documenting the expansion of human BM; supplementation with factors secreted by stromal feeder cells combined with growth promoting and growth inhibiting cytokines enabled the prolonged expansion of hematopoietic progenitors. Long term culture (several weeks) and expansion of UCB and PB was also achieved in a cocultivation setting utilizing a perfused fixed bed bioreactor seeded with immobilized stromal cells on porous glass carriers [72]. As mentioned above, perfusion has been suggested to facilitate HSC expansion by increasing the medium exchange rate [73]. However, it is also known that hematopoietic cells are extremely sensitive to shear forces which can limit their viability in stirred and perfused systems or at least affect gene expression including cytokine receptors [68, 69]. This aspect has prompted Liu and coworkers [74] to apply a rotating wall vessel (RWV) bioreactor (33 mL working volume) which ensured laminar flow, resulting in minimal shear stress and well-mixed culture conditions, and also avoided the formation of gradients. Culturing UCB in the RWV reactor, on average, enabled a 435.5±87.6-fold expansion of all mononucleated cells paralleled by a 33.7±15.6-fold increase of CD34+ cells within ~ 8 days. Although this result is encouraging the authors have calculated that a process scale-up to four 500-mL RWV reactors running in parallel would be mandatory to generate a clinically relevant transplant for an 80-kg patient. This calculation assumes that the process is inoculated with a single, typical UCB sample and the expansion kinetic observed in the current 33-mL reactor scale is translatable to the envisioned 2-L dimension process and prolonged cultivation time.

Given the complexity of this multiparameter system it's not surprising that, despite the large number of studies HSC growth in serum-free, cytokine-supplemented liquid suspension culture has been still modest to date. For the next generation of HSC bioprocess design, it was therefore proposed to perform dynamic system perturbations comprising extensive control of the cell-population (lineage selective removal/maintenance), media control (exchange/dilution), and selective growth factor supplementation to efficiently increase particularly the stem and progenitor population in the culture [75].

4 Embryonic Stem Cells

4.1 ESC Expansion: Providing the Raw Material for Future Therapies

Compared to tissue-derived adult stem cells, embryonic stem cells that were successfully derived from blastocyst stage embryos of several species including mice [76], primates [77], and importantly humans [13], offer the particular advantage of prolonged proliferative capacity and great versatility in the lineages that can be formed in culture. Translating these advantages into clinical benefits faces many challenges, including the efficient differentiation into a desired cell type, maintaining genetic stability during long term culture, ensuring the absence of tumorigenic ESC persisting in a therapeutic product, and scalability of existing protocols for mass generation of donor cells. By focusing on recent approaches of both, mouse and human ESC expansion and differentiation in bioreactors we will discuss the impact of above challenges on process scale-up.

To exploit the growth capability of ESC in their pluripotent state, a vital strategy for process development would depend on the expansion of a large starting population, which can be used to inoculate differentiation processes. Since ESC are anchorage-dependent and grow in typical colonies, current methods to scale-up their numbers have focused on flat surfaces or matrices [78]. For mass expansion, the simplest surface enlargement could be achieved by utilizing multilayered tissue culture flasks, so-called cell factories (produced by several manufacturers). They provide a relative large growth surface in limited space under standard tissue culture conditions facilitating adaptation of established tissue culture protocols. Their disposable nature would also facilitate GMP and clinical compliance. However, homogeneous cell distribution for the inoculation of multilayered flasks would require single cell dissociation of ESC combined with a medium formulation that ensures robust self-renewal. Mouse ESC (mESC) fulfill these requirements. They can be passaged by single cell dissociation and differentiation is largely avoided when the cells are grown on a simple gelatin matrix in the presence of leukemia inhibitory factor (LIF), an interleukin-6 family member that activates the Jak/Stat pathway. Unfortunately, this pathway fails to maintain self-renewal in human ESC [79].

Research to unravel the apparently multifactorial network of growth factors and downstream signaling that controls hESC pluripotency is in full swing. Members of the FGF family, particularly FGF-2, have been shown to support hESC self-renewal whereas the blockage of BMP-signaling by noggin or activin is required to retain their phenotype. For details on this topic please see recent reports and reviews [80, 81].

At present, serum replacement-based media (to avoid fetal calf serum; a product form Invitrogen, Carlsbad, USA) supplemented with FGF-2 are still broadly used for hESC culture. In addition, hESC have been mostly grown on a variety of feeder cell lines or on extra cellular matrices such as matrigel, fibronectin, laminin, or heparan sulfate and supplemented with conditioned media derived from the feeder cells [78, 82, 83]. To enable up-scaling of these culture platforms a clinical grade-human feeder cell line grown on microcarriers in spinner flasks was recently established [84]. These extensively characterized feeders have also been used to derive clinical-grade hESC cell lines [85], an important step toward the generation of fully controlled products for clinical trials. Large scale production of clinical- and cGMP-grade feeder conditioned medium might be a commercially vital strategy for hESC mass culture thereby limiting the need for costly growth factor supplementation even if a definitive cocktail will finally be available.

However, a notable discovery identified that hESC are capable of taking up substantial amounts of the potentially immunogenic nonhuman sialic acid Neu5Gc [86] and acquire bovine apolipoprotein B-100 [87] from feeder layers and the serum replacement medium, which contains animal compounds such as bovine serum albumin. Extensive research is therefore ongoing to replace bovine components either by recombinant human serum albumin and/or to simplify culture condition with just the essential serum components, such as sphingosine-1-phosphate and platelet-derived growth factor [78].

Notably, and in contrast to the still elusive definitive markers of hematopoietic stem cells, availability of numerous well established markers known to be expressed in pluripotent hESC strongly facilitates the mandatory development of completely defined culture media. Pluripotentiality markers that are downregulated upon hESC differentiation include surface antigens such as stage specific embryonic antigen (SSEA)-3 and SSEA-4, Trafalgar (Tra)-1-60 and Tra-1-81, and GCTM-2 as well as transcription factors including Oct4, Nanog, and Sox2 [88]. Immune cytology specific to these and other markers unraveled the heterogeneity of hESC cultures grown under most established culture condition suggesting progressive differentiation to some degree [88, 89]. The grade of culture heterogeneity, which also varies between independently derived hESC lines [90], apparently adds another level of complexity to the system, thereby imposing challenges to the sensitive issue of reproducibility in process development. Furthermore, the epigenetic stability of hESC is intensively discussed which may impact on the differentiation characteristics of, for example, genetically modified clonal sublines [91, 92]. Additional assays for quality control of hESC culture optimization comprise measuring the telomere length and, particularly, the regular analysis of the karyotypic integrity.

Processing of adherent cells strongly depends on single cell dissociation. It has implications for controlled scale-up and automation, where it is important to seed bioreactors or scaffolds with reproducible numbers of evenly distributed cells. This issue is particularly apparent in hESC culture where the majority of cells do not survive dissociation into a single cell suspension [93, 94]. Thus, hESC are still propagated as aggregates in standard tissue culture scale and colony dissociation is usually performed via manual scoring methods using plastic tips (with or without enzymatic pretreatment), scoring with more facilitated cutting machines developed by inventive colleagues [95], or commercially available “cake cutters” [96].

In addition to decrease survivability, single cell dissociation for passaging seems to interfere with the chromosomal integrity of hESC, particularly resulting in trisomias, probably reflecting the progressive adaptation of self-renewing cells to their culture conditions [94, 97]. Other authors have suggested that single cell adaptation and long term expansion are achievable in the absence of, at least macroscopic, chromosomal aberrations [98, 99]. If these findings are robust, reproducible, and cell-line independent, the approach might facilitate scalable hESC expansion, efficient generation of transgenic hESC lines (which has now been achieved in Christine Mummery’s group [100]) and the induction of differentiation from single cells via embryoid body formation in bioreactors. However, a recent study revealed that even conditions that prevent macroscopic aneuploidy of single cell-expanded hESC might result in subkaryotypic deletions and amplifications (identified by competitive genomic hybridization) over only 10 passages, reinforcing that present culture regimes remain suboptimal [101]. Notably, chromosomal abnormalities occurring after prolonged culturing are not limited to hESC; a recent report on mesenchymal stem cells shows that abnormal karyotypes can be detected if the cells are extensively passaged [102].

As mentioned in the introduction, the increased, systematic screening and application of small molecule inhibitors might provide new ideas and viable solutions to the field. In a recent report, the transient addition of the p160-Rho associated coiled-coil kinase (ROCK)-inhibitor Y-27632 to non single cell adapted cultures promoted survival of single cell dissociated hESC without affecting pluripotency [103]. Efficient single cell rescue and high plating efficiency might slow down the selection pressure that currently results in karyotypically abnormal cells upon culture adaptation. Thus, the compound might facilitate the single-cell-based expansion of normal hESC, generation of transgenic lines, and also the controlled inoculation of bioreactors with a single cell suspension for differentiation processes.

The single cell issue might also be resolved by alternative, potentially scalable culture strategies. Several groups have established suspension culture expansion of mouse ESC in stirred vessels by forming cell aggregates where differentiation is prevented by medium conditions, serial passaging and mechanical shearing [104, 105]. Further optimization of culture media that can efficiently avoid differentiation might allow translating this and other more automated approaches to hESC [106]. Modifying culture conditions in such systems, for example, by decreasing shear stress to allow larger aggregate formation and replenishing expansion medium by a respective differentiation medium, might allow switching from growth to differentiation in a one-step process.

Seeding cells onto microcarriers is another strategy to translate adherent, matrix dependent cells into easy-to-scale, fully instrumented and controlled stirred tank reactors. Taking advantage of the robust mouse ESC system, the groups of Zandstra and Cabral have provided initial evidence that microcarriers can be adapted to provide surface enlargement for murine ESC culture in suspension [37, 38, 104]. A high degree of carrier and cell agglomeration resulting in heterogeneous clumps was observed in these studies which substantially limits the degree of surface enlargement provided by the carrier and might also induce cell differentiation in the core of these clumps. However, expression of the tested ESC surface markers was largely retained and the ability to form embryoid bodies was also shown by Fok and coworkers [104]. Cabral's groups presented some degree of mESC expansion in an 8-day process utilizing stirred spinner flask with a working volume of up to 80 mL; unfortunately only a single passage was documented in these studies limiting conclusions about an extended applicability. A more general issue concerns the need for efficient removal of microcarriers from the final stem cell product before clinical application. However, this obstacle might be resolved if other hurdles such as the increased shear stress in stirred, microcarriers containing cultures is compatible with hESC expansion, a platform that has not yet been published but is currently developed in several labs (Blaine Phillips, Institute of Medical Biology, Singapore; Andre Choo, Steve Oh, Bioprocessing Technology Institute, Singapore; personal communication).

4.2 Scaling up ESC Differentiation: A Focus on Cardiomyocytes

Given the challenges in hESC culture, published studies on ESC expansion in scalable bioreactors have so far been limited to mouse ESC. In tissue culture, however, substantial progress has been made towards the directed lineage differentiation of human and primate ESC [107] as well. Improved differentiation regimens towards clinical relevant cell types include insulin producing beta-like cells [10, 28], dopaminergic neurons [108], hepatocytes [109] and other lineages [110].

Nevertheless, mESC have a threefold shorter population doubling time ([PDT; ~12–16 h [93]) compared to the ~36 h observed in hESC [85]. Raw material can therefore be generated much faster and attempts towards scalable differentiation have mostly utilized murine ESC as well. Many studies have focused on the generation of ESC-derived cardiomyocytes. This might be driven by the high demand of this cell type for pharmacological screening purposes [111], tissue engineering approaches and cell-based heart repair. In the next step, extensive numbers of well-characterized cardiomyocytes from mouse, primate and human ESC will be mandatory for the functional testing of these cells in physiologically relevant large animal models of human heart failure such as pigs and primates [15, 112].

Besides media formulations, efficiency and robustness of differentiation processes strongly relies on, first, the homogeneity of ESC cultures used for process inoculation and, second, the consistent production of homogeneous embryoid bodies (EBs). These are spherical structures which are induced to initiate spontaneous differentiation of ESC in suspension; they are key to process reproducibility [7]. The heterogeneity of pluripotent hESC cultures has been discussed extensively elsewhere [88, 89], so we will focus our discussion on the formation of homogeneous EBs.

Controlling cell aggregation and agglomeration during EB formation has a profound effect on the extent of ESC proliferation and differentiation; EB size was found to be critical for cardiomyocyte formation and other lineages in the mouse and human system [113–116]. Spatiotemporal formation of these spherical structures was extensively studied in mESC utilizing numerous different formats all aimed at controlled sphere formation. This included the nicely controlled but non-scalable hanging-drop technique [117], cell-encapsulation in alginate beads [114], rotating-suspension culture in a 10 mL volume [118], stirred spinner flask cultures, and controlled reactors with up to 250 mL culture volume [119–121].

Recently, we have shown stirring-controlled EB formation and mESC differentiation in a 2-L instrumented and controlled bioreactor scale, thereby enabling the production of more than 1.2 billion cardiomyocytes in a single run [7]. This cell expansion approaches the 1–2 billion functional cardiomyocytes which are irretrievably lost in a patient's heart upon infarction, a number that could readily be provided by the bioreactor approach if translatable to hESC. A coefficient of 6.4 cardiomyocytes being generated per input ESC (CM/ESC) was found in our bioreactor approach utilizing a genetically engineered mouse ESC line that facilitates enrichment of pure cardiomyocytes.

In a follow-up study, applying multiple steps of process modification particularly applying lower medium throughput and continues perfusion feeding (in contrast to batch-feeding performed in our previous work [7]), this value was even improved to

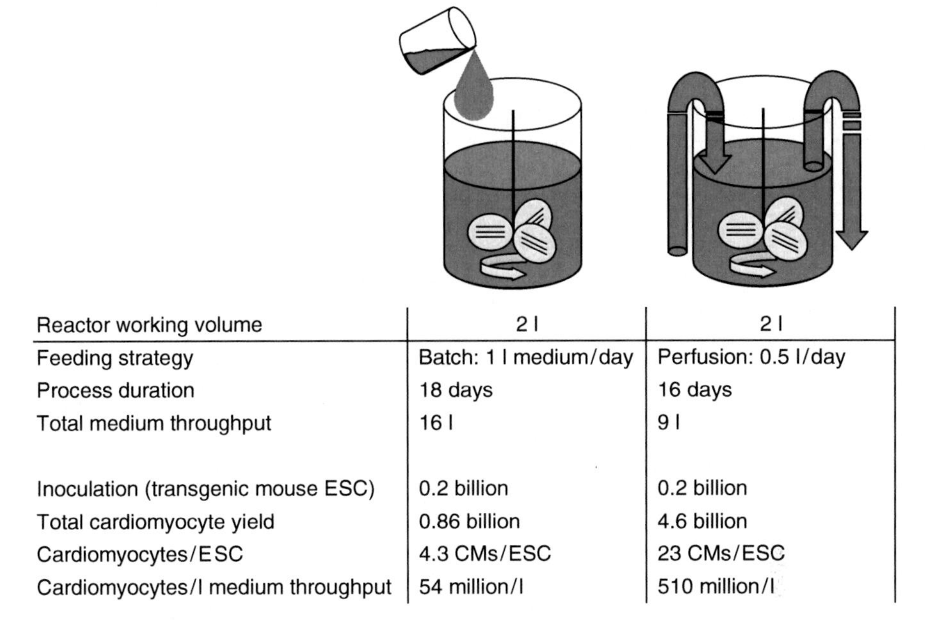

Reactor working volume	2 l	2 l
Feeding strategy	Batch: 1 l medium/day	Perfusion: 0.5 l/day
Process duration	18 days	16 days
Total medium throughput	16 l	9 l
Inoculation (transgenic mouse ESC)	0.2 billion	0.2 billion
Total cardiomyocyte yield	0.86 billion	4.6 billion
Cardiomyocytes/ESC	4.3 CMs/ESC	23 CMs/ESC
Cardiomyocytes/l medium throughput	54 million/l	510 million/l

Fig. 1 Process optimization potential. Multiple steps of process modification, particularly perfusion feeding and reduced medium throughput, resulted in a fivefold increase in cardiomyocyte yield from a transgenic mouse ESC-line in a fully controlled 2-L stirred reactor. The efficiency in cardiomyocyte generation per liter total medium throughput even increase by almost 10-fold [7, 122]

23 CM/ESC [122] thereby underscoring the enormous process optimization potential (Fig. 1). More homogeneous culture conditions achieved via continuous feeding might support a better control of ESC differentiation. In vitro differentiation of ESC is notoriously variable due to the ongoing changes in cell density paralleled by the occurrence of differentiating cell lineages and thus changes in cell physiology, cell–cell interactions, growth factor secretion, etc. Inhibition of ESC differentiation following a noncontinuous, daily medium exchange was described by Viswanathan et al. [123] as cell-secreted factors were diluted. As outlined above, it has also been reported for hematopoietic cell cultures that the consumption and release of a variety of growth factors can affect the cell type(s) generated in a process [75]. Continuous feeding strategies ensure optimal process uniformity with respect to pH, pO_2, and concentration of metabolites while manual medium exchange, at least transiently, encounters alternating pH and gassing conditions. For example, high oxygen tension has been suggested to inhibit cardiac differentiation. In a study by Bauwens and coworkers [121] a controlled, perfusion fed system at a 250-mL scale was employed. Notably, the same cell line and similar differentiation and selection conditions as in the study by Niebruegge [122] were used, but EBs were formed from encapsulated ESC. Highest cardiomyocyte yield was archived under hypoxic conditions (4% oxygen tension) resulting in a CM/ESC-coefficient of 3.77 and a drastically lower value of 2.56 CM/ESC was found at normoxia. However, the significantly higher

CM/ESC coefficient of 23 described by us was achieved at an oxygen tension of 40%; whether hypoxic conditions would further increase this value in the controlled 2-L bioreactor setting applied by us requires further experimental evidence.

Another benchmark value which is key to the commercial viability of future cell replacement therapies, is the number of cardiomyocytes that can be generated per liter (of a potentially expensive) culture medium. Under optimized conditions >500 million cardiomyocytes per liter medium were generated in our optimized 18-day differentiation and enrichment process [122].

A first demonstration of the translation efficiency of hESC into cardiomyocytes was recently provided by a monolayer differentiation protocol (sequential addition of activinA followed by BMP4) yielding three CM/hESC [26]. However, scalability and economic feasibility for the mass-production of cardiomyocytes by this growth factor-dependent, two-dimensional monolayer approach needs to be determined.

Aiming at efficient cardiomygenic differentiation of hESC in suspension, we have recently converted a coculture based protocol for directed cardiomyocyte generation into a scalable suspension process (Fig. 2), using a serum-free medium conditioned

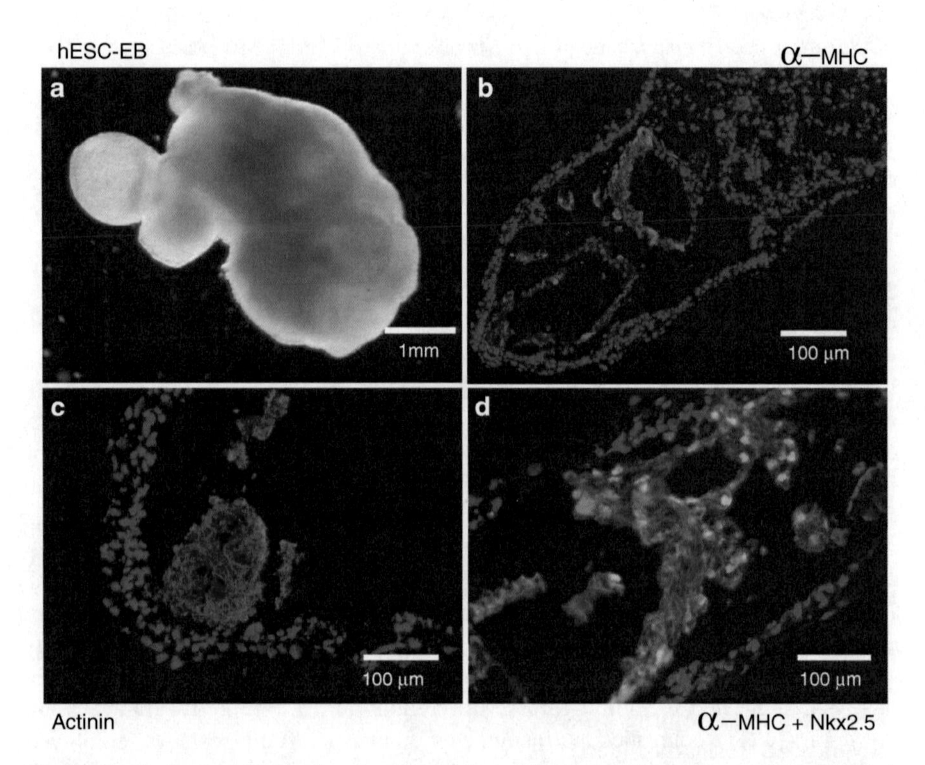

Fig. 2a–d Cardiomycyte formation from hESC in scalable suspension culture. **a** A typical cystic embryoid body after about 12 days of differentiation in a serum free medium supporting cardiomyogenesis. **b** Immune histology of EB-sections specific to cardiac markers (*in red*) alpha Myosin Heavy Chain (alpha-MHC). **c,d** Actinin (**c**) and double-staining to alpha-MHC and Nkx2.5 (transcription factor, nuclear stain *in green* (**d**)) show the formation of cardiomyocyte-clusters in cysts [85, 124, 125]

by and endoderm-like cell line END2 (END2-CM; [124]). By screening small molecule inhibitors in this system, we have identified SB203580, a specific p38 MAP kinase inhibitor, as a potent, dose dependent promoter of cardiomyogenesis. SB203580 at an optimized concentration, induced >20% of hESC to become cardiomyocytes. A parallel increase in total cell number yield approximately 2.5-fold more cardiomyocytes compared to differentiation in END2-CM alone. Besides ascorbic acid, SB203580 is one of the first molecules to act as an efficient enhancer of hESC cardiac differentiation; other factors such as DMSO and retinoic acid, known inducers of mESC cardiomyogenesis, caused no significant improvement [15].

By systematically deconstructing the cardiomyocyte inducing activity of the "xenogenic" END2-CM we have found that the common media supplement insulin can have a dramatic inhibitory effect on the formation of cardiomyocytes [125]. The insulin effect, which was also triggered by the growth factor IGF1, was mediated through activation of the PI3/Akt pathway downstream of the insulin/IGF1 receptors during early steps of differentiation. Notably, this observation might also explain the varying compliance of serum batches for cardiac differentiation. The study further identified a small molecule, the prostaglandin member PGI2, as accumulating in END2-CM and enhancing cardiomyogenesis when added into a novel, insulin-free synthetic medium at optimized concentrations. Finally, combining SB203580 with the synthetic medium yielded a fully defined, cGMP-compliant medium, which enabled efficient hESC differentiation in suspension. In a second study we found that insulin redirects differentiation of hESC from mesendoderm to neuroectoderm [126].

One major difference between mouse and human ESC that is still hampering the systematic up-scaling of differentiation is the inability of the latter to reaggregate and form EBs once dissociated to single cells [127]. High expression levels of the cell adhesion molecule E-cadherin [113] seem to underlie the aggregation of mESC, and EB formation is focused on controlling the excessive fusion tendency interfering with differentiation. In contrast, although the majority of undifferentiated hESC also express E-cadherin [128], essentially all cells die when seeded in single cell suspension. This phenotype is seemingly independent of the cell line, the dissociation method, the culture medium and the seeding density [127, 129]. Consequently, most of the present differentiation studies rely on either enzymatic whole colony lifting (thereby separating hESC from the feeder layer) or other enzymatic and/or mechanical scoring techniques aimed at providing preformed hESC-aggregates of various size for EB formation in suspension [127, 129].

These hurdles might explain the limited number of studies on hESC differentiation scale-up. Gerecht-Nir and coworkers have used small cell clumps to inoculate RWV termed slow turning lateral vessels, or high aspect rotating vessels to control floating EB formation [115]. However, scalability of these specialized reactors might be limited. A first step towards hESC differentiation in impeller-stirred systems was published by Cameron and coworkers [130] employing a 250-mL spinner flask system, while another study translated the encapsulation approach of EBs in agarose from mouse to human ESC [114]. However, all of these studies depend on the cumbersome, difficult-to-control, and hardly scalable preformation of hESC clumps before process inoculation. The only published strategies that seem to enable hESC-derived EB formation directly from single cell suspensions is seeding on three-dimensional porous alginate scaffolds [116] or the forced aggregation by centrifugation in round-bottom or V-shaped 96-well dishes [127, 129] which has recently been scaled to a 384-well format by custom-made silicon wafer-based microfabrication [131]. These studies indicate that the dissociation procedure, per se, is not irrevocably inducing hESC death but suggest that constraining physical cell–cell or cell–matrix interaction combined with chemical cues (from the substrate surface and/or the medium) are necessary to rescue single hESC. While both methods (porous alginate scaffolds and multi well dishes) are not straightforward for large-scale inoculation of stirred bioreactors, the underlying mechanism might be exploitable in future.

In summary, in vitro differentiation of ESC is a complex, continuously changing, and thus highly variable process. However, recent findings by us and others in controlled bioreactors indicate that reproducible and efficient production of differentiated lineages such as cardiomycytes is achievable. Translating highly controlled single cell inoculation and EB formation to hESC cultures and utilizing the recently developed fully synthetic differentiation media is another step towards this goal.

4.3 Enrichment of Differentiated Cell Types: The Need for Purity and Safety

Many of the envisioned hESC therapeutic as well as in vitro screening applications will require pure populations of a desired cell type such as cardiomyocytes that are devoid of any other lineage, in particular, residual, undifferentiated hESC [132]. A purification strategy is therefore essential and has proven to be effective for enrichment of hematopoietic stem cells from bone marrow and differentiated hESC populations. In the case of cardiomyocytes, however, there is no unique cardiac-specific surface marker that can be used for cellular isolation. Recently we have demonstrated that the surface marker CD166/Alcam which is specific to a transient population of heart-tube stage embryonic cardiomyocytes [133], is useful for isolating cells homologous to human embryonic cardiomyocytes from differentiated hESC populations (MASC; [134]). Using a sterile, magnet-assisted cell sorting system, we took advantage of this marker to produce cardiomyocyte populations that are greater than 60% pure from wild-type hESC.

Furthermore, Choo et al. [135] have demonstrated the ability to kill undifferentiated hESC using a cytotoxicity monoclonal antibody thereby eliminating teratoma formation in vivo in a SCID mouse model. The combination of positive and negative selection strategies will greatly facilitate in the enrichment of cardiomyocytes, which, until recently, was limited to improved differentiation strategies and hardly reproducible, selective dissociation protocols combined with Percoll gradient centrifugation [26, 136].

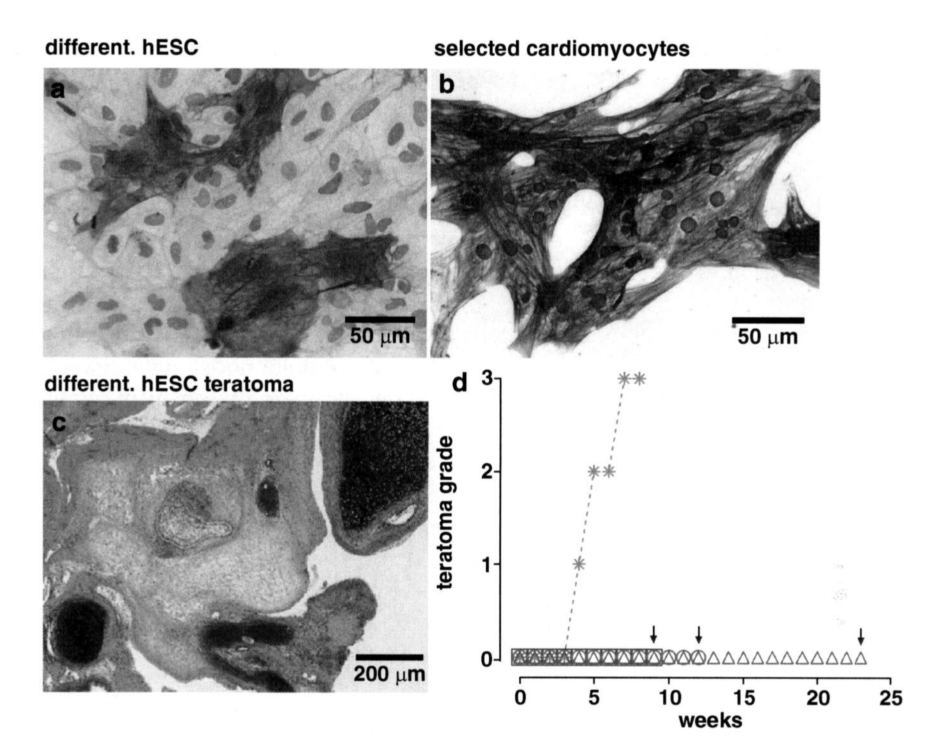

Fig. 3a–d Enrichment and safety of suspension-derived hESC-cardiomyocytes. Dissociated, differentiated embroid bodies (EBs) were seeded to generate a cell layer that contains a proportion of cardomyocytes presented *in brown* (**a**) (DAB stain specific to the cardiac marker alpha Myosin Heavy Chain). Antibiotic enrichment of cardiomyocytes (derived from a transgenic hESC-line in suspension) followed by cell seeding resulted in an essentially pure cardiomyocyte population (**b**). Injection of differentiated but not antibiotic-treated EBs as non-dissociated clumps resulted in teratoma formation is a SCID-hindlimb model (HE-stain of teratoma section in (**c**) within about 3–7 weeks (*red, dotted line* in (**d**)). In contrast, no teratoma formation was observed from an equivalent number of antibiotic-enriched cardiomyocyte-clumps injected in the same model when mice were analyzed after 9, 12, and 23 weeks (less time point tested) [142]

Previously, Field and his colleagues conceived a simple but ingenious genetic selection strategy for mouse ESC-derived cardiomyocytes. Introducing a transgene comprising the murine α-MHC promoter driving cardiomyocyte-specific expression of an antibiotic resistance gene enabled the enrichment of >99% pure cardiomyocyte populations [137]. This selection scheme was consequently applied to enrich for other cell lineages, including neural precursors and insulin-producing cells [138, 139], and adapted to mass production of cardiomyocytes in suspension culture [7, 118, 120, 122]. Aiming at the derivation of cardiomyocyte-subtypes, alternative constructs, such as the myosin light chain 2v (MLC2v) promoter in combination with a GFP-expression cassette followed by fluorescence based cell sorting (FACS), have also been utilized [140]. By generating stable transgenic lines using lentiviral vectors this strategy has been translated to hESC.

GFP expression under the transcriptional control of the human MLC2v promoter appeared to be cardiomyocyte-specific [141]. After FACS sorting, >93% of the isolated cells stained positive for cardiac-specific proteins and formed stable myocardial cell grafts for up to 4 weeks (the latest time point tested) following in vivo cell transplantation into immune suppressed Sprague–Dawley rats. The study provides the first proof-of-concept for the genetic lineage selection strategy to work in hESC. Although no teratoma formation was observed in this study, the animal model as well as the short follow-up time might not be useful to appraise this risk.

By applying the antibiotic-based lineage enrichment strategy introduced by Fields group to hESC, we have recently generated multiple transgenic hESC lines (via electroporation) and achieved >99% cardiomyocytes purity from differentiated hESC cultures [142] (Fig. 3) More importantly, applying a sensitive biosafety model for teratoma formation in SCID mice [143, 144] no teratomas were found for up to 23 weeks after the injection of antibiotic-selected cardiomyocytes clumps. In contrast, the injection of long term differentiated but not antibiotic treated EBs resulted in teratoma formation with high incidence [142]. These findings strongly underscore the necessity of efficient selection techniques and comprehensive long term safety studies in appropriate animal models. The therapeutic application of transgenic hESC lines might comprise yet another regulatory hurdle to clinical trials. However, where the genomic integration site of the transgene is well defined this technology clearly provides another level of safety in hESC-derived grafts. Whether other selection techniques will achieve the same level of scalability, purity, cell vitality, and safety remains to be demonstrated.

5 Bioprocessing of ESC- and Tissue-Derived Mesenchymal and Neural Stem and Progenitor Cells

Batch differentiation of an expanded ESC population which might be combined with a consecutive enrichment procedure is one possibility for the generation of specific progenies. The strategy is particularly useful if the differentiated cell type has no or only a limited proliferation potential such as cardiomyocytes [145, 146].

An alternative scenario is to generate intermediate cell types from ESC that are still capable of extended proliferation but are lineage-committed progenitors. Such intermediate stem- or progenitor type cells can also be derived from some adult tissues. However, the reproducible derivation of intermediate type stem cells from clinical-grade hESC might provide an invariable source of consistently uniform cells for therapeutic applications, thereby overcoming serious limitations imposed by the heterogeneity of donor tissue-derived cells.

Multipotent stem cells provide an expandable cell source that can either be used to produce more differentiated progenies or might serve directly for therapeutic or screening approaches. Recent studies on hESC- or adult tissue-derived mesenchymal stem cells and neural stem cells provide examples for this approach.

5.1 Neural Stem and Progenitor Cells

Cho and coworkers [108] have generated relative homogeneous spherical neural masses (SNM) from hESC colonies. SNM have a neural precursor phenotype and can be passaged long term in suspension culture without losing their differentiation capability. Finally, SNM have been directed into differentiated cultures consisting of 77% neurons. The vast majority, 86%, of these neurons comprise dopaminergic neurons, indicating a relative high purity of this desired cell type for Parkinson's treatment. At present SNM passaging requires mechanical handling and has not yet been scaled to bioreactors.

Neural stem cells (NSC) may also be isolated from both embryonic and adult tissue from the central nervous system (CNS). They are defined as tissue specific progenitor cells which undergo self-renewal in vitro and can be differentiated into all major cell types of the nervous system including oligodendrocytes, neurons and astrocytes [147]. NSC were thought to be particularly useful for the generation of dopaminergic neurons in vitro but the efficient differentiation towards this phenotype has been proven to be difficult. Hypoxic culture conditions appear to induce this process for human-derived tissue which forms dopamine neurons even less efficiently than NSC derived from mice [148]. A detailed description of multipotent neural stem and progenitor cell characteristics, their isolation from various sources and their envisioned therapeutic application is outside the scope of this publication. The interested reader is referred to a recent review by Hall, Li and Brundin [149]. However, the propagation of NSC in aggregates termed neurospheres is a paradigm for the expansion of pluripotent stem cells in bioreactors as outlined below.

Following the discovery of NSC in 1992 [150, 151] the group of Kallos and Behie has established and optimized scale-up of NSC cultures by controlling neurosphere size via hydrodynamic shear in stirred suspension culture [152, 153]. The process was scaled up to 500 mL culture volume in an instrumented bioreactor (temperature, pH, pO_2 control) enabling the generation of up to 1.2×10^6 cells/mL mouse NSC without interfering with the cells multipotentiality [154].

Notably, human neural precursor cells (hNPC) isolated from multiple fetal brain regions have recently also been expanded in stirred bioreactors aiming to provide tissue for neurodegenerative disorder treatments. In an initial study, reactor-expanded cells differentiated primarily into astrocytes after transplantation into the striatum or substantia nigra regions, and no behavioral improvement in a parkinsonian rat model was observed [155]. In a second study, telencephalic hNPC have been differentiated in highly enriched GABAergic cells following expansion in spinner flasks in 125 mL volume. Functional assessment in a rodent model of Huntington's disease revealed a significant behavioral improvement in motor and memory deficits following transplantation with differentiated GABAergic cells, whereas expanded but undifferentiated hNPC did not [156]. These recent studies on hNPC apparently suggest that stem cell differentiation into a desired cell type in vitro is mandatory for specific organ repair rather than to reliance on tissue specific differentiation of pluri- or multipotent stem cells following transplantation into a damaged organ. Next, it will be interesting to see at which scale primary hNPC can be expanded under optimized condition in vitro without transformation and loss of differentiation properties. Finally, functional testing of GABAergic cells in primate models will be mandatory before entering clinical trials.

5.2 Mesenchymal Stem and Progenitor Cells

Mesenchymal stem cells (MSC; also known as multipotent mesenchymal stromal cells) comprise another cell type that has originally been derived from mammalian tissue but in vitro expandable MSC-like cells were recently also generated from hESC [157].

Following the pioneering work by Owen and Friendenstein on bone marrow stromal cells 20 years ago [158], MSC have also been isolated as plastic adherent, fibroblast-like cells from multiple other sources including placenta, adipose tissue, cord blood and liver (see recent review by Brooke et al. [159]).

The original stem cell term has been thought to be inadequate by many investigators as it has not been possible to grow human MSC indefinitely in culture while maintaining their multipotent properties. Currently, there is also no in vivo assay that can be used to define the repopulation ability of these cells analogous to existing assays for hematopoietic stem cells. The anatomical location and phenotype of MSC has also not yet been well defined in vivo. However, when isolated by plastic adherence and expanded, ex vivo human MSC have been shown to differentiate into mesodermal lineages including chondrocytes, adipocytes and osteocytes [160]. In addition to the in vitro differentiation potential the International Society for Cellular Therapy (ISCT) has recently proposed that MSC should be defined based on a panel of antibodies specific to CD105, CD73, and CD90 (>95% of the population should be positive) and CD45, CD34, CD14 and CD19 (<2% of the population should be positive), but notably none of these antigens are unique to MSC [161].

Nevertheless, MSC comprise an attractive cell type for therapeutic applications given their potential for organ repair, ease and reproducibility of isolation, some level of in vitro expandability, and immunosuppressive and/or immunoprivileged properties [162, 163], which particularly favor this cell type for the generation of allogeneic "off the shelf" stem cell products. In preclinical studies of tissue repair MSC have been shown to improve the function of the heart, brain, liver, and joint and they are currently tested for the regeneration of these and other organs in clinical trials as well as for immunological disorders and solid organ transplantation, the later being recently reviewed elsewhere [159]. However, it may become apparent that MSC exert many if not all effects via paracrine mechanisms, that is secreting factors and supplying the necessary environment for host tissue to repair itself recently noted by Brooke et al. [159].

Consequently, Timmers and coworker have infused a medium conditioned by hESC-derived MSC into the coronary vasculature of pig hearts in a myocardial infarct model [164]. This was associated with a 60% reduction of infarct size and marked improvement of systolic and diastolic cardiac performance. Development of large scale cGMP-compliant processes is currently underway to establish the production of MSC-conditioned medium in sufficient quantities for clinical trials (Andre Choo and Steve Oh Bioprocessing Technology Institute, Singapore, personal communication).

Aiming at scalable MSC expansion, recent bioreactor studies have applied perfusion of human MSC embedded in three-dimensional scaffolds [165, 166]. These studies have shown that shear stress is an important biomechanical parameter in regulating MSC growth, and increased cell expansion was observed at lower perfusion rates [165]. Other culture systems, including static cultures, stirred reactors and rotated vessel reactors, which all impose highly differential shear conditions, consequently resulted in differential growth and differentiation properties of adult human bone marrow-derived MSC when cell proliferation and multilineage differentiation towards osteoblasts, chondrocytes, and adipocytes was analyzed [167, 168]. However, as with hematopoietic and other stem cell types, donor cell variability, variations in MSC isolation procedures, and a large number of cell culture variables makes direct comparison of results presented in independent studies problematic. Using hESC-derived MSC isolated under reproducible conditions and applying a meaningful side-by-side comparison of reactor systems might increase the knowledge on favorable culture conditions in future studies.

6 Conclusion and Outlook

In a recent assay on the future of stem cell biotechnology, Ann Parson [5] stated that only time will tell if “RegenMed 2.0” (Stem Cell based Therapy) will prevail or whether it will go the way of “RegenMed 1.0” (Gene Therapy based Regeneration). Unlimited availability of stem cells, the building stones of RegenMed2 in reproducible quality and at commercially viable conditions will be of fundamental importance to success.

Engineering has already provided bioreactors that can accommodate all major needs for large scale mammalian cell production. Sophisticated techniques to meet special demands posed by stem cells are continuously under development. Miniaturization has allowed scaling down (!) of bioreactor systems to a ~30 mL working volume which still allows full instrumentation and thus measurement and computational control and adaptation of key culture conditions (pH, pO_2, continuous medium supply, etc.) in multiple parallel bioreactors, thereby speeding up process development under conditions which in principle apply to 10- to 100-fold larger systems (for example from Dasgip, Juelich, Germany).

Another trend is the development of disposable bioreactors such as simple or more sophisticated spinner flasks some of which are readily equipped with active culture aeration modules (to enable increased cell densities) and ports for simplified sample collection. Establishing initial cGMP-compliant small scale processes based on disposable reactors to feed cells into phase1 clinical trials will benefit the field. Such step-by-step strategy providing stem cell products as a personalized treatment seems to be a more feasible approach, at present. Shooting for the ultimate goal, a "one-fits-all of-the-shelf" (organ but not recipient specific) stem cell product that is generated in multi-liter tanks and stocked frozen until usage is apparently not yet enabled.

Basic research still needs to define complex, interwoven networks of molecular mechanisms controlling stem cell maintenance, genomic stability, and differentiation. Systematic high-throughput technologies like "omics" approaches (gen-, transcript-, prote-, metabol-omics etc.) as well as continuous progress in developmental biology and tumor cell biology will help to understand these fundamental questions; stem cell research will vice versa feed back into these research disciplines. These findings combined with systematic screens for small molecular effectors to control identified signaling pathways will finally lead towards commercially viable process and progressive increase in production scales.

Definitions

- Bioreactor: a system that simulates physiological environments for the creation, physical conditioning, and testing of cells, tissues, precursors, support structures, and organs in vitro
- Teratoma: benign tumors containing cells from various differentiated tissues
- Stem cells are defined as being self-renewing, pluri- or multipotent, and clonogenic cells
- Clonogenic cells are single stem cells that are able to generate a line of genetically identical cells, thereby maintaining their self-renewal and differentiation potential
- Pancytopenia: shortage of all types of blood cells
- Embryoid bodies: spherical structures which are induced to initiate spontaneous differentiation of ESC in suspension

Acknowledgements I thank Blaine Phillips, William Rust, Birgit Andree, Harmeet Singh, Zhou WeiZhuang (Institute of Medical Biology, Singapore) and Andre Choo (Bioprocessing Technology Institute, Singapore) for helpful comments and a critical review of this manuscript.

References

1. Mathers JP (1998) Laboratory scaleup of cell cultures (0.5–50 liters). Methods Cell Biol 57:219–227
2. Griffiths B (2001) Scale-up of suspension and anchorage-dependent animal cells. Mol Biotechnol 17(3):225–238
3. Warnock JN, Al-Rubeai M (2006) Bioreactor systems for the production of biopharmaceuticals from animal cells. Biotechnol Appl Biochem 45(Pt 1):1–12
4. Yang JD, Lu C, Stasny B, Henley J, Guinto W, Gonzalez C, Gleason J, Fung M, Collopy B, Benjamino M, Gangi J, Hanson M, Ille E (2007) Fed-batch bioreactor process scale-up from 3-L to 2,500-L scale for monoclonal antibody production from cell culture. Biotechnol Bioeng 98(1):141–154
5. Parson AB (2008) Stem cell biotech: seeking a piece of the action. Cell 132(4):511–513
6. Kajstura J, Leri A, Finato N, Di Loreto C, Beltrami CA, Anversa P (1998) Myocyte proliferation in end-stage cardiac failure in humans. Proc Natl Acad Sci U S A 95(15):8801
7. Schroeder M, Niebruegge S, Werner A, Willbold E, Burg M, Ruediger M, Field LJ, Lehmann J, Zweigerdt R (2005) Embryonic stem cell differentiation and lineage selection in a stirred bench scale bioreactor with automated process control. Biotechnol Bioeng 92(7):920

8. Murry CE, Reinecke H, Pabon HE (2006) Regeneration gaps: observations on stem cells and cardiac repair. J Am Coll Cardiol 47(9):1777
9. Emamaullee JA, Shapiro AM (2007) Factors influencing the loss of beta-cell mass in islet transplantation. Cell Transplant 16(1):1–8
10. Docherty K, Bernardo AS, Vallier L (2007) Embryonic stem cell therapy for diabetes mellitus. Semin Cell Dev Biol 18(6):827–838
11. Ballen K, Broxmeyer HE, McCullough J, Piaciabello W, Rebulla P, Verfaillie CM, Wagner JE (2001) Current status of cord blood banking and transplantation in the United States and Europe. Biol Blood Marrow Transplant 7(12):635–645
12. McNiece I (2007) Delivering cellular therapies: lessons learned from ex vivo culture and clinical applications of hematopoietic cells. Semin Cell Dev Biol 18(6):839–845
13. Thomson JA, Itskovitz-Eldor J, Shapiro SS, Waknitz MA, Swiergiel JJ, Marshall VS, Jones JM (1998) Embryonic stem cell lines derived from human blastocysts. Science 282(5391):1145–1147. Erratum in: Science 1998 Dec 4;282(5395):1827
14. Murry CE, Field LJ, Menasche P (2005) Cell-based cardiac repair: reflections at the 10-year point. Circulation 112(20):3174–3183
15. Zweigerdt R (2007) The art of cobbling a running pump–will human embryonic stem cells mend broken hearts? Semin Cell Dev Biol 18(6):794–804
16. Rosenzweig A (2006) Cardiac cell therapy—mixed results from mixed cells. N Engl J Med 355(12):1274–1277
17. Schwartz RS (2006) The politics and promise of stem-cell research. N Engl J Med 355(12):1189–1191
18. Arnesen H, Lunde K, Aakhus S, Forfang K (2007) Cell therapy in myocardial infarction. Lancet 369(9580):2142–2143
19. Orlic D, Kajstura J, Chimenti S et al. (2001) Bone marrow cells regenerate infarcted myocardium. Nature 410(6829):701–705
20. Kawada H, Fujita J, Kinjo K, Matsuzaki Y, Tsuma M, Miyatake H, Muguruma Y, Tsuboi K, Itabashi Y, Ikeda Y, Ogawa S, Okano H, Hotta T, Ando K, Fukuda K (2004) Nonhematopoietic mesenchymal stem cells can be mobilized and differentiate into cardiomyocytes after myocardial infarction. Blood 104(12):3581
21. Miyahara Y, Nagaya N, Kataoka M, Yanagawa B, Tanaka K, Hao H, Ishino K, Ishida H, Shimizu T, Kangawa K, Sano S, Okano T, Kitamura S, Mori H (2006) Monolayered mesenchymal stem cells repair scarred myocardium after myocardial infarction. Nat Med 12(4):459–465
22. Murry CE, Soonpaa MH, Reinecke H, Nakajima H, Nakajima HO, Rubart M et al. (2004) Haematopoietic stem cells do not transdifferentiate into cardiac myocytes in myocardial infarcts. Nature 428(6983):664–668
23. Balsam LB, Wagers AJ, Christensen JL, Kofidis T, Weissman IL, Robbins RC (2004) Haematopoietic stem cells adopt mature haematopoietic fates in ischaemic myocardium. Nature 428(6983):668–673
24. Breitbach M, Bostani T, Roell W, Xia Y, Dewald O, Nygren JM, Fries JW, Tiemann K, Bohlen H, Hescheler J, Welz A, Bloch W, Jacobsen SE, Fleischmann BK (2007) Potential risks of bone marrow cell transplantation into infarcted hearts. Blood 110(4):1362–1369
25. Rubin LL (2008) Stem cells and drug discovery: the beginning of a new era? Cell 132(4):549–552
26. Laflamme MA, Chen KY, Naumova AV et al. (2007) Cardiomyocytes derived from human embryonic stem cells in pro-survival factors enhance function of infarcted rat hearts. Nat Biotechnol 25(9):1015–1024
27. van Laake LW, Passier R, Monshouwer-Kloots J et al. (2007) Human embryonic stem cell-derived cardiomyocytes survive and mature in the mouse heart and transiently improve function after myocardial infarction. Stem Cell Res 1:9–24
28. Phillips BW, Hentze H, Rust WL, Chen QP, Chipperfield H, Tan EK, Abraham S, Sadasivam A, Soong PL, Wang ST, Lim R, Sun W, Colman A, Dunn NR (2007) Directed differentiation of human embryonic stem cells into the pancreatic endocrine lineage. Stem Cells Dev 16(4):561–578
29. Schugar RC, Robbins PD, Deasy BM (2008) Small molecules in stem cell self-renewal and differentiation. Gene Ther 15(2):126–135

30. Fong WJ, Tan HL, Choo A, Oh SK (2005) Perfusion cultures of human embryonic stem cells. Bioprocess Biosyst Eng 27(6):381–387
31. van Wezel AL (1976) The large-scale cultivation of diploid cell strains in microcarrier culture. Improvement of microcarriers. Dev Biol Stand 37:143–147
32. Röder B, Zühlke A, Widdecke H, Klein J (1993) Synthesis and application of new microcarriers for animal cell culture. Part II. Application of polystyrene microcarriers. J Biomater Sci Polym Ed 5(1–2):79–88
33. Lim HS, Han BK, Kim JH, Peshwa MV, Hu WS (1992) Spatial distribution of mammalian cells grown on macroporous microcarriers with improved attachment kinetics. Biotechnol Prog 8(6):486–493
34. Koller MR, Papoutsakis ET (1995) Cell adhesion in animal cell culture: physiological and fluid-mechanical implications. Bioprocess Technol 20:61–110
35. Banu N, Rosenzweig M, Kim H, Bagley J, Pykett M (2001) Cytokine-augmented culture of haematopoietic progenitor cells in a novel three-dimensional cell growth matrix. Cytokine 13(6):349–358
36. Wang TY, Brennan JK, Wu JH (1995) Multilineal hematopoiesis in a three-dimensional murine long-term bone marrow culture. Exp Hematol 23(1):26–32
37. Fernandes AM, Fernandes TG, Diogo MM, da Silva CL, Henrique D, Cabral JM (2007) Mouse embryonic stem cell expansion in a microcarrier-based stirred culture system. J Biotechnol 132(2):227–236
38. Abranches E, Bekman E, Henrique D, Cabral JM (2007) Expansion of mouse embryonic stem cells on microcarriers. Biotechnol Bioeng 96(6):1211–1221
39. Panoskaltsis N, Mantalaris A, Wu JH (2005) Engineering a mimicry of bone marrow tissue ex vivo. J Biosci Bioeng 100(1):28–35
40. Sardonini, CA, Wu, YJ (1993) Expansion and differentiation of human hematopoietic cells from static cultures though smallscale bioreactors. Biotechnol Prog 9(2):131–137
41. Hammond TG, Hammond JM (2001) Optimized suspension culture: the rotating-wall vessel. Am J Physiol Renal Physiol 281(1):F12–F25
42. Singh H, Teoh SH, Low HT, Hutmacher DW (2005) Flow modelling within a scaffold under the influence of uni-axial and bi-axial bioreactor rotation. J Biotechnol 119(2):181–196
43. Hutmacher DW, Singh H (2008) Computational fluid dynamics for improved bioreactor design and 3D culture. Trends Biotechnol 26(4):166–172
44. Orkin SH, Zon LI (2008) Hematopoiesis: an evolving paradigm for stem cell biology. Cell 132(4):631–644
45. Thomas ED, Lochte HL Jr, Lu WC, Ferrebee JW (1957) Intravenous infusion of bone marrow in patients receiving radiation and chemotherapy. N Engl J Med 257(11):491–496
46. Sheridan WP, Begley CG, Juttner CA, Szer J, To LB, Maher D, McGrath KM, Morstyn G, Fox RM (1992) Effect of peripheral-blood progenitor cells mobilised by filgrastim (G-CSF) on platelet recovery after high-dose chemotherapy. Lancet 339(8794):640–644
47. Gluckman E, Broxmeyer HA, Auerbach AD, Friedman HS, Douglas GW, Devergie A, Esperou H, Thierry D, Socie G, Lehn P et-al. (1989) Hematopoietic reconstitution in a patient with Fanconi's anemia by means of umbilical-cord blood from an HLA-identical sibling. N Engl J Med 321(17):1174–1178
48. Kay HEM (1965) How many cell-generations? Lancet 56:418–419
49. Prchal JT, Prchal JF, Belickova M, Chen S, Guan Y, Gartland GL, Cooper MD (1996) Clonal stability of blood cell lineages indicated by X-chromosomal transcriptional polymorphism. J Exp Med 183(2):561–567
50. Lansdorp PM (2008) Telomeres, stem cells, and hematology. Blood 111(4):1759–1766
51. Ito K, Hirao A, Arai F, Takubo K, Matsuoka S, Miyamoto K, Ohmura M, Naka K, Hosokawa K, Ikeda Y, Suda T (2006) Reactive oxygen species act through p38 MAPK to limit the lifespan of hematopoietic stem cells. Nat Med 12(4):446–451
52. Andrews RG, Singer JW, Bernstein ID (1989) Precursors of colony-forming cells in humans can be distinguished from colony-forming cells by expression of CD33 and CD34 antigen and light scatter properties. J Exp Med 169(5):1721–1731

53. Krause DS, Fackler MJ, Civin CI, May WS (1996) CD34: structure, biology, and clinical utility. Blood 87(1):1–13
54. Coulombel L (2004) Identification of hematopoietic stem/progenitor cells: strength and drawbacks of functional assays. Oncogene 23(43):7210–7222.
55. Morrison SJ, Spradling AC (2008) Stem cells and niches: mechanisms that promote stem cell maintenance throughout life. Cell 132(4):598–611
56. Dexter TM, Allen TD, Lajtha LG (1977) Conditions controlling the proliferation of haemopoietic stem cells in vitro. J Cell Physiol 91:335–344
57. Mayani H, Gutierrez-Rodriguez M, Espinoza L, Lopez-Chalini E, Huerta-Zepeda A, Flores E, Sanchez-Valle E, Luna-Bautista F, Valencia I, Ramirez OT (1998) Kinetics of hematopoiesis in Dextertype long-term cultures established from human umbilical cord blood cells. Stem Cells 16:127–135
58. Nolta JA, Thiemann FT, Arakawa-Hoyt J, Dao MA, Barsky LW, Moore KA, Lemischka IR, Crooks GM (2002) The AFT024 stromal cell line supports long-term ex vivo maintenance of engrafting multipotent human hematopoietic progenitors. Leukemia 16:352–361
59. Zhang CC, Kaba M, Ge G, Xie K, Tong W, Hug C, Lodish HF (2006) Angiopoietin-like proteins stimulate ex vivo expansion of hematopoietic stem cells. Nat Med 12:240–245
60. Heike T, Nakahata T (2002) Ex vivo expansion of hematopoietic stem cells by cytokines. Biochim Biophys Acta 1592:313–321
61. Noll, T, Jelinek, N, Schmidt, S, Biselli, M, Wandrey, C (2002) Cultivation of hematopoietic stem and progenitor cells: biochemical engineering aspects. Adv Biochem Eng Biotechnol 74:111–128
62. Koller MR, anchel I, Brott DA, Palsson Bø (1996) Donor-to-donor variability in the expansion potential of human bone marrow cells is reduced by accessory cells but not by soluble growth factors. Exp Hematol 24(13):1484–1493
63. Kohler T, Plettig R, Wetzstein W, Schaffer B, Ordemann R, Nagels HO, Ehninger G, Bornhauser M (1999) Defining optimum conditions for the ex vivo expansion of human umbilical cord blood cells. Influences of progenitor enrichment, interference with feeder layers, early-acting cytokines and agitation of culture vessels. Stem Cells 17:19–24
64. Xu R, Medchill M, Reems JA (2000) Serum supplement, inoculum density, and accessory cell effects are dependant on the cytokine combination selected to expand human HPCs ex vivo. Transfusion 40:1299–1307
65. Majka M, Janowska-Wieczorek A, Ratajczak J, Ehrenman K, Pietrzkowski Z, Kowalska MA, Gewirtz AM, Emerson SG, Ratajczak MZ (2001) Numerous growth factors, cytokines, and chemokines are secreted by human CD34(+) cells, myeloblasts, erythroblasts, and megakaryoblasts and regulate normal hematopoiesis in an autocrine/paracrine manner. Blood 97:3075–3085
66. Madlambayan GJ, Rogers I, Kirouac DC, Yamanaka N, Mazurier F, Doedens M, Casper RF, Dick JE, Zandstra PW (2005) Dynamic changes in cellular and microenvironmental composition can be controlled to elicit in vitro human hematopoietic stem cell expansion. Exp Hematol 33(10):1229–1239. Erratum in: Exp Hematol 2006 Jan;34(1):122
67. Madlambayan GJ, Rogers I, Purpura KA, Ito C, Yu M, Kirouac D, Casper RF, Zandstra PW (2006) Clinically relevant expansion of hematopoietic stem cells with conserved function in a single-use, closed-system bioprocess. Biol Blood Marrow Transplant 12(10):1020–1030
68. Nielsen LK (1999) Bioreactors for hematopoietic cell culture. Annu Rev Biomed Eng 1:129–152
69. Cabrita GJ, Ferreira BS, da Silva CL, Goncalves R, Almeida- Porada G, Cabral JM (2003) Hematopoietic stem cells: from the bone to the bioreactor. Trends Biotechnol 21(5):233–240
70. Collins PC, Nielsen LK, Patel SD, Papoutsakis ET, Miller WM (1998) Characterization of hematopoietic cell expansion, oxygen uptake, and glycolysis in a controlled, stirred-tank bioreactor system. Biotechnol Prog 14(3):466–472
71. Kim BS (1998) Production of human hematopoietic progenitors in a clinical-scale stirred suspension bioreactor. Biotechnol Lett 20(6):595–601
72. Meissner P, Schroder B, Herfurth C, Biselli M (1999) Development of a fixed bed bioreactor for the expansion of human hematopoietic progenitor cells. Cytotechnology 30:227–234
73. Koller MR, Emerson SG, Palsson BO (1993) Large-scale expansion of human stem and progenitor cells from bone mar row mononuclear cells in continuous perfusion cultures. Blood 82(2):378–384

74. Liu Y, Liu T, Fan X, Ma X, Cui Z (2006) Ex vivo expansion of hematopoietic stem cells derived from umbilical cord blood in rotating wall vessel. J Biotechnol 124(3):592–601
75. Kirouac DC, Zandstra PW (2006) Understanding cellular networks to improve hematopoietic stem cell expansion cultures. Curr Opin Biotechnol 17(5):538–547
76. Evans MJ, Kaufman MH (1981) Establishment in culture of pluripotential cells from mouse embryos. Nature 292(5819):154–156
77. Thomson JA, Kalishman J, Golos TG, Durning M, Harris CP, Becker RA, Hearn JP (1995) Isolation of a primate embryonic stem cell line. Proc Natl Acad Sci U S A 92(17):7844–7848
78. Oh SK, Choo AB (2006) Human embryonic stem cells: technological challenges towards therapy. Clin Exp Pharmacol Physiol 33(5–6):489–495
79. Dahéron L, Opitz SL, Zaehres H, Lensch MW, Andrews PW, Itskovitz-Eldor J, Daley GQ (2004) LIF/STAT3 signaling fails to maintain self-renewal of human embryonic stem cells. Stem Cells 22(5):770–778
80. Liu N, Lu M, Tian X, Han Z (2007) Molecular mechanisms involved in self-renewal and pluripotency of embryonic stem cells. J Cell Physiol 211(2):279–286
81. Babaie Y, Herwig R, Greber B, Brink TC, Wruck W, Groth D, Lehrach H, Burdon T, Adjaye J (2007) Analysis of Oct4-dependent transcriptional networks regulating self-renewal and pluripotency in human embryonic stem cells. Stem Cells 25(2):500–510
82. Ilic D (2006) Culture of human embryonic stem cells and the extracellular matrix microenvironment. Regen Med 1(1):95–101
83. Sasaki N, Okishio K, Ui-Tei K, Saigo K, Kinoshita-Toyoda A, Toyoda H, Nishimura T, Suda Y, Hayasaka M, Hanaoka K, Hitoshi S, Ikenaka K, Nishihara S (2008) Heparan sulfate regulates self-renewal and pluripotency of embryonic stem cells. J Biol Chem 283(6):3594–3606
84. Phillips BW, Lim RY, Tan TT, Rust WL, Crook JM (2008) Efficient expansion of clinical-grade human fibroblasts on microcarriers: cells suitable for ex vivo expansion of clinical-grade hESCs. J Biotechnol 134(1–2):79–87
85. Crook JM, Peura TT, Kravets L, Bosman AG, Buzzard JJ, Horne R, Hentze H, Dunn NR, Zweigerdt R, Chua F, Upshall A, Colman A (2007) The generation of six clinical-grade human embryonic stem cell lines. Cell Stem Cell 1(5):490–494
86. Martin MJ, Muotri A, Gage F, Varki A (2005) Human embryonic stem cells express an immunogenic nonhuman sialic acid. Nat Med 11(2):228–232
87. Hisamatsu-Sakamoto M, Sakamoto N, Rosenberg AS (2008) Embryonic stem cells cultured in serum-free medium acquire bovine apolipoprotein B-100 from feeder cell layers and serum replacement medium. Stem Cells 26(1):72–78
88. Sperger JM, Chen X, Draper JS, Antosiewicz JE, Chon CH, Jones SB, Brooks JD, Andrews PW, Brown PO, Thomson JA (2003) Gene expression patterns in human embryonic stem cells and human pluripotent germ cell tumors. Proc Natl Acad Sci U S A 100(23):13350–13355
89. Enver T, Soneji S, Joshi C, Brown J, Iborra F, Orntoft T, Thykjaer T, Maltby E, Smith K, Dawud RA, Jones M, Matin M, Gokhale P, Draper J, Andrews PW (2005) Cellular differentiation hierarchies in normal and culture-adapted human embryonic stem cells. Hum Mol Genet 14(21):3129–3140
90. International Stem Cell Initiative, Adewumi O, Aflatoonian B, Ahrlund-Richter L, Amit M, Andrews PW, Beighton G, Bello PA, Benvenisty N, Berry LS, Bevan S, Blum B, Brooking J, Chen KG, Choo AB, Churchill GA, Corbel M, Damjanov I, Draper JS, Dvorak P, Emanuelsson K, Fleck RA, Ford A, Gertow K, Gertsenstein M, Gokhale PJ, Hamilton RS, Hampl A, Healy LE, Hovatta O, Hyllner J, Imreh MP, Itskovitz-Eldor J, Jackson J, Johnson JL, Jones M, Kee K, King BL, Knowles BB, Lako M, Lebrin F, Mallon BS, Manning D, Mayshar Y, McKay RD, Michalska AE, Mikkola M, Mileikovsky M, Minger SL, Moore HD, Mummery CL, Nagy A, Nakatsuji N, O'Brien CM, Oh SK, Olsson C, Otonkoski T, Park KY, Passier R, Patel H, Patel M, Pedersen R, Pera MF, Piekarczyk MS, Pera RA, Reubinoff BE, Robins AJ, Rossant J, Rugg-Gunn P, Schulz TC, Semb H, Sherrer ES, Siemen H, Stacey GN, Stojkovic M, Suemori H, Szatkiewicz J, Turetsky T, Tuuri T, van den Brink S, Vintersten K, Vuoristo S, Ward D, Weaver TA, Young LA, Zhang W (2007) Characterization of human embryonic stem cell lines by the International Stem Cell Initiative. Nat Biotechnol 25(7):803–816

91. Rugg-Gunn PJ, Ferguson-Smith AC, Pedersen RA (2005) Epigenetic status of human embryonic stem cells. Nat Genet 37(6):585–587
92. Hall LL, Byron M, Butler J, Becker KA, Nelson A, Amit M, Itskovitz-Eldor J, Stein J, Stein G, Ware C, Lawrence JB (2008) X-inactivation reveals epigenetic anomalies in most hESC but identifies sublines that initiate as expected. J Cell Physiol 216(2):445-452
93. Ginis I, Luo Y, Miura T, Thies S, Brandenberger R, Gerecht-Nir S, Amit M, Hoke A, Carpenter MK, Itskovitz-Eldor J, Rao MS (2004) Differences between human and mouse embryonic stem cells. Dev Biol 269(2):360–380
94. Draper JS, Smith K, Gokhale P, Moore HD, Maltby E, Johnson J, Meisner L, Zwaka TP, Thomson JA, Andrews PW (2004) Recurrent gain of chromosomes 17q and 12 in cultured human embryonic stem cells. Nat Biotechnol 22(1):53–54
95. Joannides A, Fiore-Heriche C, Westmore K, Caldwell M, Compston A, Allen N, Chandran S (2006) Automated mechanical passaging: a novel and efficient method for human embryonic stem cell expansion. Stem Cells 24(2):230–235
96. http://www.invitrogen.com/downloads/stempro-ez-brochure.pdf
97. Baker DE, Harrison NJ, Maltby E, Smith K, Moore HD, Shaw PJ, Heath PR, Holden H, Andrews PW (2007) Adaptation to culture of human embryonic stem cells and oncogenesis in vivo. Nat Biotechnol 25(2):207–215
98. Hasegawa K, Fujioka T, Nakamura Y, Nakatsuji N, Suemori H (2006) A method for the selection of human embryonic stem cell sublines with high replating efficiency after single-cell dissociation. Stem Cells 24(12):2649–2660
99. Ellerstrom C, Strehl R, Noaksson K, Hyllner J, Semb H (2007) Facilitated expansion of human embryonic stem cells by single cell enzymatic dissociation. Stem Cells 25(7):1690–1696
100. Braam SR, Denning C, van den Brink S, Kats P, Hochstenbach R, Passier R, Mummery CL (2008) Improved genetic manipulation of human embryonic stem cells. Nat Methods 5(5):389–392
101. Thomson A, Wojtacha D, Hewitt Z, Priddle H, Sottile V, Di Domenico A, Fletcher J, Waterfall M, Corrales NL, Ansell R, McWhir J (2008) Human embryonic stem cells passaged using enzymatic methods retain a normal karyotype and express CD30. Cloning Stem Cells 10(1):89–106
102. Rubio D, Garcia S, Paz MF, De la Cueva T, Lopez-Fernandez LA, Lloyd AC, Garcia-Castro J, Bernad A (2008) Molecular characterization of spontaneous mesenchymal stem cell transformation. PLoS ONE 3(1):e1398
103. Watanabe K, Ueno M, Kamiya D, Nishiyama A, Matsumura M, Wataya T, Takahashi JB, Nishikawa S, Nishikawa S, Muguruma K, Sasai Y (2007) A ROCK inhibitor permits survival of dissociated human embryonic stem cells. Nat Biotechnol 25(6):681–686
104. Fok EY, Zandstra PW (2005) Shear-controlled single-step mouse embryonic stem cell expansion and EB-based differentiation. Stem Cells 23(9):1333–42. Ebub 2005 Aug 4
105. zur Nieden NI, Cormier JT, Rancourt DE, Kallos MS (2007) Embryonic stem cells remain highly pluripotent following long term expansion as aggregates in suspension bioreactors. J Biotechnol 129(3):421–432
106. Terstegge S, Laufenberg I, Pochert J, Schenk S, Itskovitz-Eldor J, Endl E, Brustle O (2007) Automated maintenance of embryonic stem cell cultures. Biotechnol Bioeng 96(1): 195–201
107. Horn PA, Tani K, Martin U, Niemann H (2006) Nonhuman primates: embryonic stem cells and transgenesis. Cloning Stem Cells 8(3):124–129
108. Cho MS, Lee YE, Kim JY, Chung S, Cho YH, Kim DS, Kang SM, Lee H, Kim MH, Kim JH, Leem JW, Oh SK, Choi YM, Hwang DY, Chang JW, Kim DW (2008) Highly efficient and large-scale generation of functional dopamine neurons from human embryonic stem cells. Proc Natl Acad Sci U S A 105(9):3392–3397
109. Hay DC, Zhao D, Fletcher J, Hewitt ZA, McLean D, Urruticoechea-Uriguen A, Black JR, Elcombe C, Ross JA, Wolf R, Cui W (2008) Efficient differentiation of hepatocytes from human embryonic stem cells exhibiting markers recapitulating liver development in vivo. Stem Cells 26(4):894–902

110. Murry CE, Keller G (2008) Differentiation of embryonic stem cells to clinically relevant populations: lessons from embryonic development. Cell 132(4):661–680
111. Meyer T, Sartipy P, Blind F, Leisgen C, Guenther E (2007) New cell models and assays in cardiac safety profiling. Expert Opin Drug Metab Toxicol 3(4):507–517
112. Schwanke K, Wunderlich S, Reppel M, Winkler ME, Matzkies M, Groos S, Itskovitz-Eldor J, Simon AR, Hescheler J, Haverich A, Martin U (2006) Generation and characterization of functional cardiomyocytes from rhesus monkey embryonic stem cells. Stem Cells 24(6): 1423–1432
113. Dang SM, Kyba M, Perlingeiro R, Daley GQ, Zandstra PW (2002) Efficiency of embryoid body formation and hematopoietic development from embryonic stem cells in different culture systems. Biotechnol Bioeng 78(4):442–453
114. Dang SM, Gerecht-Nir S, Chen J, Itskovitz-Eldor J, Zandstra PW (2004) Controlled, scalable embryonic stem cell differentiation culture. Stem Cells 22(3):275–282
115. Gerecht-Nir S, Cohen S, Itskovitz-Eldor J (2004) Bioreactor cultivation enhances the efficiency of human embryoid body (hEB) formation and differentiation. Biotechnol Bioeng 86(5):493–502
116. Gerecht-Nir S, Cohen S, Ziskind A, Itskovitz-Eldor J (2004) Three-dimensional porous alginate scaffolds provide a conducive environment for generation of well-vascularized embryoid bodies from human embryonic stem cells. Biotechnol Bioeng 88(3):313–320
117. Wobus AM, Wallukat G, Hescheler J (1991) Pluripotent mouse embryonic stem cells are able to differentiate into cardiomyocytes expressing chronotropic responses to adrenergic and cholinergic agents and Ca2+ channel blockers. Differentiation 48(3):173–182
118. Zweigerdt R, Burg M, Willbold E, Abts H, Ruediger M (2003) Generation of confluent cardiomyocyte monolayers derived from embryonic stem cells in suspension: a cell source for new therapies and screening strategies. Cytotherapy 5(5):399–413
119. Wartenberg M, Gunther J, Hescheler J, Sauer H (1998) The embryoid body as a novel in vitro assay system for antiangiogenic agents. Lab Invest 78(10):1301–1314
120. Zandstra PW, Bauwens C, Yin T et-al. (2003) Scalable production of embryonic stem cell-derived cardiomyocytes. Tissue Eng 9(4):767–778
121. Bauwens C, Yin T, Dang S, Peerani R, Zandstra PW (2005) Development of a perfusion fed bioreactor for embryonic stem cell-derived cardiomyocyte generation: oxygen-mediated enhancement of cardiomyocyte output. Biotechnol Bioeng 90(4):452-461
122. Niebruegge S, Nehring A, B¨ar H, Schroeder M, Zweigerdt R, Lehmann J Cardiomyocyte production in mass suspension culture: embryonic stem cells as a source for great amounts of functional cardiomyocytes. Tissue Eng Part A 14(10):1591–601
123. Viswanathan S, Benatar T, Mileikovsky M, Lauffenburger DA, Nagy A, Zandstra PW (2003) Supplementation-dependent differences in the rates of embryonic stem cell self-renewal, differentiation and apoptosis. Biotechnol Bioeng 84(5):505
124. Graichen R, Xu XQ, Braam SR, Balakrishnan T, Norfiza S, Sieh S, Soo SY, Tham SC, Mummery CL, Colman A, Zweigerdt R, Davidson BP (2008) Enhanced cardiomyogenesis of human embryonic stem cells by a small molecular inhibitor of p38 MAPK. Differentiation 76:357–370
125. Xu XQ, Graichen R, Soo SY, Balakrishnan T, Norfiza S, Sieh S, Tham SC, Freund C, Moore J, Mummery C, Colman A, Zweigerdt R, Davidson BP (2008) Chemically defined medium supporting differentiation of human embryonic stem cells to cardiomyocytes. Differentiation [Epub ahead of print]
126. Freund C, Ward-van Oostwaard D, Monshouwer-Kloots J, van den Brink S, van Rooijen M, Xu X, Zweigerdt R, Mummery C, Passier R (2008) Insulin redirects differentiation from cardiogenic mesoderm and endoderm to neuroectoderm in differentiating human embryonic stem cells. Stem Cells 26(3):724–733
127. Ng ES, Davis RP, Azzola L, Stanley EG, Elefanty AG (2005) Forced aggregation of defined numbers of human embryonic stem cells into embryoid bodies fosters robust, reproducible hematopoietic differentiation. Blood 106(5):1601–1603

128. Costa M, Dottori M, Ng E, Hawes SM, Sourris K, Jamshidi P, Pera MF, Elefanty AG, Stanley EG (2005) The hESC line Envy expresses high levels of GFP in all differentiated progeny. Nat Methods 2(4):259–260
129. Burridge PW, Anderson D, Priddle H, Barbadillo Munoz MD, Chamberlain S, Allegrucci C, Young LE, Denning C (2007) Improved human embryonic stem cell embryoid body homogeneity and cardiomyocyte differentiation from a novel V-96 plate aggregation system highlights interline variability. Stem Cells 25(4):929–938
130. Cameron CM, Hu WS, Kaufman DS (2006) Improved development of human embryonic stem cell-derived embryoid bodies by stirred vessel cultivation. Biotechnol Bioeng 94:938–948
131. Ungrin MD, Joshi C, Nica A, Bauwens C, Zandstra PW (2008) Reproducible, ultra high-throughput formation of multicellular organization from single cell suspension-derived human embryonic stem cell aggregates. PLoS One 3(2):e1565
132. Hentze H, Graichen R, Colman A (2007) Cell therapy and the safety of embryonic stem cell-derived grafts. Trends Biotechnol 25(1):24–32
133. Hirata H, Murakami Y, Miyamoto Y, Tosaka M, Inoue K, Nagahashi A, Jakt LM, Asahara T, Iwata H, Sawa Y, Kawamata S (2006) ALCAM (CD166) is a surface marker for early murine cardiomyocytes. Cells Tissues Organ 184(3–4):172–180
134. Rust WL, Balakrishnan T, Zweigerdt R Cardiomyocyte enrichment from human embryonic stem cell cultures by selection of cells with surface expression of CD166. Regenerative Medicine, in press
135. Choo AB, Tan HL, Ang SN, Fong WJ, Chin A, Lo J, Zheng L, Hentze H, Philp RJ, Oh SK, Yap M (2008) Selection against undifferentiated human embryonic stem cells by a cytotoxic antibody recognizing podocalyxin-like protein-1. Stem Cells 26(6):1454–1463
136. Xu C, Police S, Rao N, Carpenter MK (2002) Characterization and enrichment of cardiomyocytes derived from human embryonic stem cells. Circ Res 91(6):501–508
137. Klug MG, Soonpaa MH, Koh GY, Field LJ (1996) Genetically selected cardiomyocytes from differentiating embronic stem cells form stable intracardiac grafts. J Clin Invest 98(1): 216–224
138. Li M, Pevny L, Lovell-Badge R, Smith A (1998) Generation of purified neural precursors from embryonic stem cells by lineage selection. Curr Biol 8(17):971–974
139. Soria B, Roche E, Berna G, Leon-Quinto T, Reig JA, Martin F (2000) Insulin-secreting cells derived from embryonic stem cells normalize glycemia in streptozotocin-induced diabetic mice. Diabetes 49(2):157–162
140. Muller M, Fleischmann BK, Selbert S, Ji GJ, Endl E, Middeler G, Muller OJ, Schlenke P, Frese S, Wobus AM, Hescheler J, Katus HA, Franz WM (2000) Selection of ventricular-like cardiomyocytes from ES cells in vitro. FASEB J 14(15):2540–2548
141. Huber I, Itzhaki I, Caspi O, Arbel G, Tzukerman M, Gepstein A, Habib M, Yankelson L, Kehat I, Gepstein L (2007) Identification and selection of cardiomyocytes during human embryonic stem cell differentiation. FASEB J 21:2551–2563
142. Xu XQ, Zweigerdt R, Soo SY, Ngoh ZX, Tham SC,Wang ST et al. Highly enriched cardiomyocytes from human embryonic stem cells. Cytotherapy 10(4):376–89
143. Lawrenz B, Schiller H, Willbold E, Ruediger M, Muhs A, Esser S (2004) Highly sensitive biosafety model for stem-cell-derived grafts. Cytotherapy 6(3):212–222
144. Soong PL, Wang ST, Putti TC, Phillips B, Dunn RD, Hentze H Transplantation into SCID mice: relevance of transplantation site, cell number, and cell dissociation, submitted for publication
145. Field LJ (2004) Modulation of the cardiomyocyte cell cycle in genetically altered animals. Ann N Y Acad Sci 1015:160–170
146. Dai W, Field LJ, Rubart M, Reuter S, Hale SL, Zweigerdt R, Graichen RE, Kay GL, Jyrala AJ, Colman A, Davidson BP, Pera M, Kloner RA (2007) Survival and maturation of human embryonic stem cell-derived cardiomyocytes in rat hearts. J Mol Cell Cardiol 43(4): 504–516
147. Storch A, Sabolek M, Milosevic J, Schwarz SC, Schwarz J (2004) Midbrain-derived neural stem cells: from basic science to therapeutic approaches. Cell Tissue Res 318(1):15–22

148. Storch A, Paul G, Csete M, Boehm BO, Carvey PM, Kupsch A, Schwarz J (2001) Long-term proliferation and dopaminergic differentiation of human mesencephalic neural precursor cells. Exp Neurol 170(2):317–325
149. Hall VJ, Li JY, Brundin P (2007) Restorative cell therapy for Parkinson's disease: a quest for the perfect cell. Semin Cell Dev Biol 18(6):859–869
150. Reynolds BA, Weiss S (1992) Generation of neurons and astrocytes from isolated cells of the adult mammalian central nervous system. Science 255:1707–1710
151. Stemple DL, Anderson DJ (1992) Isolation of a stem cell for neurons and glia from the mammalian neural crest. Cell 71(6):973–985
152. Kallos MS, Behie LA (1999) Inoculation and growth conditions for high-cell-density expansion of mammalian neural stem cells in suspension bioreactors. Biotechnol Bioeng 63(4): 473–483
153. Kallos MS, Behie LA, Vescovi AL (1999) Extended serial passaging of mammalian neural stem cells in suspension bioreactors. Biotechnol Bioeng 65(5):589–599
154. Gilbertson JA, Sen A, Behie LA, Kallos MS (2006) Scaled-up production of mammalian neural precursor cell aggregates in computer-controlled suspension bioreactors. Biotechnol Bioeng 94(4):783–792
155. Mukhida K, Baghbaderani BA, Hong M, Lewington M, Phillips T, McLeod M, Sen A, Behie LA, Mendez I (2008) Survival, differentiation, and migration of bioreactor-expanded human neural precursor cells in a model of Parkinson disease in rats. Neurosurg Focus 24(3–4):E8
156. McLeod MC, Kobayashi NR, Sen A, Baghbaderani BA, Sadi D, Ulalia R, Behie LA, Mendez I Behavioral restoration following transplantation of bioreactor-expanded neural precursor cells in a rodent model of Huntington's disease. Submitted
157. Lian Q, Lye E, Suan Yeo K, Khia Way Tan E, Salto-Tellez M, Liu TM, Palanisamy N, El Oakley RM, Lee EH, Lim B, Lim SK (2007) Derivation of clinically compliant MSCs from CD105+, CD24- differentiated human ESCs. Stem Cells 25(2):425–436
158. Owen M, Friedenstein AJ (1988) Stromal stem cells: marrow-derived osteogenic precursors. CIBA Found Symp 136:42–60
159. Brooke G, Cook M, Blair C, Han R, Heazlewood C, Jones B, Kambouris M, Kollar K, McTaggart S, Pelekanos R, Rice A, Rossetti T, Atkinson K (2007) Therapeutic applications of mesenchymal stromal cells. Semin Cell Dev Biol 18(6):846–858
160. Pittenger MF, Mackay AM, Beck SC, Jaiswal RK, Douglas R, Mosca JD, Moorman MA, Simonetti DW, Craig S, Marshak DR (1999) Multilineage potential of adult human mesenchymal stem cells. Science 284(5411):143–147
161. Dominici M, Le Blanc K, Mueller I, Slaper-Cortenbach I, Marini F, Krause D, Deans R, Keating A, Prockop Dj, Horwitz E (2006) Minimal criteria for defining multipotent mesenchymal stromal cells. The International Society for Cellular Therapy position statement. Cytotherapy 8(4):315–317
162. Le Blanc K, Ringdén O (2007) Immunomodulation by mesenchymal stem cells and clinical experience. J Intern Med 262(5):509–525
163. Ren G, Zhang L, Zhao X, Xu G, Zhang Y, Roberts AI, Zhao RC, Shi Y (2008) Mesenchymal stem cell-mediated immunosuppression occurs via concerted action of chemokines and nitric oxide. Cell Stem Cell 2(2):141–150
164. Timmers L, Lim SK, Arslan F, Armstrong JS, Hoefer IE, Doevendans PA, Piek JJ, Oakley RME, Andre Choo A, Lee CN, MBBS, Pasterkamp G, de Kleijn DPV (2008) Reduction of myocardial infarct size by human mesenchymal stem cell conditioned medium. Stem Cell Res 1:129–137
165. Zhao F, Chella R, Ma T (2007) Effects of shear stress on 3-D human mesenchymal stem cell construct development in a perfusion bioreactor system: experiments and hydrodynamic modeling. Biotechnol Bioeng 96(3):584–595
166. Zhao F, Pathi P, Grayson W, Xing Q, Locke BR, Ma T (2005) Effects of oxygen transport on 3D human mesenchymal stem cell metabolic activity in perfusion and static cultures: experiments and mathematical model. Biotechnol Prog 21(4):1269–1280

167. Chen X, Xu H, Wan C, McCaigue M, Li G (2006) Bioreactor expansion of human adult bone marrow-derived mesenchymal stem cells. Stem Cells 24(9):2052–2059
168. Wang TW, Wu HC, Wang HY, Lin FH, Sun JS (2008) Regulation of adult human mesenchymal stem cells into osteogenic and chondrogenic lineages by different bioreactor systems. J Biomed Mater Res A Apr 2. [Epub ahead of print]

Index